Gender, Culture and Northern Fisheries

Joanna Kafarowski – Editor

CCI Press – 2009

Library and Archives Canada Cataloguing in Publication

Gender, culture and northern fisheries / edited by Joanna Kafarowski.

(Occasional publications series ; no. 62)
Includes bibliographical references.
ISBN 978-1-896445-46-5

1. Women in fisheries--Canada, Northern--Case studies. 2. Women in fisheries--Alaska--Case studies. 3. Women in fisheries--Europe, Northern--Case studies. 4. Subsistence fishing--Canada, Northern--Case studies. 5. Subsistence fishing--Alaska--Case studies. 6. Subsistence fishing--Europe, Northern--Case studies. 7. Fisheries--Canada, Northern--Case studies. 8. Fisheries--Europe, Northern--Case studies. I. Kafarowski, Joanna, 1962- II. Series: Occasional publication series (Canadian Circumpolar Institute) ; no. 62

HD6073.F652G45 2009   338.3'727082   C2009-905937-1

Keywords: arctic, fisheries, gender roles, women,

© 2009 Canadian Circumpolar Institute (CCI) Press

All rights reserved.
No part of this publication may be reproduced, stored in a retrieval system, or transmitted in any form or by any means—electronic, mechanical, photocopying, recording, or otherwise without the express written permission of the copyright owner/s. CCI Press is a registered publisher with access© the Canadian Copyright Licensing Agency (Publisher Number 3524).

Cover design by art design printing, inc.
Cover images courtesy CCI and Chapter 2 authors.

Printed in Canada (Edmonton, Alberta) by art design printing inc.

See acknowledgements section of this volume for support of this publication.

ISBN 1-896445-46-5; ISSN 0068-0303
© 2009 Canadian Circumpolar Institute (CCI) Press
Occasional Publication No. 62

TABLE OF CONTENTS

Foreword v

Introduction vii

SECTION ONE:
GENDERED PARTICIPATION IN SUBSISTENCE AND COMMERCIAL ACTIVITIES

Chapter One
    Chercher Les Poissons: Gender Roles in an Aleut Indigenous
    Commercial Economy 3
    Katherine Reedy-Maschner

Chapter Two
    "Without Fish We Would No Longer Exist": The Changing Role
    of Women in Southeast Alaska's Subsistence Salmon Harvest 29
    Virginia Mulle and Sine Anahita

Chapter Three
    "It Used to be Women's Work": Gender and Subsistence
    Fishing on the Hudson Bay Coast 47
    Martina Tyrrell

Chapter Four
    Are Living Fish Better Than Dead Fillets? The Invisibility and
    Power of Icelandic Women in Aquaculture and the Fishery Economy 67
    Anna Karlsdóttir

Chapter Five
    Everyone Goes Fishing: Gender and Procurement in the Canadian Arctic 85
    Kerrie-Ann Shannon

Chapter Six
    Gender, Knowledge, and Environmental Change Related to
    Humpback Whitefish in Interior Alaska 109
    Melissa Robinson, Phyllis Morrow and Darlene Northway

SECTION TWO:
GOVERNANCE PRACTICES

Chapter Seven
"I Have Always Wanted to go Fishing": Challenging Gender and Gender Perceptions in the Quota-Oriented Small-Scale Fishery of Finnmark, Norway 131
Siri Gerrard

Chapter Eight
"It's Our Land Too": Inuit Women's Involvement and Representation in Arctic Fisheries in Canada 153
Joanna Kafarowski

Chapter Nine
Gender Equality and Governance in Arctic Swedish Fisheries and Reindeer Herding 171
Maria Udén

Chapter Ten
Beyond the Pale: Locating Sea Sami Women Outside the Official Fisheries Discourse in Northern Norway 183
Elina Helander-Renvall

Chapter Eleven
Women in Sámi Fisheries in Norway—Positions and Policies 201
Elisabeth Angell

Chapter Twelve
Gender, Human Security and Northern Fisheries 219
Gunhild Hoogensen

List of acronyms 237

Contributors 239

FOREWORD

This book provides the reader with a current accounting of the role of women in a variety of northern subsistence and industrial fisheries, both aboriginal and non-aboriginal, rural- and urban-based, in Alaska, Arctic Canada, Iceland, Norway and Sweden. Fishing often makes an important contribution to food security in northern regions, where agriculture is impossible or marginal at best, as well providing important occupational and economic diversification in small and often remote communities. In such locations the high cost and often low nutritional value of imported foods can be offset by fishing, hunting and gathering activities that contribute significantly to peoples' socio-economic circumstances and health. Indeed, these health benefits may extend beyond peoples' physical health alone, because customary food production, distribution, consumption and enjoyment sustains peoples' connection to their environment in meaningful ways that enrich their mental well-being and the integrity and vitality of their communities.

In some societies, fishing is regarded as womens' work, but in far more cases fishing is considered to be mens' work. The authors draw attention to the generally under-recognized role of women in fisheries' harvesting and formal decision-making. The conventional recognition of the primary role of men in fish harvesting often results in mens' knowledge being the principal (or only) source of important local knowledge considered by fisheries' managers and decision-makers. The resulting under-representation of womens' knowledge may compromise the quality of management decision-making, suggesting the desirability of including knowledge obtained by women more especially during the processing and food-preparation phases of product use.

However, changes associated with modernization affect all societies, and even the strictly-gendered roles of men and women engaged in fishing similarly change. Nevertheless, in situations where customary food production, sharing and consumption remain highly valued for a variety of culturally-important reasons, the likelihood remains high that fishing, in one form or another, will almost certainly persist—even as changes continue to modify the manner, frequency and strictly economic importance of fishery practices in these societies.

This volume draws attention to the need for a more critical understanding of the emphasis often placed on hunting and associated male dominance in food production in northern societies. Whereas the representation of men as hunters (and fishers) and women as gatherers and food-preparers is all too commonly encountered in the literature, this collection of studies argues that fishing as an activity may be much more ambiguous and nuanced than previously considered, and increasingly so as modernization further alters customary social roles and attitudes.

Today (and almost certainly continuing into the future), the occupational opportunities available to more highly-educated rural residents offer a wider range of choices with respect to work, place of residence, and lifestyle, suggesting that it is unwise to seek to predict how the changing roles of women in fisheries will appear in the future. This volume tests a number of assumptions and prior conclusions in respect to gender and fisheries, and indeed, of gender relations more generally, and in so doing provides useful information and insights that inform current understandings of these northern societies and social identities, as well as very likely stimulating future research.

Milton M.R. Freeman
Edmonton, Alberta

INTRODUCTION

That women are actively engaged in fisheries work around the world is without question. From crabmeat processors in Tabasco, Mexico to fish traders in Lake Victoria, Tanzania to participants in the shrimp aquaculture industry in coastal states of India, women are significantly involved in fisheries both directly and indirectly (Munk-Madsen 1998; Newell and Ommer 1999; Kumar 2004; Ahmed 2005; Neis *et al.* 2005). This degree of participation is reflected in the vast array of ongoing community-based, fisheries projects and the development of relevant programs and policies that are currently being conducted internationally.

In Europe, the AKTEA[1] European Network: Women in Fisheries and Aquaculture was initiated in 2005 to recognize women's contributions to fisheries; to increase the visibility of women's roles in fisheries; to participate in decision-making on matters affecting women's roles; to exchange ideas and experiences and to work towards the political and institutional acceptance of women's organizations in fisheries. Primarily focusing on Western Europe, AKTEA also coordinates conferences that enable academics, politicians and fishers to share experiences. In its Green Paper on the *Future of the Common Fisheries Policy* published in 2001, the European Commission acknowledged the need for the Union to ensure that the role played by women in the fisheries sector be recognized and enhanced. Extensive research has been supported by the European Union including a major 2002 study on the role of women in fisheries in thirteen European countries. Following the acceptance of this report, Member States in the European Parliament adopted a resolution on "women's networks: fishing, farming and diversification" committing to launching "the necessary actions to secure greater legal and social recognition of the work of women in the fisheries sector."[2] The International Collective in Support of Fishworkers (ICFS) with offices located in Belgium and India draws its mandate from the *International Conference of Fisheries and their Supporters* (ICFWS) held in Rome in 1984. It is an international non-governmental organization that works toward the establishment of equitable and sustainable fisheries particularly in the small-scale, artisanal fisheries. It implemented the Women in Fisheries program that supports research and policy development and publishes reports, monographs and other publications on women in fisheries (including *Yemaya* the ICSF Newsletter on Gender). In 1977, the WorldFish Center (a research centre associated with the Consultative Group on International Agricultural Research) was formed to focus on living aquatic resources in the developing world. The Center hosts the biennial *World Fisheries Forum* which includes a Gender and Fisheries component attracting international participants. This symposium "recognized

---

[1] In Greek mythology, Aktea is one of the nereids or sea nymphs who stayed close to the shore.
[2] Resolution (2004/2263(INI) was adopted on December 15, 2005. The full text is available at http://www.europarl.eu.int

the breadth and depth of changes needed to create a gender-sensitive fisheries sector and make real improvements in the lives of those involved" (Williams *et al.* 2006:3). In North America, sustained, long-term programs and organizations focusing on women and fisheries have not been as well-established or have formed to address a particular issue and then disbanded. For example, following the collapse of the cod fisheries in the late 1980s and the subsequent closure of those fisheries in Atlantic Canada, a variety of groups including Women's FishNet formed throughout the region to give women a greater voice in decision-making and a forum to respond to the introduction of an unpopular compensation package through linking with other groups facing similar issues. Spearheaded by researchers at Memorial University of Newfoundland and Labrador in Canada, this group has now broadened its focus to address related community and health issues. Despite the success of these organizations and initiatives, to date no group exists or has been created that addresses the regional needs of women involved in fisheries across the circumpolar north. Some women in this region belong to groups such as AKTEA or Women's FishNet while others may have specific issues addressed through women's organizations such as Femina Borealis in northern Europe.

An extensive literature exists that documents the contributions of women and men in fisheries, and research on gender relations in fisheries has dramatically expanded since the late 1970s (Porter 1985; Bavington *et al.* 2004; Grzetic 2004; Neis *et al.* 2005). In the circumpolar north, this literature has been dominated by Norwegian, Canadian, Icelandic and American scholars, some of whom are represented in these pages. Earlier work focused on women's involvement in, and contributions to, fishery-based households and communities (Munk-Madsen and Larsen 1989; Nadel-Klein and Davis 1988; Gerard 1995) as well as on the gendered division of labour. More recent literature adopts a gender-based analysis of the linkages between fisheries and globalization including how industrial restructuring relates to international markets and changing dynamics in local communities (Power 2005). According to MacDonald:

> Fisheries have long provided interesting vantage points from which to explore processes of capital accumulation and relations of class and gender. The community basis of most fisheries highlights relationships that might otherwise be lost in a more geographically dispersed industry. Interactions between the gender division of labour in wage work, family production and domestic work are more visible in a context where household members are integrated in one way or another into the same industry…Linking the experiences of fisheries communities worldwide will contribute to an understanding of globalization in general, its gendered nature and its failure as a basis for sustainable development in human or ecological terms (2005:18).

The gender and fisheries literature has primarily focused on community or nation-specific studies and this has resulted in a significant gap in research on areas including the Arctic. Why is adopting a regional approach to research on the roles and experiences of women in fisheries relevant or even necessary? A regional approach facilitates the comparing and contrasting of experiences across geopolitical boundaries. In the case of the circumpolar North, this is critical in regards to the experiences of Indigenous peoples in general, and Indigenous women in particular. It is clear that in any country with a significant Indigenous population, national comparisons between Indigenous and non-Indigenous peoples is meaningless. Statistics regarding health, education, income level and unemployment, amongst other factors, vary significantly. While each Indigenous group represents a separate nation and identity, it can be argued that sharing common cultural values, similar attitudes toward natural resources (including fisheries) and the challenges of navigating between traditional and Western-based knowledge systems, would be far more valuable. This is evident in the chapters that follow.

Contributors to the book were solicited through academic institutions throughout the Circumpolar North; Indigenous organizations, various relevant listservers addressing fisheries and/or Arctic issues and professional contacts. Although Indigenous involvement was strongly encouraged, few Indigenous individuals submitted manuscripts. All contributors were affiliated with universities or research institutes at the time of submission and the book primarily reflects these interests.

The first section of the volume focuses on gendered participation in subsistence and commercial activities related to the fisheries, including case studies from Nunavut in Canada, Alaska in the United States, and Iceland. Both Reedy-Maschner, and Mulle and Anahita investigate the roles of Indigenous women in Alaska. In 'Chercher les Poissons,' Reedy-Maschner explores how the sexual division of labour influences fishing practices and relationships both at sea and in Aleut coastal communities. Mulle and Anahita's chapter, entitled, 'Without Fish, we Would no Longer Exist,' addresses shifting patterns in the involvement of urban Native women Southeast Alaska in contemporary subsistence activities. Tyrrell and Shannon examine two different aspects of Inuit women's economic participation in fisheries in Nunavut. In 'It Used to be Women's Work: Gender and Subsistence Fishing on the Hudson Bay Coast,' Tyrrell considers gendered perceptions regarding the Arctic char fishery while in "Everyone Goes Fishing: Gender and Procurement in the Canadian Arctic," Shannon focuses on differing participation of community members in fishing derbies as examples of vital procurement activities. In 'Are Living Fish Better Than Dead Fillets? The invisibility and power of Icelandic women in aquaculture and the fishery economy,' Karlsdottir presents women's views on fisheries with an emphasis on gender and regional development. The last chapter of the first section, 'Gender, Knowledge and Environmental Change Related to Humpback Whitefish in Interior Alaska,' by Robinson, Morrow and Northway, investigates gendered knowledge in fisheries in the eastern interior of Alaska and how ongoing environmental change necessitates that the differing knowledge(s) of both women and men are critical for the sustainable management of natural resources.

The second section of the volume presents case studies in Norway, Sweden and Canada that explore differing governance practices. Gerrard's chapter, '"I Have Always Wanted to go Fishing": Challenging Gender and Gender Perceptions in the Quota-Oriented Small-Scale Fishery of Finnmark, Norway,' applies a gender-based analysis to the impact of the quota system and other contemporary Norwegian fisheries policies on fishers in Finnmark, Norway. Two other chapters in this section focus on Sami fisheries in northern Norway. Helander-Renvall explores how Sea Sami women reclaimed power within the fisheries sector despite marginalization by Norwegian fisheries policies. In 'Women in Sami Fisheries in Norway—Positions and Policies,' Angell considers Sami women's shifting involvement in fisheries due to the support and involvement of the Sami Parliament. Úden's chapter entitled 'Gender, Equality and Governance in Arctic Swedish Fisheries and Reindeer Herding' is a comparative study regarding governance issues affecting Swedish women in both fisheries and reindeer herding. Kafarowski reviews Inuit women's access to and participation in decision-making processes in fisheries in Nunavut, Canada. The concluding chapter of the book, 'Gender, Human Security and

Northern Fisheries,' by Hoogenson, provides a preliminary exploration of the relevance of the human security concept to gender and fisheries.

The editor acknowledges and thanks each contributor for her commitment to this project.

Joanna Kafarowski
Editor

References

Ahmed, S. (2005). *Flowing upstream: Empowering women through water management.* New Delhi: India.

Bavington, D., B. Grzetic and B. Neis (2004). Feminist Political Ecology of Fishing Down: Reflections from Newfoundland and Labrador. *Studies in Political Economy* 73: 159-182.

Gerrard, S. (1995). When women take the lead: Changing conditions for women's activities, roles, and knowledge in north Norwegian fishing communities. *Social Science Information* 34(4): 593-631.

Grzetic, B. (2004). *Women Fishes These Days.* Halifax: Fernwood Publishing.

Kumar, K. (2004). *Gender Agenda: Women in Fisheries.* Chennai, India: International Collective in Support of Fishworkers.

MacAlister E. and Partners Ltd. (2002). *The role of women in the fisheries sector. European Commission Directorate General For Fisheries Final Report.* Hampshire: England.

MacDonald, M. (2005). 'Lessons and linkages: Building a framework for analyzing the relationships between gender, globalization and the fisheries,' pp. 18-27 in B. Neis, M. Binkley, S. Gerard and M. C. Maneschy, eds., *Changing Tides — Gender, Fisheries and globalization.* Halifax: Fernwood Publishing.

Munk-Madsen, E. and M. Larsen (1989). *Gender myths with consequences: An analysis of the division of labour between men and women in the industrial fishery on land and at sea.* NFTR Report No. 1, Tromsø, Norges Fiskerihogskole.

Munk-Madsen, E. (1998). The Norwegian fishing quota system: Another patriarchal construction? *Society and Natural Resources* 11: 229-240.

Nadel-Klein, J. and D. Davis, eds. (1988). *To Work and to Weep: Women in fisheries economies.* St. John's: Institute of Social and Economic Research, Memorial University of Newfoundland.

Neis, B., M. Binkley, S. Gerard and M.C. Maneschy (2005). *Changing Tides. Gender, Fisheries and Globalization.* Halifax: Fernwood Publishing.

Newell, D. and R. Ommer, eds. (1999). *Fishing Places, Fishing People. Traditions and Issues in Canadian small-scale fisheries.* Toronto: University of Toronto Press.

Porter, M. (1985). She was the skipper of the shore crew: Notes on the history of the sexual division of labour. *Labour/Le Travail* 15: 105-23.

Power, N.G. (2005). *What do they call a fisherman? Men, gender and restructuring in the Newfoundland fishery.* St. John's: Institute of Social and Economic Research, Memorial University of Newfoundland.

Williams, M., M. Nandeesha and P. Choo (2006). 'Changing traditions: A summary report on the first global look at the gender dimensions of fisheries,' pp. 1-6 in M. Williams, M. Nandeesha and P. Choo, eds., *Global Symposium on gender and fisheries. Seventh Asian Fisheries Forum 1-2 December 2004.* Penang, Malaysia: WorldFish Center.

SECTION ONE

GENDERED PARTICIPATION IN SUBSISTENCE AND COMMERCIAL ACTIVITIES

CHAPTER ONE

Chercher Les Poissons: Gender Roles in an Aleut Indigenous Commercial Economy

Katherine Reedy-Maschner

Abstract: The term 'fisher,' used to include women when describing fishing occupations, is gaining popularity; but this trend of pushing women into a broader category with men, even where a strong sexual division of labour remains, obscures actual practices and relationships. In an Alaskan Native coastal society where subsistence and commercial fishing practices are combined, Aleut women inhabit the roles of fishermen, fishermen's wives, mothers, daughters, or girlfriends with pride, and are variably involved in fishing, processing, politics, business and family life. As men dominate the fishing arena, women fulfill crucial roles on land as they partially dictate sharing patterns which determine subsistence harvests, and as they choose mates which affirm male status roles. Their choices speak volumes, shaping dynamics on the water and in the communities. This is ever more apparent if we follow the fish from catch to pantry to table. The identity of being Aleut is tied to fishing, which is most visibly practised by men, but it is the behaviours and attentions of women that give the whole system its meaning.

Introduction

Research on women in fishing economies (*e.g.,* Chapman 1987; Ellis 1977; King 1992-1993; Nadel-Klein and Davis 1988; Nadel-Klein 2003; Sinclair and Felt 1992; Skaptadóttir 1996) or in industrialized commercial fishing (*e.g.,* Allison *et al.* 1989; Binkley and Thiessen 1988; Fields 1997; Fricke 1973; Thiessen *et al.* 1992) emphasize women's participation as challenging the stereotypical image of this male-dominated arena of rugged individualists. This literature promotes gender equality or equal participation at sea, and yet we cannot assume that equality is the goal for all women. In fact, a focus on gender equality may obscure an examination of the equal representation of men's and women's roles in fishing. Further, most works on commercial fishing in Northwestern North America and Alaska neglect indigenous women, even though they share in Alaska's largest industry.

In coastal communities and (more often) in circles distant from fishing communities, the term 'fisher' has entered the social and political jargon as more politically correct and gender-inclusive. Many women who fish reject this term (Allison *et al.* 1989; Fields 1997) and many Alaskan fishermen, understand a 'fisher' to be a "furry animal related to the marten" (Lord 1997:xi). Neutering 'fishermen' for these women/writers/fishermen is considered to be largely irrelevant since fishing is most certainly a male-dominated, male-centred way of life. Aleuts only use the term 'fisherman,' and it is understood that this includes women who fish. Aleutian commercial fishing is a male-driven cultural system within which status structures shape individual and community identity and underpin social relations (Reedy-Maschner 2004). Both men's and women's roles are complex and variously experienced and are very much tied to fishing. Simply put, Aleut men and women do not have, nor do they want, equality in fishing.

This chapter seeks to strengthen the roles of men and women through the case of Aleut sea fishermen. Through presenting a history of economic and social relations in Aleut society I will show how the changing roles of men have shaped women's roles, and that women are often proactive in driving these changes. I will demonstrate that one must follow the fish, 'chercher les poissons,' in order to understand the nature of these relationships in the modern context. Almost every fish passes from men's hands through women's hands into extensive networks of sharing, and tracing the fish from catch to pantry to stomach reveals key aspects of Aleut gender relations and cultural obligations.

Over the past several decades, feminist anthropologists have led ongoing debates over models of dominance versus complementary roles between men and women. They reacted to a dearth of women's lives being expressed in ethnography and theory and examined women through many disparate experiences and cultural meanings (Sacks 1983; Leacock 1981; MacCormack and Strathern 1981; Ortner 1974; Rosaldo and Lamphere 1974). These anthropologists focused more heavily on the cultural constructions of biological understandings of male and female than I am comfortable with, but they added historical, racial, class-based, economic and political dimensions to gender relations. Cross-culturally, men are frequently in the public sphere and women in the private sphere, but this does not always privilege one role as dominant over or subservient to another. A few studies have challenged the perceived universal ideology of male dominance, finding instead that social systems of prestige emphasize personal autonomy and egalitarianism for all men and women, creating an ideology of gender equality (*e.g.,* Lepowsky 1993; *see also* Brettell and Sargent 2000; Leacock 1981; Sacks 1979). For many societies, models of gender complementarity have surfaced, since women often feed the family on a regular basis while the men have varying degrees of success in hunting, and the hunted meat is brought into the domestic space where it is prepared and shared. For example, in her work with Iñupiat whalers, Bodenhorn (1990) argues that since wives ritually attract the animals, they are regarded as hunters in an interdependent relationship with their husbands.

Analyses of gender roles often focus on either subsistence or commercial economies and dichotomies of domestic vs. public take centre stage for each. In fishing societies, especially where fishing is industrialized, there is often a strong sexual division of labour even though women's roles and social status in fishing communities are highly variable (Acheson 1981; Ellis 1977; Nadel-Klein 2003). Nadel-Klein and Davis (1988) note that the distinction between subsistence and commercial fisheries has very different consequences in relation to women's participation. The literature on women's roles is far richer regarding women's subsistence contributions, and Nadel-Klein and Davis assert that while there are many passing references to women fishing commercially, or more often as fish processors and marketers, these lack any meaningful analyses. Women's roles have also been considered to be undermined by capitalist development, which is said to technologically and socially discriminate against women. This pushes women into subsistence production for the family while men become wage earners. This can sometimes remove both men and women from the home (Moore 1988).

The Aleut present an interesting case for these gender models. In the first instance, the "penetration of capitalism into subsistence economies" (Moore 1988:74) was embraced by the Aleut population, a conscious choice for participation by people who recognized its potential as the key to their collective survival, even though it is not an easy industry and policy changes have had uneven consequences for the villagers. Further, Aleut society has a commercial–subsistence orientation in which there is no tidy distinction between where commercial activities end and subsistence activities begin, or vice versa. This has profound significance when it comes to defining Aleut social organization and placing it within the broader literature on fishing societies. A great deal of hunter-gatherer literature emphasizes that, while these societies may engage in commercial activities, cash is often converted back into purchasing equipment for subsistence pursuits and thus, subsistence is paramount (*e.g.*, Bodenhorn 1989; Fienup-Riordan 2000; Goldsmith 1979; Rasing 1994). The Aleut, on the other hand, pour their cash income back into commercial pursuits, which are performed jointly with subsistence harvesting, a practice which also amounts to the enhancement of the status symbols of fishing vessels. Many Aleut women are also wage earners, who sometimes provide alternative incomes supplementing their husbands' income when fishing is disappointing. Aleut women may also hold crucial village-based jobs in government, economic planning and policy formation that maintain fisheries practices. The 'universal' male nature of political power is thus diffused by divisions between the land and sea, women and men.

Although decades old and critiqued many times, Ortner and Whitehead's (1981) work develops what is perhaps the most useful model through which to examine Aleut social relations. Even though they relegate biology and reproductive anatomy to mere symbols and cultural constructs, they contend that men and women are products of social and cultural processes, and that identity is often linked to gendered symbols or practices (MacCormack and Strathern 1981). Through the notion of 'prestige structures,' which they assert are the combined culturally salient sets of positions and the mechanisms in which to arrive at those positions, they show how prestige structures shape cultural notions of gender and sexuality. In the process of 'becoming' (in this case, for example, a fisherman), there are criteria that one must fulfill, which then alter his or her perceptions of self and society. Ortner and Whitehead outline how masculine ideals are constructed as Llewelyn-Davies shows that Maasai men transform themselves from 'propertyless to propertied' by gaining wives and cattle, further glamorized through ritual, which sets a standard of masculinity towards which all men strive (Llewelyn-Davies 1981; Ortner and Whitehead 1981:5). In a survey of 'brideservice' societies, Collier and Rosaldo found that "men are glorified as hunters and killers, but women are not by that fact glorified as mothers and 'lifegivers'" (Collier and Rosaldo 1981; Ortner and Whitehead 1981:6). These cases show that men are often defined by prestige categories that have little to do with women, but women's roles are often centred around men, yet do not form complementary systems. In most hunter-gatherer societies, hunting is often the major source of male prestige, and men define themselves by their skill as hunters (*e.g.*, Fienup-Riordan 1983; Lee 1993; Lee and DeVore 1968; Radcliffe-Brown 1922). The central activity of the Aleut is male-oriented and female activities support those of the males. Thus, prestige structures and supporting ideologies can provide the motivation and explanation for behaviour. Status, prestige, kinship and marriage organizations are critical to understanding gender constructions. Women and men have different status structures where different forms of masculinity and femininity are locally constructed, which are then variously valued and expressed within the culture.

Most Aleut women do not fish on a regular basis, but women are essential to maintaining rights of access in the outward political arena. The Aleut are a marginalized people within the non-Native and Alaska Native communities and they fish the most controversial salmon fishery in the state. This fishery has been treated as non-traditional, fished almost exclusively by seasonal transients from Washington, in media and on political stages (Gay ADN 2/26/2004, 3/2/2004, 4/28/2004; Hensel 1996:169-172; Johnson 1997:6; Knowles 2000a, 2000b; Loy ADN 1/31/2001, 2/2/2001, 1/13/2002; Ruskin ADN 7/20/2000). It is considered expendable by the majority of residents of the state because it is a mixed stock intercept fishery in which commercially harvested fish are bound for rivers in western Alaska and Asia and the primarily subsistence-based societies residing there. Aleuts define themselves as commercial fishermen (with subsistence as a vital part of that) and they engage fully in the market economy. This has contributed to their misrepresentation within Alaska because Aleuts are often perceived as not 'acting Native' (Reedy-Maschner 2004). This requires constant political engagement with resource managers and politicians, and women conduct much of this work. Women manage family businesses, the household, the processing and storage of fish and game and they often hold land-based jobs to supplement the family income in lean years. They have also created economic development schemes to market local salmon. Here, I emphasize through examples that, while men fish, own boats and permits, manage crews and bring to shore the fish and the prestige attached to these practices, none of this would be successful without the local–global efforts of women and their affirmation of men's status, which, in turn, affirms their own. Men understand this, but take it for granted. Therefore, Aleut men's and women's roles can best be defined as variable in which some relationships can be considered male-dominant, and some can be considered complementary, but most often, relationships and divisions of labour cannot be clearly delineated. Women can and do fish and hunt, men can and do process fish and game, and both manage households, businesses, and politics.

The Aleut Hunter-Gatherer-Fisherman

The Aleut of the lower Alaska Peninsula, Aleutian Islands and Pribilof Islands are a commercial fishing society which engages simultaneously in for-profit enterprises in fisheries and subsistence harvesting of wild fish and game. The majority of Aleut families rely on the sea for a living as commercial seiners, gillnetters, longliners, trawlers and pot fishermen. An archaeological record reveals their 10,000-year relationship to a marine ecosystem (Dumond 1987; Laughlin 1963, 1980; Laughlin and Aigner 1975; Maschner et al. 1997; Maschner 1999a, b, 2000; Maschner and Hoffman 2003; McCartney 1984) to which the living Aleut link their maritime sociocultural identity. Today, there are 13 communities in the Aleutian and Pribilof Islands.[1] This chapter focuses on the Aleut of the lower Alaska Peninsula/Eastern Aleutian Islands region who reside in the villages of False Pass, King Cove, Sand Point, and Nelson Lagoon. Collectively, these villages

---

[1] Combined, these have a total population of approximately 3,500. Many Aleuts also reside in Anchorage (Morgan 1976) as well as the Pacific Northwest, where approximately 600 Aleuts live in Washington and Oregon and are members of the Northwest Aleut Association.

have a permanent population of 1,891 of which approximately 70% are Native Aleut (Census 2000). This population increases three-fold during peak fishing seasons. Commercial fishing occurs virtually year-round with salmon fished in the summer, crab in the fall and winter, cod and pollock in the winter, halibut in winter and spring, and herring in the spring. Aleut fishermen dominate the local salmon fleets, fishing in regional waters designated as 'Area M' by the Alaska Board of Fisheries.

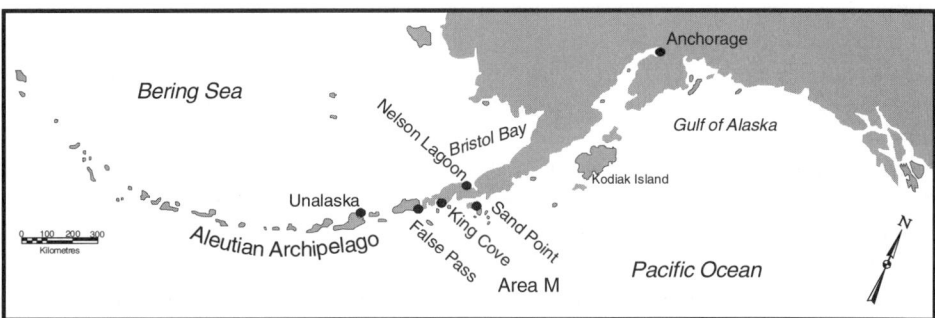

Figure 1. *Aleutian villages and Area M.*[2]

Although the Aleut are considered to be a hunter-gatherer society, they defy dominant definitions of hunter-gatherer peoples, which a large literature has tended to place outside global economic, political and social processes (*debated in* Burch and Ellanna 1994; Kelly 1995; Leacock and Lee 1982; Myers 1988). Although they have lived in their Native homeland for millennia, the Aleut are western and industrialized with both autonomous and government-dependent political and economic status. Villages are kin-based, yet foreign-owned canneries and white fishermen are ever-present. These Aleut are wealthy relative to most North American and global indigenous peoples (and many earn more than university professors). For example, the median household income in King Cove was averaged at $45,893 for the year 2000 (Census 2000). (However, a fisherman's income depends entirely on the amount of fish caught, the market value, number and duration of fishing openings, and regulations affecting fishing). They prefer to eat wild fish, caribou, birds and clams, but for any given meal, store-bought pork chops or pizza might be on the menu. Few speak the Aleut language and there is no everyday traditional dress or adornment, unless you count fishermen's raingear. Most homes have cable or satellite television, and they worry over Iraq and al-Qaeda or muse over television episodes.

The Aleut have formed a number of non-governmental organizations to develop multi-community responses to state, federal, and environmental challenges to their commercial fisheries. A few Aleut leaders travel to Washington, D.C. on a regular basis, and participate in the Arctic Council and other international meetings, to protect their livelihood. The fisheries are an international enterprise, heavily regulated and policed by state and federal bodies, and the Aleut continuously engage with these entities. Fishermen pilot

---

[2] Area M is the Alaska Board of Fisheries designation for the waters to the North and South of the Alaska Peninsula surrounding the villages of Nelson Lagoon, False Pass, King Cove, and Sand Point.

boats through the dangerous waters of the Pacific Ocean and Bering Sea, and yet there is a sense that the outside world is potentially more threatening.

In everyday life, 'traditional' ways of living combine fairly seamlessly with the modern American orientation. The Aleut do not spend much time trying to reconcile two worlds; this is simply one world for them, which sometimes is easier to navigate than at other times. Commercial fishing is a volatile industry worldwide, with unpredictable runs and risky boating conditions ever-present, but the Aleut have extended layers of difficulty to deal with. Because of their commercialized life, they are often portrayed in both media and politics as non-Native thieves of the fish intended for other Native Alaskans. The *Anchorage Daily News* has published numerous articles that obscure or ignore their indigenous identity (*e.g.,* Gay 2004a ,b,c; Loy 2001a, b, 2002; Ruskin 2000). Environmentalists challenge the Aleut fisheries each time they campaign to add another northern marine species to the Endangered Species List. Fish farms in Chile and Norway have displaced traditional Aleut markets for wild Alaska salmon. Thus, the political status of their fisheries is contentious, and the Aleut must engage individually and collectively with policymakers on a year-round basis in order to continue their livelihood. Fishing is a community identity, an Aleut identity. Women participate in the Aleut 'fishermen' identity, fully entrenched in this fishing society and culture even though some have never even pulled a fish out of the water. They keep the local industry alive by attending important meetings while their husbands are out at sea.

Historical Processes:
Aleut Gender Roles from Traditional Society to Commercialization

Rank and Power

Most ethnohistoric writings agree that men's and women's roles in traditional Aleutian society were sharply defined (Black 1984; Jochelson 1933; Jones 1976; Lantis 1970; Liapunova 1996; Townsend 1983; Veniaminov 1984). However, Robert-Lamblin's paper on traditional Aleut women's roles proposes that an Aleut woman, much more so than an Inuit woman, "enjoyed real independence *vis-à-vis* men because she could at all seasons and close to her dwelling secure enough food for her consumption" (1982a:199). Robert-Lamblin writes that the environmental constraints of the High Arctic required Inuit men and women to conform to rules of labour as interdependent units in order to survive, but that the Aleut had more favourable ecological conditions such that women did not always 'need' men to provide for them.[3] If anything, men were perhaps lost without women. In one extreme case in 1802, 200 Aleut men died from paralytic shellfish poisoning after eating toxic mussels in Peril Strait, Southeast Alaska (Lantis 1984). It was the women who tested the shellfish by touching one to their tongue and waiting to see if it went numb before eating them and, in their absence, the men did not know that the shellfish were not safe to eat.[4]

As a ranked society in which hereditary classes of nobles, commoners and slaves formed villages, women and men had internal rankings based upon kin organization and

---

[3] Ecological models of social relations in the High Arctic have been reviewed and challenged by Bodenhorn (1990, 1997).

[4] Clams and mussels are now sent to Anchorage for testing each year.

family resources (Lantis 1970). Marriage between villages was often with other nobility and functioned to form alliances between them, creating power through extensive family ties (Townsend 1983; Veniaminov 1840:II, *c.f.* Lantis 1970). Nobility could simply take wives from other men, abduct them from other villages (which often instigated war), or demote a wife to slave status depending upon her own position relative to the community (Tolstykh in Jochelson 1933:12). Polygamy and polyandry were common among those who could afford to care for the additional family members, yet few women had more than two husbands and few men had more than two wives, although some of higher rank reportedly had more than six wives (Veniaminov 1984).

Girls were trained to prepare and store fish and game, make clothing and kayak skin covers, weave baskets and mats, and collect edible plants; young boys were trained in navigation, kayak operation, hunting and warfare (Robert-Lamblin 1982a), and to "endure everything possible" while learning survival, military and hunting skills (Veniaminov 1984:191-192). Robert-Lamblin found evidence that during puberty, menses and childbirth, women had great powers to contaminate or influence hunting and fishing success, and were generally feared by men (Jochelson 1933; Veniaminov 1984). She believed that women controlled the 'group's equilibrium' through these 'supernatural powers' and the "indirect but conscious «control» that she could exert on hunting activities," but "the fact remains that the predominant role and prestige went to the Aleut hunter as supplier of meat, the food held in highest esteem, and skins necessary for clothing" (Robert-Lamblin 1982a:201; Dall 1870).

Colonial Shaping of Economic and Social Relations

The colonization process in the Aleutian Islands, beginning in 1741, was a gendered and racialized undertaking. Immediately following the arrival of Russian fur hunters to the Aleutians, Aleut men were conscripted as labour, often forcefully removed from their villages and taken to prime sea otter and fur seal hunting grounds. Although men came to derive high status from these hunts, the activities removed men from their villages and they subsequently lost control of their homes and families (Lantis 1970). This exclusively male invasion by Russians also seized Aleut wives and daughters for their own gratification, producing a Creole generation. After piecing together a genealogy for an Aleut woman in which her female ancestors were Aleut and her male ancestors were all Russian and Scandinavian (a product of the turn-of-the-twentieth-century cod industry), she asked, "Where did all the Aleut men go?" In combination with disease and a smallpox epidemic, which resulted in heavy losses, it is clear that Russians had a huge impact on shifting marital patterns. Russian Orthodox priests suppressed polygamy and polyandry in the early nineteenth century, and they began arranging marriages as they saw fit. Arrangements occurred between the same families, for example, two sisters of one village might marry two brothers of another[5] (Robert-Lamblin 1982b). Nonetheless, almost every living Aleut today can trace their ancestors to a Russian, European or Scandinavian man and an Aleut woman; thus, it appears that Aleut men simply did not have many opportunities to reproduce during this time. Russians pulled out of Alaska *en masse* after

---

[5] This is still common, although by choice, not arrangement. In King Cove, for example, four brothers of one family are married to four sisters from another.

the sale of the territory to the United States in 1867, which perhaps tipped the balance back in favour of Aleut or Creole men.

Access to New Industries

After the American purchase of Alaska in 1867, sea otter hunts continued until they reached the brink of extinction, and were eventually outlawed in 1911. The cod industry became the new boom (and bust) in the Aleutians in which many Aleut men were successful, and for which there was an influx of Scandinavian fishermen (Shields 2001). The advent of the cod industry was followed shortly thereafter by the development of a salmon industry, which has been the most consistent North Pacific fishery for the past century. Women worked alongside men in salting cod and salmon. Present-day King Cove, False Pass, Sand Point, and Nelson Lagoon were established around canneries at the beginning of the twentieth century from several nearby villages and dwindling cod stations. Initially, the canneries were supplied with company-owned salmon traps, which required little labour to operate, and they employed small fleets of fishermen, both outsiders and Aleuts, to fish other areas where there were no traps. Aleut men chose to be fishermen instead of work in the cannery where there was a steadier income. They leased boats from the cannery or fished by setting nets or beach seining from small skiffs. Canneries also financed boat purchases, giving Aleut men a more significant role. When fish traps were outlawed in 1959 following a territory-wide fisheries crisis,[6] fish were required to be caught using only boats, which made the canneries solely dependent upon fleets of fishermen.

From the 1940s to the 1970s, local Aleut women were the main cannery workers, and many worked full-time and raised large families simultaneously. When fishing became a lucrative business for their husbands, and their identity as fishermen became solidified, many of these women retired, although several widows or those women whose husbands did not fish full time, stayed on. The labour force shifted toward a staff of largely Filipino, Mexican, or American youth and these jobs became seen as less desirable. Women today reflect on their cannery work with dignity, but always couch this in a description of how much the canneries have changed and how it is no longer a good job. A few young Aleut women still work in the canneries, although it is meant to be as temporary as possible. Instead, many Aleut women hold jobs in other capacities, which afford them some community status, but they are not as valued as fishing jobs community-wide. Still, Aleut women maintain a degree of independence from men's work, and a single mother can be very successful within her village through employment opportunities in local political organizations.

As canneries diversified into herring and crab processing, the number of outsider fishermen in the region increased, and they tended to have larger, more efficient boats. These men were also seen as taking over the salmon fisheries across the state. Following statehood in 1959, the proposed solution to outside advantage and overfishing was the created exclusive access rights to salmon fishing through the allocation of permits. Limited Entry allocated a fixed number of fishing permits per district that were distributed to those fishermen who could demonstrate their prior participation in commercial fishing and their economic dependency. The plan was an attempt to give power and a sense of

---

[6] Fish traps were located in the path of migrating salmon which were corralled into a pot in the centre of the trap. They were then brailed out of the pot and onto a tendering vessel. The state built requirements for maintaining adequate escapement into Alaska's constitution.

ownership to local fishermen. While this is true for those Aleut fishermen who received permits, this also resulted in the exclusion of many who could not demonstrate their fishing history and a stratification within the villages based upon levels of access occurred. Aleutian fishermen can be divided into those with no access, crew access only, or those with permits in one, two or all three set gillnetting, drift gillnetting, and purse seining salmon fisheries. Social stratification roughly corresponds to this access such that permitted fishermen also control the political arenas.

Limited Entry further circumscribed fishing for future generations. Permits are individually owned, and can be loaned to relatives, inherited, or sold. Typically, permits are handed down from fathers to sons. A fisherman may not own more than one permit per gear type per fishing area. Aleutian salmon permits have sold recently for tens of thousands to hundreds of thousands of dollars, although their value rises and declines sharply depending on the previous season's success. Many fishermen used their additional permits to finance boat purchases, or have sold permits to outsiders in lean years, which has again resulted in an increase in non-resident fishermen. This practice is heavily criticized by those residents who do not want to lose control of their fisheries (Langdon 1980).

Thus, access to boats, permits, crew labour, and revenue is controlled by the Limited Entry permit plan, but Limited Entry also affects access to other social and political resources. Those who are fully within the fishing franchise as captains and their families, and those who participate at different levels as crew or in support of fishing, experience fishing and life on land quite differently. In this way, the status of being a fisherman is a limited resource. Variability in one's ability to access the fisheries, and especially to attain the status of captain, results in societal, financial, and gender inequalities. Relationships between members of Aleut communities are analyzed within this framework.

Gender Relations in a Limited Entry World

Although Russian expansion resulted in the marginalization of Aleut men with regards to marital and social relations, prosperity in fishing in the twentieth century has resulted in increased numbers of men in the Aleutians. Every Aleut claims a mixed heritage, and Aleut men today dominate their villages, although I hesitate to speculate on the specific demographic mechanism for this shift. This shift in demography, combined with constraints placed upon men's ability to attain the status of fisherman, has created particular concerns for modern gender relations.

The following conversation with men in the King Cove Harbor House, which is a men's gathering space at the harbour's edge where all important fisheries business and gossip is carried out, illustrates some of the difficulties men have faced in a humorous way. My husband and I had been talking to a group of fishermen about ethnohistorical accounts of Aleut men going on distant raids. One fisherman said of his ancestors:

> Those guys had to paddle all the way to Kodiak, kill a few Eskimos, and then paddle all the way back, just to impress women. Today all you need is a big boat.

These fishermen immediately shifted the conversation from a discussion centred around war expeditions to one centred around impressing women. They began describing how hard it was to impress women 'back then' and how one has to impress women today. The primary reason for the war raids, they said, was 'stealing women.' One fisherman, whose wife is from Kodiak, added with a wink and a chuckle, "We still do that." In Nelson Lagoon, I was told stories about a living elder who, in the 1950s, travelled down the coast and around the tip of the Alaska Peninsula to Sanak Island in a dory "just to get himself a woman," and another man who never made it to Sanak but got 'weathered in' at False Pass, and there, he "found one that would do." Marriageable women appear to have been scarce resources, and men travelled great distances to find wives. Ethnohistorical accounts coincide with these modern stories. In the past, Aleut men raided other Aleut villages as well as distant villages in Bristol Bay and Kodiak and captured women for themselves (Golder 1963a; Townsend 1980, 1983; Veniaminov 1984). The Aleut also have stories of men fighting, performing feats of strength, and maintaining the status of a great hunter in order to win wives (Dall 1870; Golder 1963b).

Access to and skills in fishing and hunting continue to be the primary concerns for Aleut males. To impress women and each other, fishermen exaggerate the sizes of their catches, their bravery in dangerous boating conditions, daring rescues, and how they have outsmarted Fish and Game authorities. Wealth from fishing adds to their overall sense of prestige, yet the wealth is poured back into vessel upgrades or top of the line equipment, which enhances these symbols of empowerment. Despite the number of wealthy, outside fishermen in the region, Aleut women prefer to marry or partner with Aleut men, but they seek out men who meet prestige criteria in relation to fishing, or sons of established fishermen who are guaranteed a future in the industry (add to this the claim of several young women who insisted that Aleut men are very well endowed). Although non-Aleut husbands and wives from Anchorage or outside of Alaska are found in all villages, their mating preferences are markedly different from some parts of Native Alaska where women are increasingly marrying non-Native men and leaving the villages (Fienup-Riordan 1990; Hamilton and Seyfrit 1994a, b). Fishing, then, is necessary for male survival in terms of negotiating their place in the hierarchy of fishing, demonstrating their skills, and attracting women. The very first detail one woman proudly announced to me in describing her new boyfriend was, "Kate, you won't believe it! He owns four boats!"

Demographically, the Aleutians East Borough has the highest percentage of men and boys relative to women and girls in the State of Alaska. In an overall population of 2,697 people, 64.9% are male and only 35.1% are female (Census 2000). Transient fishermen tip the scales even further with each fishing season. For the four villages examined here (n=1891), 60.3% are male and 39.7% are female. For men and women 18 and older (n=1480), 62.8% are men and 37.2% are women (Census 2000). This excess of men and shortage of women has exaggerated factors affecting gender relations and the prestige structures of men and women. There are more men than available female partners, more men than there are fishing permits, more men than there are crewing opportunities, and more men than there are non-fishing wage jobs. From the women's perspective, there is a scarcity of the 'right kind of man' for them to marry (although this does not preclude sexual relationships). Knowing the uncertainties of fishing, a few stated that they intend to marry outside the villages and move away. Several women seek out transient white fishermen because they are "related to everyone in town!" For young men, it is more difficult to bring non-local women to the village and persuade them to stay, so they can be torn between finding a mate or staying in the village to fish. In 2000, the ratio of men to women over the age of 18 was 1.69:1. Thus, there are shortages all around. There are fewer women relative to men, but there are also relatively fewer sought-after choices for women. Thus, the Native preference, coupled with the women's preference for fishermen, can generate difficulties and frustrations between men and women.

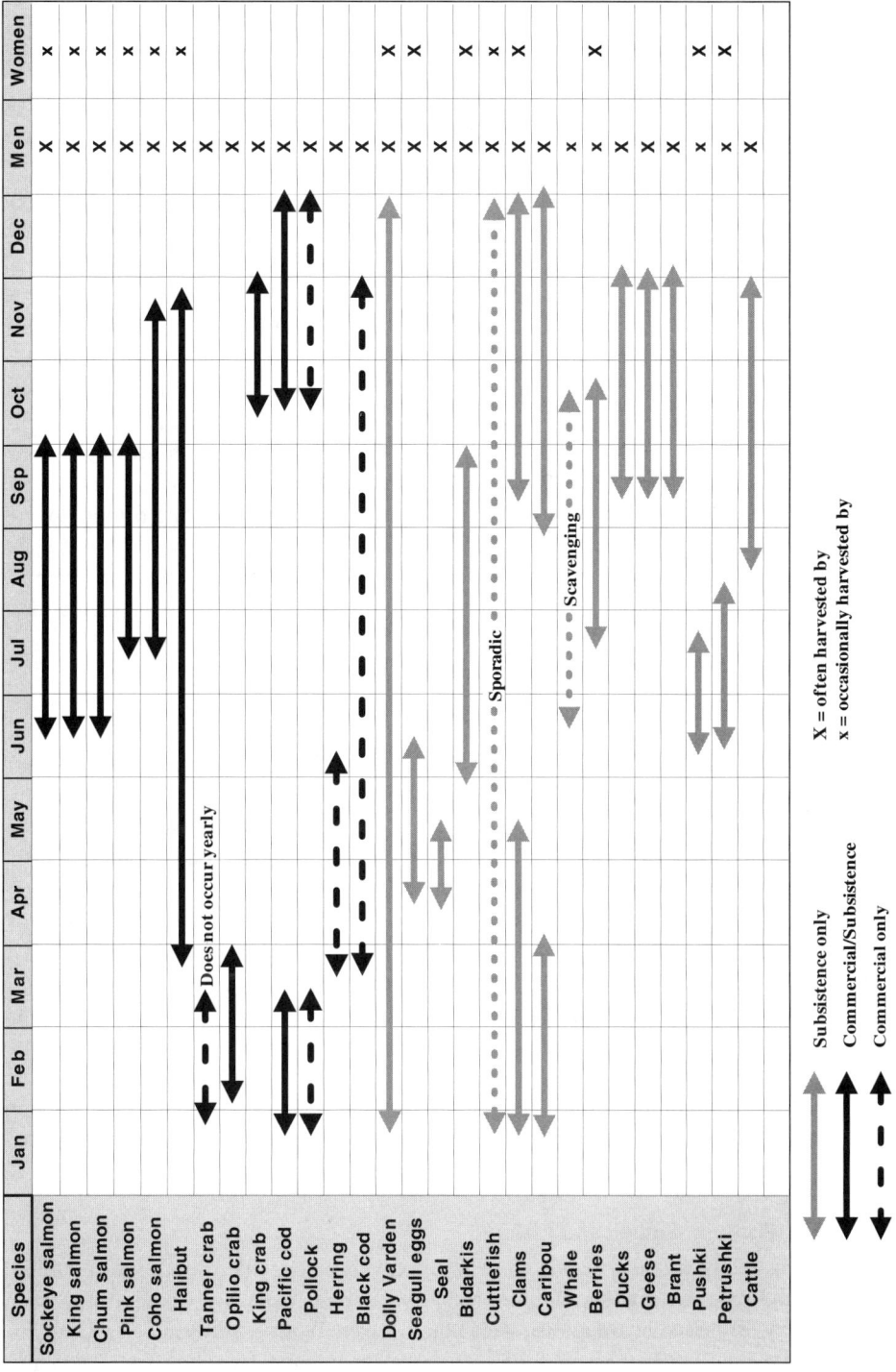

Figure 2. *Approximate schedule of seasonal harvests, mostly based upon King Cove, False Pass and Sand Point.*

Subsistence and Commercial Fishing

The most distinguishing feature of Aleut subsistence practices is how closely they are linked to commercial fishing. The locations of villages do not provide easy access to a broad range of foods, and although the beaches and uplands are hunted, the volume comes from travel aboard commercial vessels. Subsistence collecting, harvesting, hunting and eating is a daily affair for all Aleuts. Subsistence is locally called 'home pack,' which covers every fish, bird, intertidal or land species processed and consumed. Most hunting parties for terrestrial species such as caribou or geese consist of men or teenage boys roaring out of town on 4-wheelers in full camouflauge with guns strapped to their backs, legal seasons be damned. Women can more typically be found scouring the beaches for bidarkis (black katy chitons), sea urchins, and clams. As one woman often said looking out her window at the beach, "There's bidarkis smiling at us down there." Women usually are not part of the acquisition of fish and game or the primary butchering of waterfowl and game, but they do most of the processing and storing and teach these skills to their children. In calm seas, skiff-loads of families will head to nearby bays or to the outer islands to collect food. Wives may accompany their husbands fishing in order to collect along the beaches, but men will also collect on the beaches during fishing downtime in the absence of women. Collecting is not considered 'women's work' but is usually performed by women. The usual schedule of seasonal subsistence and commercial harvests is depicted in Figure 3, but is subject to yearly changes depending on regulations or species availability.

Women might fish for other fish species off the docks or in the creeks, but most fish for home use (primarily salmon, halibut, cod, crab) are taken from commercial catches using commercial gear by the captain and crew. Sockeye and king salmon are the preferred fish, and are usually stored in greater quantities, but can only be caught at sea. Almost all king salmon are kept for subsistence because the price offered by the canneries is often low (average was $.25/pound in 2002 for each cannery) and Aleut women would rather eat them. Fish are delivered to the cannery by individual boats or tenders and processed there, such that shore labour for salting or drying fish, often performed by women prior to canneries and the fresh fish market, is no longer in practice. Subsistence fish are separated out on a net-by-net basis at the discretion of the captain and kept in seawater until the fish can be taken home or delivered to those intended.

Alternatively, fishermen can bag the salmon while out on their boats and send them back to town with the tenders, which are the large boats that move fish and supplies from the fishing grounds to the cannery. The tendermen leave the fish in metal containers on the fish docks and radio to those for whom the fish are intended. Someone then picks them up to take home or delivers them. In Nelson Lagoon, individual fishermen always bring in fish since the cannery is located across the bay from the village.

Women in Fishing, Women and Fishing

As I have stated, fishing is a decidedly male activity, but women's knowledge of fishing practices and requirements surrounding these practices is extensive, though they may only occasionally step aboard a boat or fish themselves. Women crew on a few boats; if they do not regularly fish, they can fill roles as crew on a moment's notice if a crew member quits, gets fired, or is injured. With men gone fishing for extended periods of time, women manage the home, assume the primary parenting responsibilities, work in politics, or perform other community duties. Men fill jobs at the harbour as an extension of the

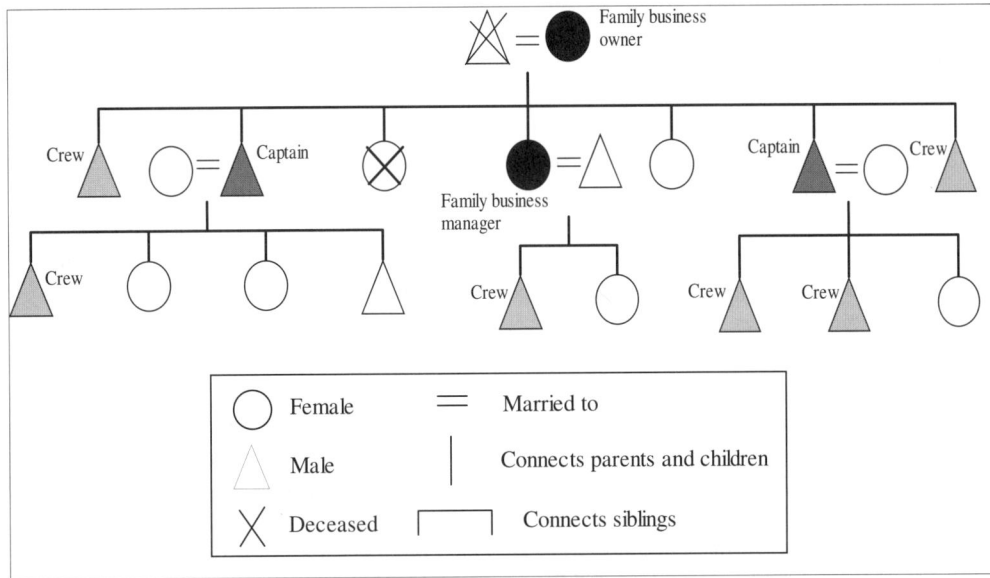

Figure 3. *Partial genealogy illustrating an Aleut family fishing corporation.*

fishermen's sector. Men maintain some independence from household duties, but they must also get along with other men on a cramped boat for long periods of time, and cook and clean for themselves. When a woman is on board, she is not automatically expected to cook for the crew.

As Robert-Lamblin outlined (1982a), in the past, 'female pollution' was a concern in matters of fishing. Until relatively recently, menstruating women were prohibited from walking the beaches because they were believed to have the power to deflect fish. One woman told me that just twenty years ago her father forbade his wife and daughters from coming aboard his boat because it would 'jinx it,' and he would only take them on the boat on the last day of the fishing season. But later in life, after he had suffered some health problems, his wife began fishing with him so she could keep watch on his condition and radio for help if necessary. Thus, women were forbidden on boats until more pragmatic considerations outweighed the dangers that women represented on board. Today, there is no lasting belief that women are polluting and there is no institutionalized segregation, but they still experience difficulties surrounding gender. There are no women captains in these Aleut fleets, however, women do fish as fishermen, not as 'women fishermen' (Allison 1988:231). On board, they function in multiple capacities: as skiffman, cook, deckhand, among many roles. A woman is usually linked to the captain as family or partner; if she is not, it is still assumed by the community that she must be.

Many vessels are named after women in the villages, and women feel connections to the boats because of this. One young woman, whose father's boat carries her name, felt responsible and even embarrassed if something went wrong with the boat, and pride if something went right. She became half-owner of the boat after her parents divorced, and would curse the "piece of shit Volvo engine" in one breath, and brag in another, buying drinks all around at the bar if her boat was the highliner (the fisherman with the largest catch) for an opener.

Becoming a Fisherman

Limited Entry bestowed permanent fishermen's status on a select number of fishermen in the 1970s, and their descendants have benefited. Becoming a captain is extremely difficult if one stands outside of this legacy because of the skyrocketing costs of buying into fishing. Sometimes it can come with ownership of a boat but no permit or, more easily, ownership of a permit through transfer (even temporary transfer) but not the boat. This means that the fisherman must find a partner, and also split his profits with that partner. Captaincy can also come 'empty handed' such that a fisherman may run a boat holding a permit through temporary transfer, but both the permit owner and the boat owner take unknown percentages of his overall income. These can develop into situations in which the captain could 'prove himself' and gain more permanence in his arrangement with the boat/permit owners, although it is rare that one can work his way into inheritance of a fishing operation outside of family. In a few cases, crew hires have taken the form of a groom service, where a man will crew for his prospective/actual father-in-law, or a permit transfer might take the form of a bride price and be given to the young man as a symbolic dowry.

In 2003, most of the salmon fishing openers coincided with terrible wind and rainstorms. On one seiner, a 15-year-old girl was crewing for a friend of her mother's with her cousin and his friend. The girl was on a slippery, icy deck, helping to pull the net in, and every time the boat rolled to her side, a sea lion leapt up trying to bite her. One crewman had already slipped off the deck and fell overboard earlier that day but thankfully was rescued right away. The storm was so severe that most boats quit early and headed for the harbour. When she returned, she told her mother the stories, saying, "Mom, I'm a fisherman now!" She knew that she had accomplished something momentous and had crossed a threshold, earning her the title of fisherman in that one experience. However, fishing is certainly not her career choice—she was earning money to attend a basketball camp out of state. Her title of fisherman is temporary. After high school, she has plans to attend college and most likely will never fish again. For the young men who were fishing alongside her, it is a different story altogether. For them, crewing bestows a perpetual status, and although they strive to get on more successful boats with highliner captains, they can probably never hope to be captains themselves since their fathers do not have the assets for them to inherit.

The opposite of these young men can be found in a case of a young man from a prominent family whose father owns a boat and two permits and is often a highliner. His father usually hires his permit-less brothers as crew and his son salmon-fished with an uncle who had no sons of his own. For him, future opportunities abound from the senior males in his family. He is already in a prime position for inheriting an operation from his father, and has the skills to run it. Since his father shows no signs of retiring, his uncle gave him the opportunity to fish the salmon season using his boat and permit. As captain, he organized his crew (which consisted of two friends of his who are in no position to inherit) and in the first opener, they were the highliners. Everyone in town was singing their praises. Young women began talking about these men, and they took a stronger interest in the captain when out at the bars. At the end of a successful season (relative to performances of other boats, but still low numbers because of few fish), his uncle, who had relocated further out the Aleutian chain, offered to sell him the operation. Thus, he had proven his worth to his uncle, father, his crew, the village, and to himself.

Being a Fisherman

Retaining the role of fisherman, once it has been attained, is not an easy task for everyone. Through several years of fieldwork, I have witnessed men (and women indirectly) raise and lower their standing in the community, changing their entire demeanour, based upon newly found or lost participation in fishing. Successful fishermen tend to already have, but also require, two things: a reliable crew and a solid marriage. Crews are valuable resources for captains at sea by their skills in fishing and on land by maintaining the captains' status as highliners by doing their boasting for them.

Almost every highliner or successful fisherman has a relatively stable marriage: he can attract a wife more easily but he also needs a wife to be successful. There is evidence that the initial distribution of permits, as well as today's ownership, corresponds to marital stability. Husbands and wives will fish together for many reasons, one of which is so they do not have to pay a crew. This is especially the case if they still owe on their boat or permit, which many of them do. They send their children to stay with non-fishing relatives (or with the willing anthropologist) if they do not take them on the boat. However, staying at home, child rearing, and supporting their husbands are valued activities, and women are not pressured to do more than that. Women also act as a kind of reliability check for men's boastings. There is evidence to suggest that men exaggerate the pounds of fish caught, and may even tack a few extra feet onto their boat lengths when talking to someone they want to impress. One wife laughed over her husband bragging to others about his catches, "but I see the fish tickets, so I know better," she said.

Long fishing openings also serve as a respite between spouses. When men are gone, women and children seem to have more freedom in household duties and in recreation. Some spouses fear bad fishing years because of the potential for spousal abuse. In good fishing years, there is a good deal of hedonistic behaviour, such that men might have been chasing women before, but become more successful in the catch. Men might be compelled to travel to Anchorage to party, and consequently cheat on their wives. This goes both ways. "Some women pass themselves around when guys are out fishing," one man stated, and I was aware of a few obvious cases of this. At the beginning of crabbing seasons, when there is an influx of fishermen touting themselves as highliners, an already sexually charged atmosphere is heightened.

In lean years, a few entire families, or more often just the male heads of household, will leave the village in search of employment. Some families reside permanently in Anchorage and return to the villages to fish and catch up with family. Occasionally, divorced couples with children will find the father leaving the village for other employment, returning only to fish. His ex-wife and children are often left in the care of her extended family. Of course, the stress of coping with lean fishing years is often blamed for the divorce in the first place.

The Business of Fishing

Fishermen's wives are sometimes land-based managers, as they manage the boat's finances, permits, and insurance. Figure 4 illustrates one such family business in which they have formed an actual corporation, owned by the mother, managed by the sister, and fished by the brothers and their sons. The mother became owner of the family business after her husband passed away. The corporation owns the boats (two seine boats, plus one tender/crabber) while the brothers own the permits. The women of this family corporation are intensely involved in managing finances, permits, insurance, and other compliances.

Gender, Culture, and Northern Fisheries – Chapter One

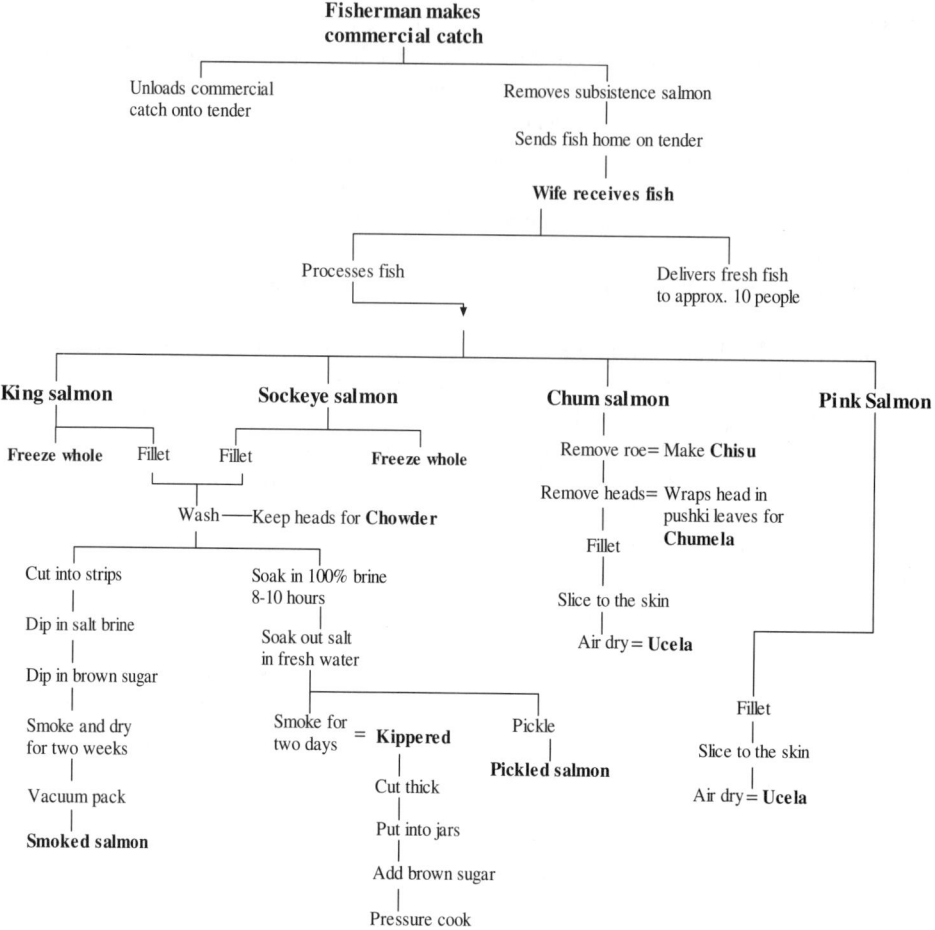

Figure 4. *Example of salmon from catch to stomach, and subsequent distribution, from a single fisherman (from Reedy-Maschner 2004:88).*[7]

Only a few extended Aleut families have incorporated as a business, and the advantages of doing this are undefined.

More often, wives remain separate from the business of fishing, and do not involve themselves with crew hires or crewshares, equipment purchases, or fishing itself, but restrict themselves to household and family management. Many men mend their own nets, although women may do this as well. Non-fishing, small local businesses are meant to provide filler income, such as childcare services, salons, taxis, hotels, bars, eateries, charter services, and fuel sales. Many of these businesses do not advertise and are operated out of private homes on irregular schedules, so you have to be 'in the know' to get a

---

[7] As a mixed stock fishery, chums, sockeyes, and kings are often caught together and can all be brought or sent home from one fishing opener. Pink salmon run later in July and August, and are usually caught separately from the other species.

haircut, rent a movie, or order a pizza. Women manage the majority of these land-based businesses. They experience ups and downs along with fishing due to extending credit in bad fishing seasons. Newcomers to the villages have noted gender patterns. As a temporary health worker stated, "Males are tied into fishing but it seems to be matriarchal. They raise the kids [together]. All other responsibilities are up to the women." Women often become the sole consistently employed member of the family and, in the classic hunter-gatherer sense, are keeping their families alive with steady work while the men wait for a fishing opener or other job.

Chercher Les Poissons: Family Organization, Production and Distribution

To this point, we have seen how commercial/subsistence fishing is organized and gendered and explored some of the requirements and complications men and women face in relation to fishing. I now turn to the organization of production and distribution of fish within the villages. For the most part, all fish pass through the hands of women, who utilize them to solidify relationships and enhance their status and that of their spouses.

Previous studies in the region distinguished between major family lineages based on surnames and male heads of household (Braund and Associates 1986). 'Dominant families' were so delineated because their members occupied formal leadership positions and because they often owned fishing assets in direct proportion to their family size. Most families are male-centric lineages, organized together as boat owners, permit holders, crewmen, parents, husbands and wives, and children in a cohesive network. Variably throughout the year, they are salmon seiners, tenders, crabbers, and cod, pollock, and halibut fishermen.

Even though some elder men claim to be 'retired,' they can be found in the Harbor House everyday, which is the men's lounge at the harbour where fishing business is carried out, and they may still fish. These families rely on each other on land and at sea. A variety of fish species almost always move from husbands to wives and children first, and then are used to solidify other relationships.

Although most lineages are organized around male relatives who fish, in one village, five female heads of household, who are also sisters from the most extensive lineage, are strongly influential community-wide, and their husbands' positions,' although highly respected and influential, are enhanced and maintained by the status of their wives. Through this 'sorority,' the related families are organized around fishing and sharing. Even though one of these sisters is deceased, her offspring operate within the network almost as if she were still living. Living members of these sorority-linked families including spouses comprise approximately 25% of the village, and as the village's largest family, it is also the one whose members hold many political positions and command the greatest respect. This sorority encompasses so many living Aleuts into its network that it cannot be dismissed as an anomaly in a male-organized lineage society. One sister stands out as the head of the family, with access to almost every relative and their goods and services, though she does not have direct access to fishing through her husband or sons. Her status is ascribed in that she was born into an extensive lineage but it is also achieved because she has had a long life as a community health provider and is well liked. Most immediate relatives to her have close relationships and a steady flow of fish and food sharing, care, and communication between them. Some relatives have weaker associations with less flow between them, indicating that there are differences between the biological relationship and the social relationship. The flow of actual foods (salmon, bidarkis, cuttlefish, crab, for example) most often coincides with the social relationship. In order to get access to someone's labour, material wealth or sharing, an individual often goes through their social relationships with the women. This family has tremendous fishing assets with numerous boats and permits between them. The sisters receive the majority of fish from their fishing relatives and redistribute finished products such as kippered salmon or *chisu* (caviar). One of the sisters is also a pivotal figure in Anchorage because she and her husband own a house there and spend their winters there. Most relatives who pass through for whatever reason stay in their house, whether they are home or not.

Those family members who are connected to the network, but who do not have opportunities to contribute many goods and services back are more dependent upon their biological relationship. One brother's wife is deceased and he does not fish commercially, however, he does accept fish from his sisters and friends, and will fish with rod and reel in town. This family network also demonstrates that fish and fish products do not always move in one direction, from supplier to receiver, but that since there are several vessels fishing in the network, fish move in more unsystematic ways.

Limited Entry restricts the number of boat owners and permit holders, relegating most men in the village to crewmen status. This situation greatly affects an individual's ability to provide subsistence resources to his household or to other households that depend on him. Crewmen must negotiate with their captains for the amount of fish they can keep for their own and their relatives' subsistence use. Generally, captains allow crewmen to keep as much as they need to, although there were a few reported cases of stinginess when it came time to deliver the fish to the cannery. Generally, however, the lower the price the canneries are paying for fish, the more people bring home.

Thus, decisions people make regarding the proportion of fish to remove from the commercial catch are based upon knowledge of past sharing quantities, assessment of current needs and numbers of people they share with, their wife's or mother's knowledge of who should get fish and how much, and the price of fish offered by the cannery. Women keep close tallies on numbers of fish needed for each household in their sharing network, including their own, and will tell their husbands, fathers, brothers or uncles how many more fish and which species they need to bring home for themselves and for elders, widows, widowers, and other relations in their sharing network. In this way, they are crucial to the status of the fishermen and themselves. The production of fish within larger, wealthier households is impressive in quantity and quality, to say the least. Figure 4 shows an example of just how many products can result from the salmon season, which in this case, represents the work of a single wife during the entire June to September salmon season. This model is not representative of every fisherman's catch, but tends to reflect the activities of the relatively wealthy who have a steady, guaranteed supply of salmon, perhaps one-third of the village population. Salmon processing and distribution charts would appear differently depending upon who is fishing, with whom one fishes, and the season. In many cases, heavy subsistence demands are placed upon individual crewmen.

Figure 5 illustrates how one crewman must supply two main households (his aunt and uncle's as well as his own) with salmon from negotiations with his captain. He is lucky in that his captain is generous and the boat he fishes on is often the highliner for each opening as well as for the whole season. His aunt, then, is the sole salmon supplier for three other households (her mother's and two daughters) from whatever her nephew is able to bring her. Quantities vary from year to year.

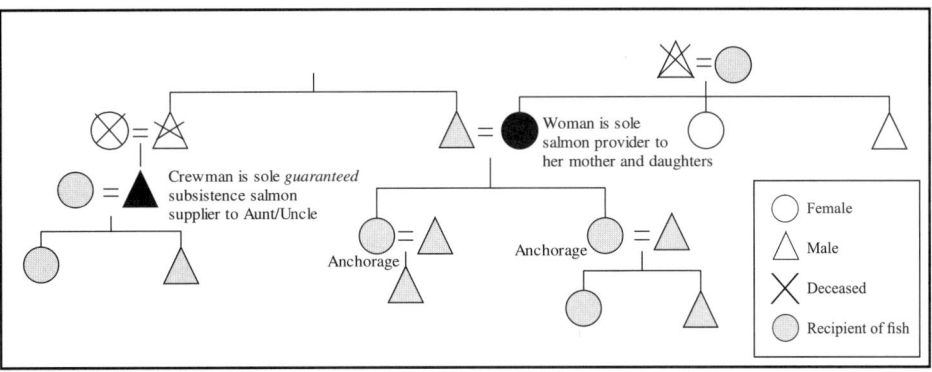

Figure 5. *Crewman and his aunt as sole subsistence providers.*

In this case, none of the other households receive salmon from anywhere else, since her daughters live in Anchorage and their husbands do not fish, and her mother is an elderly widow. This crewman is the sole guaranteed salmon source for his aunt and uncle. Although this crewman's father had initially received two salmon permits during Limited Entry and had owned a boat, all have been sold over the years under circumstances undisclosed to me, and his parents are now both deceased. This crewman has also had his share of trouble with the law, which makes his ability to supply his family with fish even more uncertain. He is not married to the mother of his children, and was released from jail on the condition that he continues fishing because he is required to support them. His uncle has not been employed as a crewman for several years due to health problems. His aunt holds a full-time job at a local store, pays the bills, and keeps food on the table when subsistence foods are not easily obtained. She uses a broad range of species beyond salmon in the sharing, broader than in wealthier homes where salmon is guaranteed in abundance. Other sources of fish species come into her household from a variety of sources, but this is inconsistent (Fig. 6).

In this case, the primary receivers are her relatives, but she also shares freely with neighbours, friends, those who drop in for a beer, and this anthropologist. Sharing is expected between some family members, but is only partially prescribed in that surpluses are generally shared with family members first, but portions of fish or finished products frequently go to several different households from one fisherman or crewman.

Men returning from the sea with a bounty of fish are also bringing in material affirmations of status, and they can earn prestige through sharing. Fish pass largely from men to women, who convert them into a variety of products, which are then shared. This system is tenuous for some, guaranteed for others, but provides the foundation of social relations. Women constantly remind the men in their lives of the necessary requirements.

Women Politicians

I now briefly turn to the formal political arenas in which the future of Aleut commercial and subsistence fishing practices stands precariously on the shoulders of a few savvy women and men. Firstly, I must return to the political climate outside the Eastern Aleut

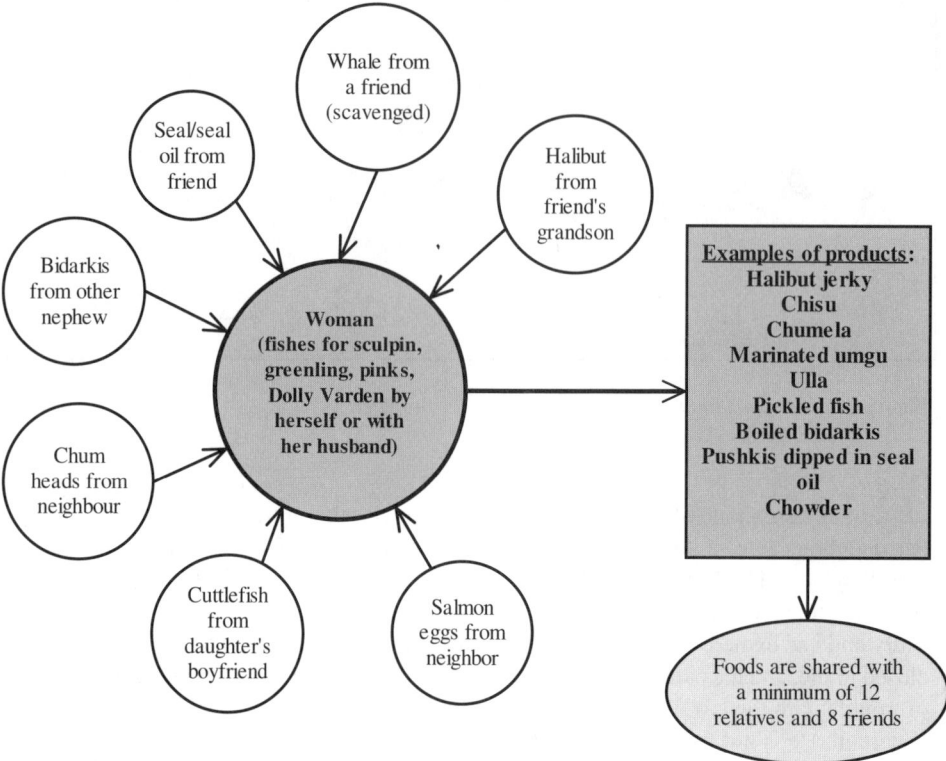

Figure 6. *Other species moving in and out of one woman's household over the course of a year.*

villages that directly affects local practices and relations. The Aleut fish a region designated 'Area M' by the Alaska State Board of Fisheries. This term has become synonymous with their summer salmon fishery, but is likewise used to describe the region and its people by both those living within Area M and those across the state who claim to have an interest in their fishery. Public opinion, media, state government, anthropologists, conservationists, and western Alaskans in particular, have portrayed these Aleut fishermen as non-Native non-residents stealing the fish that are ultimately bound for their rivers of origin to spawn in western Alaska and Asia (*e.g.*, Gay ADN 2/26/2004, 3/2/2004, 4/28/2004; Hensel 1996, 2001; Johnson 1997; Knowles 2000a,b; Loy ADN 1/31/2001, 2/2/2001, 1/13/2002; Ruskin ADN 7/20/2000). Politically motivated claims against this Aleut fishery have provided state and federal fisheries regulators with the justification to restrict, or close fishing grounds and the time allotted for fishing.

Fisheries meetings held in Anchorage during which many of these regulations are altered or imposed often coincide with opening seasons for various species, and men have typically been unwilling to sacrifice fishing time to spend the minimum of $1,000 and several days required to travel to Anchorage and to testify at these meetings. The majority of Aleut rely on a few to speak for all. Women stand at the political foreground because they have more willingness and freedom to travel to Anchorage for meetings when the men must fish. The Eastern Aleut have a plethora of political, social and economic organizations to contend with, so many that it is daunting for most people. "We have trouble getting men to talk," said a woman who manages her family's fisheries corporation. But for some of these women, there is a fearless way that they attack political disputes. "It's 'cause I don't know any better so I don't get intimidated," said one woman. Several Aleut women who engage in these struggles have, in fact, never fished from aboard a commercial vessel.

Within the villages, leaders tend to be from large, prominent families. Women control much of the supporting political structures as administrators, while men usually fill the actual posts of mayor, corporation president, and tribal council president, for example. Thus, women's power is individualized and diffuse within the villages, but they have a tremendous amount of power in state, national, and international forums. Sometimes, women from several villages will act collectively on Aleut issues. Women are the forerunners in working to set national-level policy on fisheries, management and allocation of resources, village infrastructure, and improving the value of their fish in processing and marketing. For example, the head of the federal Kodiak–Aleutians Regional Advisory Council is an Aleut woman; the Board of Directors for Aleutia, Inc., their regional marketing scheme, consists of several Aleut women; and the Aleut Marine Mammal Commission is managed principally by women. An Aleut woman led the unprecedented plan to bring fishermen from the Arctic–Yukon–Kuskokwim region to her village to learn how the fisheries operate, and hopefully find common ground in the fight over salmon rights.

The gendered political arena appears to be complementary, yet, despite their ambition and high visibility in politics, women still choose their battles based on what their husbands, fathers, uncles and grandfathers want and need. They will often defer to the fishermen's desires on particular policy issues, even if the link between fishing and the issue at hand is not clear. At the same time, political and economic challenges in fisheries are treated as Aleut concerns. Fisheries are extremely volatile and women, men and children are strongly affected when access to or the health of the fishery changes. Thus, the stakes are community-wide.

Conclusion

Although there is a division of labour such that typically men fish, hunt, hold political office and own boats and permits, and women collect bidarkis and clams, prepare and preserve fish and game, share wild foods and manage offices, men and women cross over and blur these lines in constant motion. Women are not homogeneous in any specific arena of life: some women stay home and process foods and raise children; others hold a job and process foods and manage the household; others are intensely involved in village and global politics; others own boats and/or permits and fish regularly. Men often pick bidarkis; kipper, smoke and jar salmon; share fish and game; go berry picking; and testify at resource management meetings, for example, and thus are engaging in a broad range of activities.

Most women do not want to be equal to men in fishing because the social rewards are not the same for them. Ortner and Whitehead (1981) argue that gender is a prestige structure where women's roles, activities, and products of their labour are generally accorded less prestige than their male counterparts. Black (1998) adds that species hunted by men carried symbolic significance whereas species hunted and gathered by women and children are seen as utilitarian. This model is partially true for these villages where men gain or lose status in providing fish and game, and women gain or lose status in food preparation, in the quality of the final products, and in sharing. The status of women is not, however, as vulnerable to the volatility of fishing success because of the diversity of roles and responsibilities among them. Women nonetheless maintain a close watch on the fishermen in their lives, since their identities are linked to the status of 'their' fishermen.

Being a capable fisherman is the essence of being an Aleut man. Adult men spend the majority of their time working on their boats both in the harbour and at sea, and must continuously demonstrate their abilities as fishermen. Status is maintained through continuous hard work and success. Women support their activities by providing supplies to the boat, taking care of in-home responsibilities, and fiercely fighting for their rights to continue fishing. Women also hold land-based jobs to provide money for fishing as well as for their families, especially in lean years. Occasionally they will fish with their husbands, but they often wait for him to send them fish to process and/or distribute. Men and women employ a wide range of hunting, fishing and processing techniques that they have learned throughout their lives, and although most of these chores are gender-marked, a rigid sexual division of labour is neither possible nor desirable.

Much of the feminist anthropological literature has emphasized a 'male bias' paradigm suggesting that women are often 'forced' to express themselves through the dominant male ideologies, and that women are 'muted' because of this (*reviewed in* Moore 1988:3). This is certainly a recognized, empirically documented trend for many parts of the world; however, this does not begin to capture the situation for Aleut women. Here, the identity of being Aleut is inextricably intertwined with the fishing economy, the most visible aspects of which are largely practised by male Aleuts, but through which the behaviours and attentions of women give the whole system its meaning. Fishing is recognized and affirmed by women as having status, and prestige categories are centred around this. We cannot assume that a 'male' prestige system is exclusively male, since women choose mates based upon certain criteria, and their choices speak volumes, signalling to men which behaviours and attitudes are desired. Within the villages, women are crucial to maintaining the status of their fishing spouses, kin, and themselves through their attitudes, political pursuits, and especially in supplying salmon and other products to their subsistence networks.

Acknowledgements: This research was supported by the Wyse Fieldwork Grant and the Richards Fund of the Department of Social Anthropology; the H. M. Chadwick Fund from the Department of Anglo-Saxon, Norse and Celtic; and the Ridgeway-Venn Travel Studentship, all of the University of Cambridge, UK. The bulk of this work was made possible by the National Science Foundation, Arctic Social Sciences grant OPP-0094826, and the Department of Anthropology, Idaho State University. Any opinions, findings, conclusions, or recommendations expressed in this chapter are those of the author and do not reflect the views of the National Science Foundation. I thank my Ph.D. supervisor Barbara Bodenhorn for her guidance. Special recognition goes to the people of the Aleutians East Borough, Alaska, and Peter Pan Seafoods, Inc.

References

Acheson, J. (1981). Anthropology of Fishing. *Annual Review of Anthropology* 10:275-316.

Allison, C.S. Jacobs and M. Porter (1989). *Winds of Change: Women in Northwest Commercial Fishing.* Seattle: University of Washington Press.

Allison, C. (1988). 'Women Fishermen in the Pacific Northwest,' pp. 230-260 in J. Nadel-Klein and D.L. Davis, eds. *To Work and to Weep: Women in Fishing Economies.* St. John's: ISER, Memorial University of Newfoundland.

Binkley, M. and V. Thiessen (1988). 'Ten days a "grass widow"—forty-eight hours a wife: sexual division of labour in trawlermen's households.' *Culture* 8(2): 39-50.

Black, L. (1998). Animal world of the Aleuts. *Arctic Anthropology* 35(2): 126-35.

Black, L. (1984). Atka: An Ethnohistory of the Western Aleutian Islands. *Alaska History,* No. 24. R.A. Pierce, ed. Kingston, Ontario: The Limestone Press.

Bodenhorn, B. (1989). *"The animals come to me; they know I share": Iñupiaq kinship, changing economic relations and enduring world views on Alaska's North Slope.* Unpublished doctoral dissertation, Cambridge University.

Bodenhorn, B. (1990). "I'm Not the Great Hunter, My Wife Is": Iñupiat and anthropological models of gender. *Études/Inuit/Studies* 14(1-2): 55-74.

Bodenhorn, B. (1997). 'Person, Place and Parentage: Ecology, Identity and Social Relations on the North Slope of Alaska,' pp. 103-132 in S.A. Mousalimas, ed. *Arctic Ecology and Identity.* International Society for Trans-Oceanic Research 8. Bupest: International Society for Trans-Oceanic Research.

Braund, S. & Associates (1986). *Effects of Renewable Resource Harvest Disruptions on Community Socioeconomic and Sociocultural Systems: King Cove.* Report for the U.S. Department of the Interior, Minerals Management Service, Alaska OCS Region, Anchorage, Alaska. Social and Economic Studies Program Technical Report No. 123, 419 pp.

Brettell, C. and C. Sargent, eds. (2000). *Gender in Cross-Cultural Perspective.* Upper Saddle River, N.J.: Prentice Hall.

Burch, E. S., Jr. and L. Ellanna, eds. (1994). *Key Issues in Hunter-Gatherer Research.* Explorations in Anthropology Series. Oxford: BERG.

Census 2000. (2000). *United States Federal Census.* Electronic document: www.census.gov.

Chapman, M. (1987). Women's fishing in Oceania. *Human Ecology* 15(3): 267-88.

Collier, J. and M. Rosaldo (1981). 'Politics and gender in simple societies,' pp. 275-329 in S. Ortner and H. Whitehead, eds. *Sexual Meanings: The cultural construction of gender and sexuality.* Cambridge; New York: Cambridge University Press.

Dall, W.H. (1870). *Alaska and Its Resources.* Boston: Lee and Shepard.

Dumond, D.E. (1987). *The Eskimos and Aleuts.* London: Thames and Hudson Ltd.

Ellis, M.E. ed. (1977). *Those Who Live From the Sea: A study in Maritime Anthropology.* New York: West Publishing Company.

Fields, L.L. (1997). *The Entangling Net: Alaska's Commercial Fishing Women Tell Their Lives.* Urbana: University of Illinois Press.

Fienup-Riordan, A. (1983). *The Nelson Island Eskimo: Social Structure and Ritual Distribution.* Anchorage: Alaska Pacific University Press.

Fienup-Riordan, A. (1990). *Eskimo Essays: Yup'ik Lives and How We See Them.* New Brunswick, NJ: Rutgers University Press.

Fienup-Riordan, A. (2000). *Hunting Tradition in a Changing World: Yup'ik Lives in Alaska.* Rutgers University Press.

Fricke, P., ed. (1973). *Seafarer and Community.* London: Croom Helm.

Gay, J. (2004a). Fish Board ruling ignites Area M fervor. *Anchorage Daily News.* February 26.

Gay, J. (2004b). Area M fishing isn't over yet, fishermen say. *Anchorage Daily News.* March 2.

Gay, J. (2004c). Area M testimony spills over. *Anchorage Daily News.* April 28.

Golder, F.A. (1963a). Aleutian Stories. *Journal of American Folklore* 18(70): 215-222.

Golder, F.A. (1963b). The songs and stories of the Aleuts, with translations from Veniaminov. *Journal of American Folklore* 20(76): 132-142.

Goldsmith, S. (1979). *Man-in-the-Arctic Series Documentation.* University of Alaska, Anchorage: Institute for Social and Economic Research, 341 pp.

Hamilton, L. and C. Seyfrit (1994a). Coming out of the Country: Community Size and Gender Balance among Alaska Natives. *Arctic Anthropology* 31(1): 16-25.

Hamilton, L. and C. Seyfrit (1994b). Female Flight? Gender Balance and Outmigration by Native Alaska Villagers. *Arctic Medical Research* 53 (Supplement 2): 189-193.

Hensel, C. (1996). *Telling Our Selves: Ethnicity and Discourse in Southwestern Alaska.* Oxford.

Hensel, C. (2001). Yup'ik identity and subsistence discourse: Social resources in interaction. *Études/Inuit/Studies* 25(1-2): 217-227.

Jochelson, W. (1933). *History, Ethnology, and Anthropology of the Aleut.* The Netherlands: Oosterhout, N.B. Carnegie Institution of Washington. Publication 432.

Johnson, C. (1997). 'The Role of Indigenous Peoples in Forming Environmental Policies,' pp. 1-2 in E.A. Smith and J. McCarter, eds. *Contested Arctic: indigenous people, industrial states, and the circumpolar environment.* Seattle: University of Washington Press.

Jones, D.M. (1976). *Aleuts in Transition: a Comparison of Two Villages.* Seattle: University of Washington Press.

Kelly, R. (1995). *The Foraging Spectrum: Diversity in Hunter-Gatherer Lifeways.* Smithsonian Institution Press.

King, M.H. (1992-3). A partnership of equals: women in Scottish east coast fishing communities. *Folk life* 31:17-35.

Knowles, (former) Governor T. (2000a). *Declaration of Disaster.* 7/19/00. Electronic document: www.state.ak.us.

Knowles, (former) Governor T. (2000b). *Letter to Alaska Board of Fisheries Chairman Dan Coffey.* 8/9/00. Electronic document: www.state.ak.us.

Langdon, S.J. (1980). *Transfer Patterns in Alaskan Limited Entry Fisheries. Final Report for the Limited Entry Study Group of the Alaska State Legislature.*

Lantis, M. (1970). The Aleut Social System, 1750 to 1810, from early historic sources. Pp: 139-172. In *Ethnohistory in Southwestern Alaska and the Southern Yukon: Method and Content.* M. Lantis, ed. Studies in Anthropology 7. Lexington, KY: University Press of Kentucky.

Lantis, M. (1984). 'Aleut,' pp. 161-184 in D. Damas, ed. *Handbook of North American Indians, vol. 5: Arctic.* Washington, DC: Smithsonian Institution Press.

Laughlin, W.S. (1963). Eskimos and Aleuts: their origins and evolution. *Science* 142 (3591): 633-645.

Laughlin, W.S. (1980). *Aleuts: Survivors of the Bering Land Bridge.* New York: Holt, Rinehart and Winston.

Laughlin, W.S and J.S. Aigner (1975). 'Aleut adaptation and evolution,' pp. 181-201 in W. Fitzhugh, ed. *Prehistoric Maritime Adaptations of the Circumpolar Zone.* Chicago: Aldine.

Leacock, E. (1981). *Myths of Male Dominance: Collected articles on women cross culturally.* New York: Monthly Review Press.

Leacock, E. and R. Lee, eds. (1982). *Politics and History in Band Societies.* Cambridge University Press.

Lee, R.. (1993). *The Dobe Ju/'hoansi.* Fort Worth, TX: Harcourt Brace.

Lee, R. and I. DeVore, eds. (1968). *Man the Hunter.* Chicago: Aldine.

Lepowsky, M. (1993). *Fruit of the Motherland: Gender in an egalitarian society.* New York: Columbia University Press.

Liapunova, R.G. (1996). *Essays on the Ethnography of the Aleuts (at the End of the Eighteenth and the First Half of the Nineteenth Century.* J. Shelest, trans. W. Workman and L. Black, eds. Rasmuson Library Historical Translation Series, Vol. 9. Fairbanks: University of Alaska Press.

Llewelyn-Davies, M. (1981). 'Women, warriors, and patriarchs,' pp. 330 - 258 in S. Ortner and H. Whitehead, eds. *Sexual Meanings: the cultural construction of gender and sexuality.* New York; Cambridge: Cambridge University Press.

Lord, N. (1997). *Fishcamp: Life on an Alaskan Shore.* Washington, D.C.: Counterpoint.

Loy, W. (2001a). Area M fishing gutted. *Anchorage Daily News.* January 31.

Loy, W. (2001b). Fish board curbs fishermen in False Pass. *Anchorage Daily News.* February 2.

Loy, W. (2002). On the rocks. *Anchorage Daily News.* January 13.

MacCormack, C. and M. Strathern, eds. (1981). *Nature, Culture and Gender.* New York; Cambridge: University of Cambridge Press.

Maschner, H. (1999a). Prologue to the Prehistory of the Lower Alaska Peninsula. *Arctic Anthropology* 36(1-2): 84-102.

Maschner, H. (1999b). 'Sedentism, Settlement and Village Organization on the Lower Alaska Peninsula: A Preliminary Assessment,' pp. 56-76 in B. Billman and G. Feinman, eds. *Settlement Pattern Studies in the Americas: Fifty Years since Viru.* Washington: Smithsonian Institution Press.

Maschner, H. (2000). 'Catastrophic Change and Regional Interaction: The Southern Bering Sea in a Dynamic World System,' pp. 252-265 in M. Appelt, J. Berglund and H. C. Gulløv, eds. *Identities and Cultural Contacts in the Arctic. Proceedings from a Conference at the Danish National Museum, Copenhagen, November 30 - December 2, 1999.* Copenhagen: Danish National Museum and Danish Polar Center.

Maschner, H. and B. Hoffman (2003). The Development of Large Corporate Households along the North Pacific Rim. *Alaska Journal of Anthropology* 1(2): 41-63.

Maschner, H., J. Jordan, B. Hoffman, and T. Dochat (1997). *The Archaeology of the Lower Alaska Peninsula.* Report 4 of the Laboratory of Arctic and North Pacific Archaeology, Madison: University of Wisconsin.

McCartney, A. (1984). 'Prehistory of the Aleutian region,' pp. 119-135 in D. Damas, ed. *Handbook of North American Indians, vol. 5: Arctic.* Washington: Smithsonian Institution Press.

Moore, H. (1988). *Feminism and Anthropology.* Minneapolis: University of Minnesota Press.

Morgan, L. (1976). Anchorage and Fairbanks—the Biggest Native Villages of them all. *Alaska* 42(3): 33-37, 77-79.

Myers, F. (1988). Critical trends in the study of hunter-gatherers. *Annual Review of Anthropology* 17: 261-282.

Nadel-Klein, J. (2003). *Fishing for Heritage: Modernity and Loss along the Scottish Coast.* Oxford: Berg.

Nadel-Klein, J. and D.L. Davis, eds. (1988). *To Work and To Weep: Women in fishing economies.* Social and Economic Papers No. 18, St. John's: Institute for Social and Economic Research, Memorial University of Newfoundland.

Ortner, S. (1974). 'Is female to male as nature is to culture?,' pp. 67-88 in M. Rosaldo, and L. Lamphere, eds. *Women, Culture and Society.* Stanford: Stanford University Press.

Ortner, S. and H. Whitehead (1981). 'Introduction: Accounting for Sexual Meanings,' pp. 1-27 in S. Ortner and H. Whitehead, eds. *Sexual Meanings: The Cultural Construction of Gender and Sexuality.* New York; Cambridge: Cambridge University Press.

Radcliffe-Brown, A.R. (1922). *The Andaman Islanders.* New York: The Free Press.

Rasing, W.C.E. (1994). *'Too Many People': Order and Nonconformity in Iglulingmiut Social Process.* Reeks Recht & Samenleving No. 8. Cip-Gegevens Koninklijke Bibliotheek, Den Haag.

Reedy-Maschner, K. (2004). *Aleut Identity and Indigenous Commercial Fisheries.* Ph.D. dissertation. Department of Social Anthropology, University of Cambridge, UK.

Robert-Lamblin, J. (1982a). Woman's Role and Power Within the Traditional Aleut Society. *Folk* 24: 197-202.

Robert-Lamblin, J. (1982b). An historical and contemporary demography of Akutan, an Aleutian village. *Études/Inuit/Studies* 6(1): 99-126.

Rosaldo, M. and L. Lamphere, eds. (1974). *Women, Culture and Society.* Stanford: Stanford University Press.

Ruskin, L. (/2000). Y-K, Norton Sound runs declared state disaster. *Anchorage Daily News.* July 20.

Sacks, K. (1979). *Sisters and Wives: The Past and Future of Sexual Equality.* Contributions in Women's Studies, No. 10. Westport, CT: Greenwood Press.

Shields, Captain E. (2001). *Salt of the Sea: The Pacific Coast Cod Fishery and the Last Days of Sail.* Lopez Island, WA: Pacific Heritage Press.

Sinclair, P.R. and L.F. Felt (1992). *Separate worlds: gender and domestic labour in an isolated fishing region.* Canadian Review of Sociology and Anthropology. 29(1): 55-71.

Skaptadottir, U. (1996). Gender construction and diversity in Icelandic fishing communities. *Anthropologica* 38(2): 271-87.

Thiessen, V., A. Davis and S. Jentoft (1992). The veiled crew: an exploratory study of wives' reported and desired contributions to coastal fisheries enterprises in northern Norway and Nova Scotia. *Human organization* 51(4): 342-52.

Townsend, J.B. (1980). *Ranked Societies of the Alaskan Pacific Rim.* Senri Ethnological Studies 4: 123-156.

Townsend, J.B. (1983). 'Pre-contact Political Organization and Slavery in Aleut Societies,' pp. 120-132 in E. Tooker, ed. *The Development of Political Organization in Native North America.* Proceedings of the American Ethnological Society, 1979. Washington, D.C.: American Ethnological Society

Veniaminov, I. (1840). *Notes on the islands of the Unalaska District, Volumes I, II, III.* St. Petersburg, Russia: Russian-American Company.

Veniaminov, I. (1984). 'Notes on the islands of the Unalaska District [1840],' L. Black and R.H. Goeghega, trans. R.A. Pierce, ed. *Alaska History* 27. Kingston, Ont.: The Limestone Press.

CHAPTER TWO

"Without Fish We Would No Longer Exist": The Changing Role Of Women In Southeast Alaska's Subsistence Salmon Harvest

Virginia Mulle and Sine Anahita

Abstract: When addressing subsistence activities, the primary focus of social scientists has been the role of men. The purpose of this study is to document the role of urban Native women in contemporary subsistence activities and to examine the relationship between modernization and its impact on the traditional roles of women. The women interviewed have perceived a change in the gendered division of labour characterized by the increasing participation of women, which is primarily due to the influence of factors external to Native society and consistent with trends toward the modernization of society. These women see increased involvement as a result of more general societal changes in which women have become more active participants in their society. The women's movement, supporting the equality of women in all spheres of contemporary society, has played a significant role in the beliefs of the women interviewed toward the increasing participation of women in the contemporary salmon harvest. However, for these urban Native women, there remains a continuous struggle between the desire to practice their traditional ways of life and the pull of modernization.

Introduction

While the roles of men and women in subsistence economies have been addressed in the social science literature (Krause 1956; Oberg 1973; Klebnikov 1976; Klein 1980; de Laguna 1983; Emmons 1991; Moss 1993; Betts 1994; Dauenhauer and Dauenhauer 1994; Arnold 1997; Goldschmidt and Haas 1998), the specific roles of women have not been emphasized nor clearly articulated. When writing about and discussing subsistence activities, the primary focus has been on men, and women's roles have been embedded in the history of male activities and practices. Very little research on the subsistence salmon fisheries harvest in southeast Alaska (primarily conducted by the Alaska Department of Fish and Game, Division of Subsistence) has focused exclusively on the roles that women play in the traditional and customary use of subsistence resources. In her research in a Tlingit community in southeast Alaska, Klein (1976) found that people in the community believed that the study of Tlingit women was both welcome and needed. She wrote "an older woman [...] contended that Tlingit women have been de-emphasized in this literature and 'you can't begin to understand the Tlingits unless you know the importance of the women.'" Kan (1996) has argued that there has been a general lack of attention paid to the cultural and historical experiences of Native North American women, or that these experiences have been subsumed under the male domain (Fisher 1999).

The purpose of this research study was to begin an exploration of the roles of women in contemporary subsistence activities, and, in particular, the impact of modernization on women's traditional roles in relation to the subsistence salmon harvest. Twenty-five urban southeast Alaska Native women were interviewed in Juneau, Alaska in 2003, by the same trained interviewer.[1] Approval was obtained by the Institutional Review Board of the sponsoring institution and informed consent forms were signed by all participants. An open-ended survey questionnaire was used, which included demographic information, questions regarding the women's participation in the preparation, harvesting, processing and distribution of the salmon, as well as regarding their perceptions of change in participation patterns. The open-ended questionnaire allowed the women to tell their stories in their own words in a semi-structured format. Each interview lasted from one to three hours. Highlighting the process of change that has occurred in women's roles in contemporary subsistence activities will serve to shed further light on women's changing roles in general.

The importance of the relationship of the Tlingit people to salmon in southeast Alaska is well documented (de Laguna 1983; Emmons 1991; Dauenhauer and Dauenhauer 1994). Salmon has been the major resource around which the economy has revolved, and it has long constituted the principal item in their diet. As the most significant source of all subsistence activities, salmon has been intertwined with the political, spiritual and social relationships of the Tlingit. The earliest account of a Tlingit salmon fishery (described as 'abundant') was in 1786 by the French explorer Jean Francois Galaup de la Perouse off Yakutat Bay in Lituya Bay (Price 1990). The strong bond connecting the Tlingit to their marine environment, and more specifically to salmon, endures today.

Women's Roles in Production

In traditional subsistence economies, items for household use such as food, clothing, and shelter, are produced by members of the household. Because the extended family is the work unit, subsistence production is shared equally by men and women (Lorber 1994). Women's and men's work in subsistence economies may be different and organized by

---

[1] *Federal Register* (Volume 72, No. 87, May 7, 2007), pp. 25689-25690. Under Federal regulations, population size is a fundamental distinguishing characteristic between rural and non-rural (urban) communities. A community with a population of 2,500 or less is deemed rural, unless it possesses significant characteristics of a non-rural (urban) area, or is considered to be socially, economically, and communally part of a non-rural (urban) area. A community with a population of more than 7,000 is presumed non-rural (urban), unless it possesses significant characteristics of a rural nature. A community with a population above 2,500 but not more than 7,000 is evaluated to determine its nonrural (urban) status. *Alaska Administrative Code.* Title 5.99.015 Joint Board non-subsistence areas. Under the State of Alaska subsistence fish and game regulations, urban areas in Alaska regulatory language nonsubsistence areas, are determined by the Alaska joint board of fish and game. The joint board has identified the boundaries of non-subsistence areas. A non-subsistence area is an area or community where dependence upon subsistence is not a principal characteristic of the economy, culture, and way of life of the area or community.

gender, but among Arctic Native peoples who continue to rely on subsistence production, gendered labour is generally complementary (Williamson 2004), and the tasks expected of women and men are interdependent (Bodenhorn 1990). The complementary division of labor along gender lines can be seen in Arctic women's historical contribution of wild foodstuffs to family and community tables, such as greens, mushrooms, berries, bird eggs, roots, small game, and fish, while men typically provide large game animals, such as moose, caribou, or whales. Even in economies that rely on men's labor to provide large game, women provide critical logistical labour, organizing travel, butchering and processing kills, and distributing food (Bodenhorn 1990; Jarvenpa and Brumbach 1995).

As capitalism and industrialization expanded in the late nineteenth century, subsistence products became commoditized, and women and men began to work for wages in order to purchase the products that were formerly produced in households (Lorber 1994). Production in the circumpolar region was also affected by the expansion of capitalism and industrialization. For example, commercial fisheries in British Columbia in the late nineteenth century particularly exploited Native women and children, drawing them from their homes to work in salmon canneries at the lowest wages possible (Muszynski 1988). In mixed economies, which characterize many Alaska Native groups, subsistence production is currently mixed with wage labour, which is sharply gendered (Kleinfeld and Andrews 2006).

Women's Activities Related to Preparing to Fish and Fishing

Juneau, Angoon, Klukwan, Hoonah, Sitka and Metlakatla were among the many communities in southeast Alaska in which the twenty-five women involved in this study fished (Fig. 1). Ninety-six percent of the women participants in the study reported that they fished with others, particularly family and friends, and that they fished several times each summer. Four percent also engaged in commercial fishing. The gear used by the women fishers was primarily nets (73%), and rod and reel (45%), or a combination of both.

Prior to fishing, women engaged in shopping activities and food preparation. Seventy-five percent of the women fishers shopped for gear, bait and food, and prepared food and drink to take fishing. They got clothes organized for fishing and made arrangements with others to fish. While fishing, the women baited hooks and cooked on board. None of the women stated that they actually 'got' fish. Rather, they engaged in what could be described as women's 'helping work,' while the men were involved in 'getting the fish.' Even if not fishing themselves, women helped others get ready to fish. Similar to the preparation activities of women who did fish, non-fishers shopped for supplies and gear and prepared foodstuffs, snacks, drinks and lunches for the fishers to take with them. When friends and family returned from fishing, almost all of the fishers and non-fishers prepared food. Some women mentioned a practice in which they engaged while men were out fishing that relates to a traditional role of women in the fisheries harvest. Some "offered verbal encouragement" to their husbands before they left for fishing, and others did "not think of him while he's gone." In her ethnographic study of the hooligan[2] fishery in southeast Alaska, Betts (1994) noted that women have historically sewn net frames

---

[2] 'Hooligan,' also referred to as 'eulachon' or 'oolichan,' is a small anadromous ocean fish.

Map courtesy of: Alaska Department of Fish and Game, Division of Subsistence

Figure 1: *Map of Southeast Alaska*

and woven mesh for dip nets, however, none of the women interviewed engaged in these activities.[3]

The women fishers all go to the same area to fish. Sixty percent of the women fished in traditional areas that their family has used for generations. Historically, Native fishers in southeast Alaska fished in their clan salmon streams (Oberg 1973), waters in which resource harvest and processing took place and for which the matrilineal clans claimed ownership (Betts 1994). These salmon streams were strictly guarded and regulated by clans and clan house leaders, and any violation of the clan ownership could lead to serious conflict (Arnold 1997).

This contemporary role of women in processing fish is consistent with the traditional work of women in the salmon fisheries harvest. Men caught the salmon, and women processed and prepared the fish. Traditionally, women removed the fish heads, cleaned, split, filleted and stripped the salmon (Oberg 1973). Today, women continue to play important roles in the subsistence production of finished salmon, as they contribute their labour in processing the resource. The primary processing work in which the women engaged were cooking the salmon to eat fresh, and freezing it. Seventy-five percent of

---

[3] George Emmons, *The Tlingit Indians.* Edited with additions by Frederica de Laguna. (American Museum of Natural History, Anthropological Paper No. 70. Seattle: University of Washington Press and the American Museum of Natural History 1991).

Figure 2.
Women processing fish.

the women canned or jarred the fish; just over 50% smoked the fish, and 33% dried it. Traditional methods of processing fish were to boil, dry, and smoke it (Olson 1967; Oberg 1973; Klein 1980; Emmons 1991; Arnold 1997). In his ethnographic work, Emmons (1991) related tales of fish being smoked in southeast Alaska as far back as 1787, and Dombrowski (2001) reports a smokehouse present in Kake in 1910. Today, the most frequently reported method of processing the fish is to freeze it. Fewer women engaged in the more traditional methods of processing as no women boiled the salmon; few smoked the salmon, and even fewer dried it. The mean age range of the women who smoked and dried the salmon was the early 40s (age range was 32-57 years); very few of the younger women either smoked or dried the fish. Freezing fish through vacuum packing is a method of processing more adaptable to a modern, contemporary lifestyle, and was the preferred processing activity of the women, seemingly to the exclusion of more traditional methods.

Women also engage in other processing work including cleaning and sterilizing jars, making brine, salting and pickling the fish; cleaning racks, gathering wood and keeping the smokehouse fire going; and labelling and packaging salmon to send to others. While Arnold (1997) has reported that it was the men who traditionally presided over the smoking of fish, many of the women interviewed said they started and/or maintained fires in the smokehouse. In addition, Betts (1994) noted that women have directed the hooligan smoking process, including making adjustments to the fires.

Historically, women have engaged in processing work in the fisheries harvest, even as the commercial fishing industry in southeast Alaska modernized. Throughout the boom period of the fish cannery industry in southeast Alaska, from the 1880s through the beginning of World War I, women worked in processing the fish while men fished on the cannery boats. Working on the 'slime lines,' women cleaned the fish and arranged for its processing (Dombrowski 2001). Today, they continue to engage in similar activities in the contemporary salmon harvest.

An area in which the experience of women in this historical commercial fishing industry parallels that of working women today, is in regard to pay. With both men and women working ten-hour days on the slime line in the early 1890s, women were paid $1.00 a day, and men were paid between $1.25 and $1.50 (Arnold 1997). Johnson (1981) stated that, for doing the same processing work early in the twentieth century, women received $1.50 a day, and men received $3.00 a day. Klein (1976) has reported that the monetary value placed on fishing, the men's work, was higher than the monetary value placed on the cannery work of women. It is interesting to note that this pay gap is 20% to 23%, which is nearly identical to today's average pay gap (Weinberg 2004).

In their processing work, all women used the meat or body of the fish, and 45% used the heads and eggs. Fifty-three percent of the women threw the parts of the fish that they didn't use into the garbage, while the rest (47%) threw the parts back into the water to 'complete the natural cycle.' This was particularly interesting to note, as the traditional use pattern was to return the fish parts to the water. According to Tlingit mythology, if salmon bones were returned to the water after the flesh was utilized, the spirit of each fish would return to its underground home, regain its human form, and return again the next year. If the bones were thrown away on land, the spirit of the salmon person could not return, and if some bones did not make it back to the water, the salmon person when resurrected, might be missing a leg or arm, and in his anger would not return to the stream where he had suffered mistreatment (Arnold 1997). Another traditional use of the salmon was to make medicine. While none of the women said that they made medicine, some said that "salmon is medicine." More of the women appeared not to adhere to traditional use patterns than those who did adhere to them.

Women's Activities Related to Distribution of the Fish: Sharing

Salmon or salmon products were shared primarily with family and friends as well as with elders, widows, and single parents. One single parent stated that while she did not share salmon, others shared with her. Women who shared their salmon did not expect something back in return for the fish when it was given, but for some, there was an expectation or understanding that something would be given back at a later time. Only one woman did expect something in return for her fish. All of the women were involved in decision-making regarding sharing. Seventy percent of the women alone made the decision about who the fish would be shared with. Women also made the decision in consultation with their husband or partner, their mother, or their children. This pattern of female decision-making is consistent with Betts' (1994) report on the hooligan harvest in the Chilkat

Valley in southeast Alaska, where she found that the senior female of the producing group took charge of decision-making regarding the distribution of the resource.

Women's Activities Related to Distribution of the Fish: Exchange

Half of the women exchanged fish for beadwork, seal and other fish. Fewer fishers (40%) exchanged their salmon than did non-fishers. Only 4% of the women sold fish to friends in other communities; she was a non-fisher and her uncle made the decision regarding how the fish would be sold. As with decision-making regarding with whom the salmon or salmon products were shared, all of the women who exchanged the fish were involved in the decision-making process. Seventy-seven percent of the women alone made the decision; others stated 'women' in general, or a specific woman, such as 'gramma,' made the decision.

The women were all personally involved in the negotiation process for the exchange of the fish. Eighty-four percent of the women alone negotiated the exchange; others stated that 'women' in general engaged in the negotiation process. A review of studies that address the traditional patterns of trade, exchange and negotiation of goods in southeast Alaska Native communities indicates that women are frequently cited as having had great influence over the men (Klebnikov 1976; Klein 1976, 1980, 1995). In his work in southeast Alaska, Oberg (1973) found that frequently a 'shrewd old woman' was taken along with the men when trading in order to keep a check on exchange values. According to Klein (1995), as late as 1882, European explorers reported that 'a single woman' frequently went with the men to ensure that the male traders got a good return for their goods. There are several reports where women in fact vetoed deals that were made by the men, who then had to return to cancel the exchange they had negotiated. There are also reports of early traders who would demand that women be absent during trade negotiations, as they believed the presence of the women put the traders at a disadvantage (Krause 1956; Klein 1980). While these studies provide evidence that women were engaged in the negotiation process, they suggest that they were informed 'behind the scenes' players during the negotiation process, while the men were dominant. There is, however, some indication that women were more active participants in the process, and actually engaged in direct negotiations with early traders (Krause 1956; Klein 1976; de Laguna 1983).

The pattern discerned in this study, where women were active participants in the negotiation process, suggests a role change for women from traditional exchange patterns. Historically, women's involvement in exchange and negotiation was most frequently as a private advisor. The change to a public role today may reflect the effect of modernization on traditional gender roles where women have become more active participants in society at large, including engagement in business and commerce.

Potlatches, community dinners, and fundraisers are among the special occasions where women donate their salmon. Along the northwest coast of North America, ceremonial performances associated with declaration and confirmation of social status were called Potlatches, a term from the Chinook jargon, the lingua franca of the Northwest Coast. Ceremonial giving was a central feature of these events. Potlatches honoring the dead, funeral potlatches and 40-day parties are still an intimate part of Tlingit and Haida society and are regularly held throughout the region. Ceremonial giving of Native foods is a central feature of these events. For ceremonial occasions in Juneau and Ketchikan, Native foods are sent in from the villages. Wild foods commonly consumed during the ceremonial event, and also taken home include dried salmon, halibut, herring roe on hemlock branches and on kelp, venison, seal meat, seal oil, hooligan oil, and dried seaweed (Drucker 1955; deLaguna 1990; Turek 2006).

Eighty-eight percent of the women contributed salmon or salmon products to potlatches and 63% brought salmon to community events. In addition to donating the fish, three women served the fish at the event. Traditionally, food for the potlatches came from surplus that the women had put aside (Klein 1976). In this study, participatory activity in the potlatch differed by whether or not the woman was a fisher. The fishers brought their fish to the potlatches, and the non-fishers brought and cooked the fish, made salmon spread, and served at the potlatch. Kan (1989) found that historically, women primarily donated food to the potlatch, whereas contemporary women are involved in multiple tasks during potlatching.

Traditions Involving Women in the Fisheries Harvest

Traditional behaviours related to women in subsistence fishing harvest and hunting activities have been well documented in histories of Native southeast Alaska peoples (Kan 1989, 1996; Emmons 1991; Moss 1993; Betts 1994; Arnold 1997). These traditional practices had to do with women's behaviour and thoughts before, during, and after the hunt or harvest. For example, for four days prior to going hunting or fishing, a man could not have sexual relations with his wife (Emmons 1991). Many of the traditions surrounding women related to menstruation. It was believed that animals were repelled by menstrual blood, thus, women stayed away from both hunting and fishing implements as well as fresh salmon (Moss 1993; Betts 1994; Kan 1996; Arnold 1997). Menstruating women were considered dangerous, and it was believed that they would bring bad luck to male hunters and fishers (Kan 1989). While he was hunting or fishing, a man's wife was usually obligated to remain quiet, to do no work, to refrain from any angry act, and also possibly to fast, since her behaviour was supposed to influence that of the quarry pursued by her husband (Emmons 1991). There is also a tradition surrounding the cutting and preparation of the fish. When preparing the fish, women always cut from the vent, along the belly to the throat. If not cut in this manner, it was believed that the fish or their spirit would feel offended and would then desert the stream (Emmons 1991). Women butchered the salmon on cutting surfaces with their heads facing upstream while they themselves faced downstream, and hung the salmon on drying racks with the heads pointed upstream (Arnold 1997). Sixteen percent of the women spoken with did have some traditional knowledge regarding women and the fisheries harvest. Study participants included a mother-daughter pair wherein traditional knowledge had been passed from mother to daughter. Other women knew that they were not supposed to go near the fishing gear or boat when menstruating because it would cause bad luck, while others knew not to spit out the bones of the first fish of the season because they would be spitting away the luck of the men in their future fishing activities. One woman said that she could not think of the men while they were out fishing, and that she had to dedicate herself to processing the fish when the men returned from fishing. Interestingly, none of the literature examined revealed traditionally held beliefs about women not spitting out the bones of the first fish of the season, thinking of the men while fishing, or dedicating one's self to preparation activities once the men returned with the fish. It may be that this latter activity is more of an efficient behavioural expectation to get the fish taken care of rather than one of tradition. Also, none of the women had knowledge of the traditions associated with cutting patterns of the salmon.

The general lack of traditional knowledge concerning women and the subsistence fisheries harvest among the women interviewed may indicate a loss of traditional knowledge patterns among contemporary women. After thinking about any rules or traditions concerning fishing that relate specifically to women, a 43-year old woman said, "You know, I don't know. I'll have to ask mom." It was suggested by one woman that there are no longer 'superstitions' held concerning the disadvantage of having women present during the fisheries harvest. The transposition of the word 'superstition' for 'tradition' may note a shift to a more modern or contemporary way of thinking, and an effect of modernization on traditional ways of thinking. This word transposition may reflect an aspect of religious assimilation that occurred in Alaska, a result of the role played by Christian missionaries in shifting Alaska Native understanding of traditional practices from tradition to superstition.

Perceived Gender Role Change in Participation in Fishing

The women were asked if they thought there were more women involved in fishing today than when their mothers were growing up. If they believed that there were more women involved, they were also asked why they thought this change had occurred. Fifty-six percent of the women thought that there were more women involved in fishing today than when their mothers were growing up—a range of approximately 40 to more than 80 years ago—while 44% thought that women were less involved. Their responses differed by age, as 69% of the women under the age of 50 felt that women today were more involved in fishing than were their mothers, and 61% of women over the age of 50 believed that women today were less involved in fishing than were their mothers. A pattern emerged in that respondents thought that women were heavily involved in fishing prior to the 1940s, that participation then decreased until the 1970s when it began to increase, and that it continues to increase today. Several reasons were mentioned to explain this pattern, particularly regarding why women believed there are higher rates of participation today than in the past. The decline of fish camps, the fact that women in their mother's generation were sent to boarding schools where they were discouraged from continuing traditional practices, and that their mothers were living at a time when they were "learning to be Americans" were all cited as reasons for the decline of women's participation in fishing during the period between 1940 and 1980.

The observations of the women are supported by historical data. Since the 1880s when commercial fishing began and Native involvement in the Alaska fishery increased (especially among Native peoples in the southeast), women participated in the fisheries harvest through their involvement in the canning industry. Women worked in the commercial sector processing fish throughout the 1940s. During that time, subsistence fishing was subsumed by, or became part of commercial fishing activities and travelling to fishing camps declined. One 57-year-old woman said that her mother travelled to fish camps in the 1920s, and that the decline in that activity resulted in a decline in women's overall participation in fishing: "When my mom was growing up, they still went to fish camps to gather for the winter—when I was growing up and now, men and women don't go to camps anymore." Fish camps were semi-permanent family campsites usually occupied from June through October, depending on the timing of the salmon runs. Families moved to fish camps in the spring, fished through the summer and into the fall, and moved back to their winter villages after the last salmon runs. By the early 1900s, the U.S. Federal Government required that children be sent to schools located in the winter villages, and school boards charged twenty-five dollars for each student who didn't attend school. Families now found it difficult to leave winter villages for months at a time and less time was spent at fish camps. Women stayed home with the children so that they could attend school, and men traveled to fish camps alone. This contributed to the decline in women's travel to fish camps and the participation in the fisheries harvest. Traditional fish camps continued to be used regularly into the mid 1940s, but Natives were, by this time, intimately involved in the commercial fishery, spending more of their time fishing for wages or working in the canneries. After World War II, fish camps became even less important as the seasonal cycle revolved around commercial fishing. Fish camps were still being used to harvest subsistence salmon into the 1960s (Turek *et al*. 2007; Dombrowski 2001).

There was an overall decline in commercial fishing in southeast Alaska during and after WWII as stocks were depleted (Turek 2003). During this time, children were sent away from their homes to attend boarding schools where they could receive a 'modern' education. Part of this education included an acceptance of Western ideas with a concurrent devaluation of traditional beliefs and way of life. A woman whose mother had been raised in the 1940s said that women in her mother's generation were shipped off to boarding schools and discouraged from continuing traditional practices. Another woman mentioned that her mother, who grew up in the 1960s, had not been taught how to fish, "My mom's generation wasn't taught to go fishing, they were taught not to be Native, that included fishing."

The 1960s was a revolutionary period in the history of the United States, when the civil rights of previously marginalized societal groups were increasingly asserted and defended. As a result of the civil rights movement, women's and Native groups also began to assert their rights as fully participating members of society. In response to having been asked why she thought there were more women involved in fishing today, one woman responded that, "It was the civil rights movement. Women are engaging in more 'men-like' activities, women go fishing now and didn't then." A revitalization of interest in and emphasis on Native traditional culture took hold across the nation. Several women stated that when their mothers were growing up in the 1950s, Native people were focused more on learning contemporary Western, more modern ways of living, and that those ways did not include fishing. They did note, however, that today, "things were changing back." In addition to a revitalization of interest in traditional Native culture, women's awareness of their rights as citizens was also increasing as a result of a resurgence of interest and support for Native and women's rights. Among the urban Native women interviewed for this study, responses regarding women's roles in fishing were consistent with the general societal patterns during these time periods, particularly as they were affected by external forces of a rapidly advancing, modern society. The decline in commercial fishing, decreased participation in traditional fish camps, and mandatory attendance at boarding school led to a decline in women's participation in fishing. For many Native women in southeast Alaska, the revitalization of Native culture and the women's movement provided the impetus for both an increased awareness of the opportunities available to women in society, as well as a renewed interest in traditional Native culture.

Gender Roles and the Division of Labour

Studies that have recorded the roles of men and women in the subsistence fisheries harvest in southeast Alaska have strongly supported a traditional gendered division of labour in the harvest. From some of the earliest works such as Olson's (1967) ethnography describing the lives of the Tlingit around the turn of the nineteenth to twentieth centuries to the more recent work of Betts (1994), who recorded an account of the hooligan harvest activity in 1990-1991, one of the most consistent reports has been that of a traditional gendered division of labour in both subsistence and commercial enterprises, where men caught the fish and the women processed them (Olson 1967; Oberg 1973; Klein 1980; Emmons 1991; Arnold 1997; Dombrowski 2001). In his ethnography of the Tlingit in the early 1930s, Oberg (1973) reported that women took care to perform only those tasks that "belonged to them," and did not enter the men's sphere of work.

Only 22% of the women interviewed believed that the division of labour in the contemporary subsistence salmon harvest is strictly defined by gender roles, that gender roles today are not less defined than when their mothers were growing up, and that gender role participation in the subsistence salmon harvest has remained the same over generations. The eldest woman with whom I spoke (65 years old), did not think that women's participation in the harvest has changed, and that men still "get the fish," or "he does his role" and the women still "process the fish," or "she does her role."

Seventy-eight percent of the women felt that the division of labour in the contemporary salmon harvest is not strictly defined by gender roles, and that it is less defined today than when their mothers were growing up. The women believed that, in general, gender roles in contemporary society are not as strictly defined as they have been in the past and that it is now acceptable for women to participate in work that has traditionally been defined as men's work. One woman said that a strict gendered division of labour is "a thing of the past," and another said that women "are more independent." Other women said that "women do what they want when they want—it's okay for women to fish and process," "women are allowed to do men's jobs," and "we don't do things the same way we used to do them with the way things have changed....it's okay for women to fish." Overwhelmingly, the women credited social movements such as 'women's lib' for the changed status of women in society, both in Western society and their native Tlingit society. Dombrowski (2001) agrees that contemporary gender roles in Native societies have changed, and that women today play a more central role in the leadership of the family, and hold jobs that enhance their status.

Many of the women felt that it is more acceptable for women to fish today, because "women have more rights" and are able to "do what they want," as a result of the "women's movement." These remarks underlie the cultural shift that has taken place in society where women have been given more responsibility outside of the home, and that Native women have been part of this change. As described by one woman, they have experienced a change from "Tlingit culture" to a "modern day, urban culture." This suggests that there is a perception on the part of some Native women that the modernization of contemporary society has contributed to a shift in traditional gender roles such that there is no longer a strict division of labour in the subsistence salmon harvest in southeast Alaska.

However, 12% of the women interviewed viewed less defined gender roles as a return to traditional ways, and not a reflection of more liberal societal attitudes regarding men and women's work. They defined traditional ways as those in which men and women did the same work—men did women's work and women did men's work. Evidence from Klein's (1980) ethnographic work supports these contentions. She found that the theme of equal participation in the economy is closer to the traditional ideal than to the Euro-American ideal. Arnold (1997) states that the range of subsistence activities in which families engaged required both individual and collective efforts, and cooperation between family members.

The women were asked if there was anything else that they would like to talk about regarding the roles of women in the subsistence salmon harvest. Some women said that it is important to continue to fish because it provides a healthy diet. However, the most frequently reported response from women was that subsistence fishing is an important part of the Native way of life and it is necessary to keep alive that part of their culture. They felt that it was critical that they teach their children about fishing so that their people will never forget their traditions. According to a woman, "It is said to be the woman's responsibility to teach the kids...we should never forget and always teach kids how to fish." The teaching of children is a traditional role of the mother, as she is seen as a clanswoman and the first clan teacher for her children (Klein 1995). The importance of continuing traditional practices was paramount among these women but it is also recognized that these practices are threatened by modernization. As one woman said, "there is a clash of cultures that gets in the way of our involvement with salmon. Today more than ever, it's hard to be Native. We have non-Native expectations like work that we tend to when we should be out fishing; it's hard to be traditional in a contemporary culture."

These women have perceived a change in the gendered division of labour from the time that their mothers were young, characterized by the increasing participation of women in the harvest. This increased participation is believed to be primarily due to the influence of factors external to Native society and consistent with trends toward the modernization of society. Industrialization in southeast Alaska, particularly the introduction of the cannery industry in the late 1880s, resulted in high rates of women's participation in the fisheries harvest which continued through the early 1940s. Participation then decreased as women's travel to fish camps declined, as children were sent away from their villages and community to attend boarding schools where Euro-American ideology was dominant and Native tradition devalued, and as cultural transformations occurred in the 1960s around the time of statehood where becoming 'American' was highly valued. These events resulted in a loss of engagement in traditional cultural practices among Native peoples. With the advent of civil rights movements in the 1960s, gender role behavioural patterns began to change. The Native revitalization movement sparked a renewed interest among Native peoples in returning to and including traditional practices in their lives. However, the women's movement, supporting the equality of women in all spheres of contemporary society, appears to have played a more significant role in the beliefs of the women interviewed regarding the increasing participation of women in the contemporary subsistence salmon harvest.[4]

---

[4] Tuhiwai L.Smith, *Decolonizing Methodologies: Research and Indigenous Peoples* (New York: St. Martin's Press, 1999:154, 167-8). These findings contradict literature by Indigenous women scholars such as Smith (1999) and Anderson (2000). Smith (1999) in her work on Maori women, argues that Western feminism which has typically addressed white, middle-class women's issues cannot provide a framework necessary for Indigenous women's multifaceted challenges surrounding the consequences of capitalism as well as the intersection of race, class and gender. She concludes that Western feminism is too 'simplistic' due to its primary focus on gender, and has little impact on the lives of Indigenous women. Smith further contends that only Indigenous people themselves have the tools necessary to accurately examine the complexities of their culture. Also, Smith seems to suggest that Western feminists have betrayed their men but that Maori women will not be disloyal to members of their own ethnic group: men or women. Race and the effects of imperialism trump concerns over

This increased involvement is seen as a result of a more general change in society in which women have become more active participants. Contemporary society has experienced a breakdown of what were once considered rigid beliefs regarding gender roles, including ideas about what constitutes men's and women's work. This external societal trend appears to have had a strong effect on attitudes toward, and engagement in, traditional role activity among the Native women with whom we spoke.

For many of the women, traditional knowledge regarding the salmon harvest has not been passed down, and this may suggest a culture gap that was affected by their living in a Western-dominated society. It was revealed that the women possessed little knowledge of, nor did they engage in, traditional practices related to clan ownership of fishing areas, use patterns of the fish, and knowledge concerning the relationship of women to the salmon fisheries harvest.

This chapter has, in part, addressed the struggle of Native women in southeast Alaska to harmonize their traditional way of life in regard to the work they do in the salmon harvest with the effects of modernization. In societies where people who practice traditional ways of life reside with a dominant non-Native group, one may ask how traditional cultural practices of indigenous peoples are reconciled with modern ways. Is it possible for indigenous women to successfully harmonize their contemporary roles with their traditional role in cultural maintenance?

A Promise of Role Integration?

Two possible outcomes or models of racial/ethnic relations which have occurred when there has been an intrusion of a dominant culture into a traditional culture have been assimilation and accommodation/pluralism. The assimilation model requires that indigenous peoples are transformed by an outside dominant group into full members of the dominant society by giving up their traditional ways of life to embrace the dominant group's cultural core, or contemporary ways of life (McLemore and Remo 2004). Indigenous people are expected to accept the cultural elements of the dominant group and relinquish the cultural elements of their traditional culture, including but not limited to religion, family structure, patterns of gender relations, political and economic systems.

The accommodation/pluralism model allows ethnic group members the freedom to participate in all of society's major institutions (i.e., educational, political, economic) while simultaneously retaining or elaborating on their own ethnic heritage (Kallen 1956). The acceptance and practice of the ideology of pluralism, where the dominant culture

---

those of gender alone and leaves Smith to consider Western feminism to be of little use to the convolutions of Maori women.

K. Anderson, *A Recognition of Being: Reconstructing Native Womanhood* (Toronto: Second Story Press 2000: 63, 275-276.) Anderson's argument against Native women accepting Western feminism is based on three lines of contention: in some matrilocal Native societies, women secured and maintained power through control of economic resources; that the predominantly white, middle-class Western feminism has not only failed to address issues of race but that white women have been complicit in the oppression of Native women; and that Native women do not want to want to acquire power held by white males through rejection of the traditional Native power held within motherhood, which has typically been seen as a source of oppression by Western feminists.

accommodates to the traditional practices and beliefs of the minority group, would suggest the possibility that Native women could successfully integrate their traditional roles in cultural maintenance through their work in the contemporary salmon harvest in southeast Alaska.

In the United States, however, the pattern of ethnic/race relations generally practiced has been that of assimilation (Gordon 1964). This is evident in relations with indigenous peoples both geographically within the boundaries of the United States (American Indians and Alaska Natives) as well as with indigenous peoples outside of the geographic borders (the Philippines, Puerto Rico, Hawaii, Cuba) (Drinnon 1980).

Consistent with assimilationist theory, Alaska Natives were perceived as uncivilized inferior people who would benefit from the replacement of their traditional beliefs and activities with those of the civilized intruders. For example, forced education of Native children was practiced by taking them away from their homes and sending them to distant government boarding schools; there was an almost entire replacement of Native languages with English; and, missionaries were successful in permanently altering the sociopolitical clan structure of Alaska Native peoples. Native women today are keenly aware of this loss of structure and cultural tradition and express a strong desire to see it reintegrated into their lives and the lives of their children. However, for there to be successful integration between traditional and contemporary cultures, specifically in the inclusion of traditional gender role work with contemporary gender roles, an accommodation/pluralist strategy, where traditional cultural beliefs and practices are valued, must be incorporated into the ethos of a society. With the history and practice of assimilationist ideology and policy that has been dominant in the United States, it is unlikely that this will occur. For urban Native women then, there remains a continuous struggle between the desire to practise their traditional ways of life and the modern way of life that contemporary society has imposed. As summarized by a 30-year old Tlingit woman, "modern day TV, music takes away from working with fish—kids play rather than watch and learn—our culture has accommodated to contemporary urban culture, and we have lost."

Conclusion

The importance of the salmon subsistence harvest fishery to the people and cultures of southeast Alaska cannot be overstated. While the role of women has not been the focus in most Western historical documentation, they have always been involved in and played a significant role in fishing work. The close relationship of Tlingit to subsistence fishing was best expressed by one woman interviewed who said, "fishing and subsistence are important and a part of the Tlingit way of life. To live without fish we would surely eventually no longer exist."

The roles that women have played in fishing in southeast Alaska have changed as their traditional lifestyle has been affected by the intrusion of a Western-dominated culture. According to the women interviewed during this study, that change has been characterized by a more public role where their participation has become more evident and they have taken on roles formerly occupied only by men. These women have clearly perceived a change in the division of labour in the salmon subsistence fisheries harvest characterized by the increasing participation of women, and they attribute this change to the influence of factors external to Native society. These women see increased involvement as a result of a more general societal change in which women have become more active participants in their society. The women's movement, supporting the equality of women in all spheres of contemporary society, has played a significant role in the beliefs of the women interviewed toward the increasing participation of women in the contemporary salmon harvest. However, this change has also created a 'clash of cultures,' where traditional practices have been influenced by a more modern approach to fishing and processing the fish. For example, the time-saving method of freezing fish has replaced the more traditional practice of smoking and drying fish; women are both fishing and processing the fish rather than just doing processing work; and they have become active, and, in some cases, the sole participants in decisions regarding with whom the fish will be shared and exchanged, rather than playing a role in influencing the decisions that men make. The mere fact that women do engage in fishing, which has traditionally been the domain of men represents perhaps the most significant change in gender roles. This positive change was described by one woman when she said "I'm glad it's still going on and we can still fish—they haven't taken that from us […] like (name omitted) […] we might sneak out and fish anyway if it were illegal. Teaching our kids is a good thing. If they learn to fish they'll be set for life."

These changes have not come about without consequences. The women interviewed have recognized that along with their increased involvement in the subsistence salmon fisheries harvest, traditional practices have been lost. To these urban Native women, there remains a continuous unresolved struggle between the desire to practice their traditional ways of life and the pull of modernization.

References

Anderson, K. 2000. *A Recognition of Being: Reconstructing Native Womanhood.* Toronto: Second Story Press.

Arnold, D.F. 1997. *"Putting up Fish": Environment, Work and Culture in Tlingit Society 1780s–1940s.* Ph.D. dissertation, University of California at Los Angeles.

Betts, M.F. 1994. *The Subsistence Hooligan Fishery of the Chilkat and Chilkoot Rivers.* Technical Paper No. 213. Juneau, AK: Division of Subsistence, Alaska Department of Fish & Game.

Bodenhorn, B. (1990). "I'm Not the Great Hunter, My Wife Is": Iñupiat and anthropological models of gender. *Études/Inuit/Studies* 14(1-2): 55-74.

de Laguna, F. (1983). 'Aboriginal Tlingit Sociopolitical Organization,' pp. 71-85 in E. Tooker, ed. *The Development of Political Organization in Native North America.* Proceedings of the American Ethnological Society, 1979. Washington, D.C.: American Ethnological Society.

de Laguna, F. (1990). 'Tlingit,' pp. 203-228 in W. Shuttles, ed. *The Handbook of American Indians. vol. 7: Northwest Coast.* Washington D.C.: Smithsonian Institution Press.

Dombrowski, K. (2001). *Against Culture: Development, Politics and Religion in Indian Alaska.* Lincoln: University of Nebraska Press.

Dauenhauer, N. Marks, and R. Dauenhauer, eds. (1994). *Haa Kusteeyi Our Culture: Tlingit Life Stories.* Seattle: University of Washington Press.

Drinnon, R. (1980). *Facing West: The Metaphysics of Indian-Hating and Empire-Building.* New York: Meridian Press.

Drucker, P. (1955). Indians of the Northwest Coast. New York: McGraw-Hill.

Emmons, G.T. (1991). *The Tlingit Indians. Edited with additions by Frederica de Laguna.* American Museum of Natural History, Anthropological Paper No. 70. Seattle: University of Washington Press and the American Museum of Natural History.

Fisher, A.H. (1999). 'This I Know from the Old People:' Yakima Indian Treaty Rights as Oral Tradition. *Montana: The Magazine of Western History* 49: 2-17.

Goldschmidt, W. R. and T. Haas (1998). *Haa Aaní Our Land: Tlingit and Haida Land Rights and Use,* Thornton, T.F. (ed). Seattle: University of Washington Press,

Gordon, M.M. (1964). *Assimilation in American Life.* New York: Oxford University Press.

Jarvenpa, R. and H.J. Brumbach (1995). Ethnoarchaeology and Gender: Chipewyan Women as Hunters. *Research in Economic Anthropology* 16: 39-82.

Johnson, F. (1981). *Tlingit Fishing in Southeast Alaska/Stories by Frank Johnson.* C. Hendrickson, (ed). Ketchikan, AK: Ketchikan Indian Corp,

Kallen, H.M. (1956). *Cultural Pluralism and the American Idea: An Essay in Social Philosophy.* Philadelphia: University of Pennsylvania Press.

Kan, S. (1989). *Symbolic Immortality: The Tlingit Potlatch of the Nineteenth Century.* Washington: Smithsonian Institution Press.

Kan, S. (1996). Clan Mothers and Godmothers: Tlingit Women and Russian Orthodox Christianity, 1840-1940. *Ethnohistory* 43: 613-41.

Klebnikov, K.T. (1976). *Colonial Russian America: Kyrill T. Khlebnikov's Reports, 1817-1832.* Portland: Oregon Historical Society.

Klein, L.F. (1976). "She's One of Us, You Know": The Public Life of Tlingit Women: Traditional. Historical, and Contemporary. *Western Canadian Journal of Anthropology* 6: 164-83.

Klein, L.F. (1980). 'Contending with Colonization: Tlingit Men and Women in Change,' in M. Etienne and E. Leacock, eds. *In Women and Colonization: Anthropological Perspectives*, New York: Praeger.

Klein, L.F. (1995). 'Mother as Clanswoman: Rank and Gender in Tlingit Society,' pp. 28-45 in L. F. Klein and L.A. Ackerman, eds. *Women and Power in Native North America*, Norman: University of Oklahoma Press.

Kleinfeld, J. and J.J. Andrews. (2006). Postsecondary Education Gender Disparities among Inuit in Alaska: A Symptom of Male Malaise? *Études/Inuit/Studies* 30(1): 111-121.

Krause, A. (1956). *The Tlingit Indians: results of a trip to the northwest coast of America and the Bering Straits.* Gunther, E., trans. Seattle: University of Washington Press.

Lorber, J. (1994). *Paradoxes of Gender.* New Haven, CT: Yale University Press.

McLemore, S.D. and H.D. Remo (2004). *Racial and Ethnic Relations in America.* Boston: Allyn and Bacon.

Moss, M. (1993). Shellfish, Gender, and Status on the Northwest Coast: Reconciling Archaeological. Ethnographic and Ethnohistorical Records of the Tlingit. *American Anthropologist* 95: 631-53.

Muszynski, A. (1988). Race and Gender: Structural Determinants in the Formation of British Columbia's Salmon Cannery Labour Forces. *Canadian Journal of Sociology / Cahiers Canadiens de Sociologie* 13(1-2): 103-120.

Oberg, K. (1973). *The Social Economy of the Tlingit Indians.* Seattle: University of Washington Press.

Olson, R.L. (1967). *Social Structure and Social Life of the Tlingit in Alaska.* University of California Publications. Anthropological Records Volume 26. Berkeley: University of California Press.

Price, R.E. (1990). *The Great Father in Alaska: The Case of the Tlingit and Haida Salmon Fishery.* Douglas, AK: First Street Press.

Smith, T.L. (1999). *Decolonizing Methodologies: Research and Indigenous Peoples.* New York: St. Martin's Press.

Turek, M.F. (2003). *Personal interview.* Juneau, Alaska, August 10.

Turek, M.F. (2006). *Field Notes.* Division of Subsistence, Alaska Department of Fish & Game. Juneau, AK.

Turek, M.F., A. Paige, E. Cheney, J. Dizard, and N. Soboleff (2007). *Kake Subsistence Salmon Harvest Use Patterns.* Technical Paper No. 309. Juneau, AK: Division of Subsistence, Alaska Department of Fish & Game.

Weinberg, D.H. (2004). *Evidence for Census 2000 About Earnings by Detailed Occupations for Men and Women.* Washington, D.C.: U.S. Department of Commerce, Economics and Statistics Administration.

Williamson, K.J. (2004). 'Gender and Equity, the Arctic Way,' pp. 187-191 in N. Einarsson, et al., eds. *Arctic Human Development Report,* Akureyri, Iceland: Stefansson Arctic Institute.

CHAPTER THREE

"It Used to be Women's Work": Gender and Subsistence Fishing on the Hudson Bay Coast

Martina Tyrrell

Abstract: Throughout spring, summer and fall each year, Inuit men and women in Arviat, on the west coast of Hudson Bay, are engaged in a variety of fresh- and salt-water fishing activities. These range from mass fishing derbies to small family fishing outings to lakes and rivers. Despite the economic and nutritional value of these fishing activities, they are almost universally constructed as leisure pursuits and, therefore, accorded a low social status. During summer and early fall, Arctic char are fished from the sea. This is the most important fishery in Arviat, and is conducted in two ways—by boat, or by setting nets along the shoreline. Despite an almost complete absence of women from shoreline char fishing, it is perceived by Arviat's women to be an exclusively female form of fishing. Due to recent political and socio-economic changes in Nunavut, many women have taken up full-time formal employment, thus severely restricting the amount of time they have to engage in 'leisure' activities such as subsistence char fishing. This exclusion of women from summer char fishing has implications for gendered forms of environmental knowledge and skill, and for gendered spaces within the Arctic landscape.

Introduction

For at least the past three decades, feminist anthropology has sought to overcome long-seated biases within the discipline that privileged male knowledge, male economic activities, and male spaces. Hunter-gatherer scholarship, emphasizing 'man-the-hunter', has been forced to make room for 'woman-the-gatherer,' and to take account of the fact that, in most hunter-gatherer societies across the world, female-gathered foods account for "more than half and at times nearly all of what is eaten" (DiLeonardo 1991:7). From this starting point, anthropologists such as Frances Dahlberg (1981), Sherry Ortner (1974; 1981), and Henrietta Moore (1989) have emphasized the study of women, women's roles within the family and society, and the politics of gender. Arctic scholars have followed this trend. Barbara Bodenhorn (1990) and Carol Zane Jolles and Kaningok (1991) among others have emphasized both the importance and fluidity of gender roles within Inuit society, demonstrating the shared nature of work, the complementary roles of husbands and wives as they work for each other, and the lack of strictly adhered to divisions of labour. Rewriting anthropology from this more nuanced gendered perspective is more than an intellectual exercise. Mark Nuttall (1998) argues that effective strategies for sustainable development can only be developed when varied gender roles are taken into account, and when the potentially different impacts that men and women have on the environment is better understood. Nuttall (1998) noted that few researchers had documented the social and economic contributions made by women in Arctic communities, although ten years on, it would appear the tide has turned.

In this chapter, I explore subsistence fishing practices in one Inuit community on the west coast of Hudson Bay. Examining a range of fishing activities from fresh and salt-water derbies to summer Arctic char fishing, I contrast the social and economic contributions of women to both male and female perceptions of these contributions. In relation to the Arctic char fishery, I argue that despite perceptions to the contrary, female participation has declined, due to long-term changes to settlement patterns and more recent changes to formal employment practices. Finally, I consider the potential impact that this decline of women fishers may have on enculturation and enskilment, and on certain forms of gendered environmental knowledge.

Methodology

Since 2000, I have conducted anthropological research into the role of the sea in Inuit life on the west coast of Hudson Bay.[1] On research visits to the community of Arviat, varying in duration from two to twelve months, I have undertaken a thorough examination of Inuit perception, knowledge and use of the sea and marine animals. I have critically analyzed the sea as an inhabited place, an extension of the land in both summer and winter, and a place where knowledge, kinship and identity are enacted (Tyrrell 2005). Embedded within the role of the physical environment is the relationship Inuit maintain with marine animals—seals, beluga whales, polar bears, Arctic char and various sea birds. Knowledge of these animals and engagement with them in and after the hunt remains central to Inuit notions of identity and belonging. This chapter is based primarily on empirical data gathered between September 2002 and September 2003 during doctoral fieldwork, and is supplemented by data gathered on two subsequent research visits to Arviat in summer 2006 and fall/winter 2007.

Throughout the course of field research, the predominant methodology has been participant observation. In order to acquire a critical insight into perception, knowledge and use of the sea, I have engaged in a wide variety of sea-related activities, from hunting, fishing and travelling at sea and along the coastline, to the transformation of harvested animals into food and clothing. Through these activities, I have gained an understanding of the social, cultural and economic value of the sea to Arviarmiut.[2] The data upon which this chapter is based were gathered through my participation in fresh- and salt-water fishing derbies in 2001 and 2003, and in the sea-run char fishery during the summers of 2003 and 2006. Participant observation was supplemented by informal and formal interviews with male and female fishers, and with wildlife officers and managers within Arviat.

---

[1] Permits to conduct this research between 2001 and 2007 were granted by the Nunavut Research Institute, and I received formal support from the Arviat Hamlet Council and the Arviat Hunters' and Trappers' Association.

[2] *Arviatmiut*: The people of Arviat.

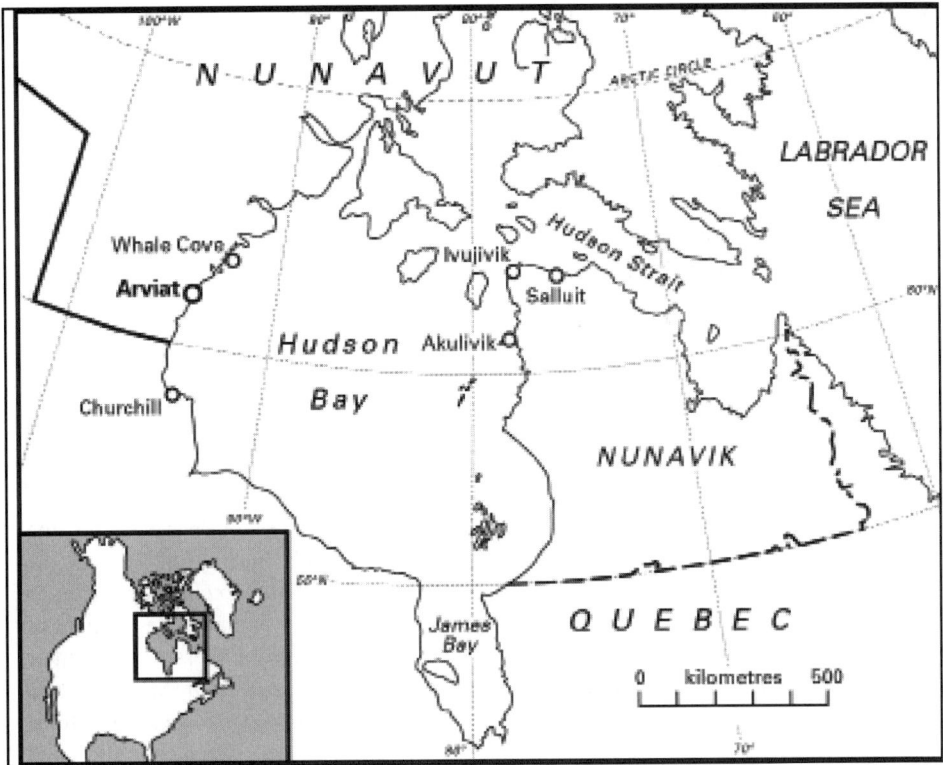

Figure 1. *Location of Arviat, on the west coast of Hudson Bay*

The Setting

Arviat is situated on the west coast of Hudson Bay. It is the most southerly community in the Kivalliq region of Nunavut, lying approximately 250 kilometres north of Churchill, Manitoba. It is one of Nunavut's most populous towns, and continues to expand rapidly due to a combination of high birth rate and in-migration. According to the 2006 Canadian census, 40% of Arviat's population of 2,060 people is under the age of 15, and the birth rate is one of the highest in all of Canada. A large school-age population, offices of various regional and Territorial Government departments and the in-migration of Inuit and *qablunaat*[3] have further added to the growth of the community.

The town can be described as having a mixed subsistence economy. Main sources of waged or salaried employment are the hamlet, stores, schools, government offices, health centre, and community and Inuit organizations. Yet much of the food, and many of the resources consumed in Arviat, still come from the land and sea. Caribou are the predominant source of country food[4] and are hunted almost year-round. Wolves, and to a lesser

---

[3] *Qablunaat:* Non-Inuit/Euro-Canadian people. Singular: *qablunaq*.

[4] Country food refers to all types of food procured from land and sea, including terrestrial and marine mammals, fish, shell-fish, birds, eggs, and berries.

extent musk-ox, are hunted in spring, and trout, grayling and whitefish are harvested from lakes and rivers. In May, the honking of migrating geese fills the air, and hunters take to the land to hunt geese and gather goose eggs. In late summer and early fall, cranberries, cloudberries and crowberries grow in rich profusion on the ground, and are gathered either for immediate consumption or for storage and use throughout winter. In winter, men set trap lines on land or sea-ice, to trap foxes for sale to southern fur markets. From the sea, the predominant resources are ringed and bearded seals and beluga whales. Seals are hunted by a variety of means throughout much of the year and are valued for their pelts and as food for humans and dogs.[5] Beluga whales are hunted during the eight weeks of their migration in summer and early fall each year. In Arviat, only beluga *maktaaq*[6] is eaten by humans, while large amounts of beluga meat are used to feed Arviat's many dog-teams. Arctic char are fished along the shoreline throughout summer, predominantly for subsistence use, but are occasionally sold to local stores, or sent south to Churchill. Polar bears are hunted each fall by both Inuit and visiting trophy hunters in a strictly managed hunt that provides income to some families through either the sale of bear pelts or the guiding of visiting *qablunaat* hunters.

In an economy where subsistence and market practices are inseparable, the line between subsistence and commercial use of resources is easily blurred. Some animals, such as foxes, wolves and, to some degree, polar bears, are hunted exclusively for their commercial value, and the pelts of these animals are sold, via the Nunavut Department of Environment's local wildlife office, to southern fur auction houses. Excess skins of other animals, such as ringed and bearded seals, are sold in a similar fashion. Many women in Arviat supplement their family income by producing garments for sale, made from either animal pelts or store-bought materials. Sometimes women make commissioned pieces, while at other times, these women attempt to sell their wares to resident or visiting *qablunaat* or at the government-run Kiluk Sewing Centre. Many of these same women also make wall-hangings which they market in the same manner. These wall-hangings are often made from a combination of store-bought duffel and caribou skin, depicting animals and traditional scenes of Inuit life. Some women carve soapstone or make dolls from seal and caribou skin. Four or five Arviat women have gained national and international renown for their carvings and dolls and these are sold at 'Native' galleries in southern Canada and abroad. Caribou antlers and soapstone are carved into ornaments and jewellery by a number of men, while a few also make elaborately decorated *ulus*[7] and snow goggles.

The production and sale of these items, combined with formal employment and government welfare payments, contribute to the subsistence activities of many families. Revenue generated by these means pays for such items as skidoos, Hondas,[8] boats, outboard motors, rifles, bullets, fishing nets, sewing equipment, store-bought clothing, as well as the store-bought foods that supplement all Arviarmiut diets (*see* Wenzel 1991 and Dahl 2000 for comparative accounts of mixed subsistence Inuit economies in Baffin Island and Greenland respectively).

---

[5] In 2006, 27 dog-teams were registered with the Arviat Dog Racing Club, each comprising, on average, 10 animals.
[6] *Maktaaq*: beluga whale skin and the layer of fatty tissue underneath.
[7] *Ulu*: A half-moon shaped woman's knife, made of metal, with a handle made of wood or bone
[8] Honda: All terrain vehicles, known universally in Arviat as Hondas.

Fishing for Food and Fun

Between 1996 and 2001, Arviarmiut harvested an average of 4,356 Arctic char each year (NWMB 2002:5). In that same period, the average number of cod taken was 869, along with 1,358 lake trout, 655 Arctic grayling, and 392 whitefish.[9] Throughout the year, these species are harvested using a variety of different fishing methods, and while most are harvested for human consumption, some are also fed to dogs, or are fished in anticipation of winning prizes at annual fishing derbies.

As with many other subsistence activities in the Arctic, fishing for subsistence cannot be disentangled from fishing for leisure. Stern (2000) has written that, as Inuit become more dependent on waged employment, subsistence activities are viewed more as forms of leisure, thus making them optional rather than essential activities. Many Inuit in waged employment say they travel on land or sea as much to escape the confines of town life as to provide valuable food and pelts for their immediate and extended families. Some part-time hunters in Arviat have told me that "getting away from it all" is often their main priority, and a successful hunt is a bonus.

However, this relegation of subsistence practices to the realm of leisure masks the fact that country food continues to make an important contribution to the economy of Inuit communities. Kuhnlein *et al.* (2003:28) have found that the 450 residents of Qikiqtarjuaq, on Baffin Island, consume on average 140 kilograms of country food each day, or a total annual consumption of over 50,000 kilograms. In Arviat, the mammals, fish, birds and other resources harvested from land and sea form the staple diet in most homes. However, because of the contention by many people in full-time employment that their subsistence practices are first and foremost leisure pursuits, the contribution they make to the economy runs the risk of being undervalued or overlooked. I shall return to this later, with regard to perceptions by and of women fishers.

Beyond subsistence being perceived as a leisure pursuit, and as a need to "get away from it all," throughout the year, a variety of organized leisure events take place within the community, and these too contribute to the subsistence economy in important ways. Tournaments, derbies and contests, such as skidoo or dog racing competitions, fishing derbies, community games, drag races, or tea and bannock making contests, all provide Arviarmiut with the opportunity to participate in friendly competition, combined with the chance to win monetary or material prizes and, particularly with regard to fishing derbies, provide food for humans and dogs.

Most of the fishing activities outlined in this chapter are considered by Arviarmiut to be leisure activities. It is interesting to note that the one fishing activity perceived to be of greatest economic value is that engaged in almost exclusively by men. As other anthropologists have shown (*see* DiLeonardo 1991), male roles and subsistence practices are often privileged within society in comparison to the subsistence contributions of women. This perception exists despite the significant material contributions of women. The various fishing activities discussed next offer fishers of both genders and all ages the opportunity to participate in sociable activities away from the confines of town, while at the same time contributing to the subsistence economy through the provision of food for humans and dogs.

---

[9] During this same period, an average of 3,339 caribou and 391 ringed seals were harvested each year (NWMB 2002:5-6).

Fishing Derbies

Each year, throughout the long dark days of winter, Arviarmiut (and women in particular) begin to speak with growing anticipation and excitement about the arrival of spring and the two fishing derbies that take place each May. As May draws near, anticipation mounts, and people get ready—buying provisions, readying sleds, preparing warm clothing—to travel inland and later, onto the sea-ice to participate in these two derbies.

The fresh-water fishing derby takes place during the first weekend of May, attracting large numbers of Arviarmiut. Entire families travel inland to various lakes in close proximity to or more distant from Arviat, in order to fish for trout, char, grayling or whitefish. Using mechanical augers or drilling by hand, holes are made in the still-thick lake ice, through which people jig for fish. During this weekend, small groups of men and women and extended families cluster together on lakes often 20 kilometres distant from Arviat, jigging for fish at their leisure. The derby is an important social occasion, and everyone has an equal chance of winning first prize for the biggest fish, which, in 2003, was a not-insubstantial $5,000.

For many, especially women, children and elders, the fresh-water derby offers the first opportunity to travel on the land following a long winter confinement in town. The derby is greatly anticipated for this reason and, for weeks leading up to the event, conversations with and amongst women are peppered with excitement at the prospect of travelling inland to fish at lakes with their families. Some entrants return to Arviat at the end of each day's fishing, while others camp on their chosen lake for all or part of the weekend, making the most of their fishing opportunities. Trout, grayling and whitefish are savoured by most Arviarmiut, and so the derby provides a welcome alternative to a winter diet heavily dependent on caribou and store-bought food. Land-locked char are considered less tasty than the sea-run equivalent, but are still enjoyed by some at this time of year. Therefore, many of the fish taken during the derby are eaten and shared on the spot, while more are taken back to Arviat for later use. Some fish is taken home to feed dog-teams, while some inevitably ends up being abandoned on the ice. Arviarmiut do not consider this abandonment a waste of fish, as other animals such as foxes and gulls will eat these discarded fish and thus benefit from what Inuit have left behind (*see* Freeman 2005 for a short discussion on Inuit perceptions of waste). For this fishing derby, many people cite the potential monetary reward as the most important aspect of the weekend, but it is clear that the fresh-water derby provides an opportunity for so much more than this. A break from the confines of town and the opportunity to get out on the land for possibly the first time in spring, combined with the prospect of a welcome change in diet, and food for ever-hungry dogs, are equally important aspects of the derby.

During the third week of May, a second fishing derby takes place on the sea-ice immediately in front of the community. This time, rock cod, a species never eaten by Arviarmiut (as they are considered ugly and unpleasant to eat) is the target. However, rock cod is an important species for feeding the town's many dog teams and, as with the fresh-water derby, those not taken home to feed dogs are left for scavenging gulls and foxes. The four-day event brings virtually the entire community out onto the sea-ice. The town has a distinct feeling of desertion throughout the weekend, as families move out onto the ice for the duration. Tents are either hauled out on large sleds or are erected directly on the ice, and extended families camp and fish together in clusters scattered across the sea-ice as far as the eye can see. Fishing for rock cod is dependent on tides so, for those who choose to return to Arviat each day, movement on and off of the sea-ice occurs in conjunction with flood and ebb tides.

Socially, the salt-water derby is perhaps more important than the fresh-water derby. As the shoreline is so close to Arviat, those lacking the means to get to the more distant lakes for the first derby, can simply walk out onto the sea-ice from their homes. Indeed, it is not uncommon to see individuals or small family groups walking out from the community onto the ice and taking up fishing positions at abandoned holes. This second derby is notable in that women vastly outnumber men. Rousselot *et al.* (1988) have noted that, across the Arctic, ice fishing is predominantly carried out by old people, women and children. In Arviat, the proximity of the sea-ice renders it accessible to people of all ages, with or without access to transport. The tools required are cheap and easy to make, and while most extended families have access to at least one mechanical auger, others drill fishing holes by hand, or occupy fishing holes that have been abandoned by other fishers. First prize for this fishing derby in 2003 was a new Honda, provided by the Arviat Dog Racing Club which organized the derby. For many, fishing for the derby prize is serious business, and I observed in both 2001 and 2003 that, as the derby progressed, fewer children appeared on the ice, and only the most eager adults stuck it out to the end.

Both of these fishing derbies provide opportunities for all Arviarmiut—men and women, young and old—to engage in fishing activities that make an important seasonal contribution to the subsistence economy by providing food for both humans and dogs. And for one lucky individual, the derby provides a windfall of either cash or an expensive and important piece of equipment such as a Honda, which provides access to prime locations for caribou hunting or char fishing. The cash prize is quite often used to buy one or more pieces of hunting equipment, and men and women who have won first prize in previous years have spent their prize money on such items as new outboard motors for boats, rifles, Hondas, etc. The good fortune of an individual derby winner is transformed into the good fortune of an entire family, where possibilities for subsistence practices and access to country food are expanded. Both derbies also provide opportunities for those who might not usually get out on the land or sea—some women, children and elders—leave the community for the first time following the long cold winter. These derbies, however, take place on only two weekends throughout the entire year, while other fishing activities occur on a more regular basis (for a more detailed examination of fishing derbies, see Shannon, this volume).

Fresh and Salt-Water Fishing

During spring and fall each year, fishing takes place on the lakes and rivers to the north, west and south of Arviat. Before ice has formed on lakes in fall, and during the process of break-up in spring, fishers make the most of the open water to set fishing nets. Once the ice has formed, fishers seek out known areas of open water, drill holes in the ice to jig for fish or, in a small number of cases, set nets under the lake ice. At the end of spring and during summer and early fall, fishing takes place on the open sea.

Year-round, a certain amount of fishing takes place in lakes and rivers for trout, char, grayling, and whitefish. The places of most importance are along the Maguse River to the north of Arviat, on Maguse Lake to the northwest, and at Dionne Lake to the west. Many other lakes and river systems are also used, but these generally are of lesser importance and are used by far fewer Arviarmiut. Almost no fishing takes place in winter, but once spring arrives, groups of men or family groups make trips inland by skidoo and sled to fish through holes in the ice, or at areas of permanently open water. Freshwater fishing declines in summer as people concentrate on the greater availability of sea-run char, but once char begin their fall migration up Maguse River, nets are set along the river and in Maguse Lake for these anadromous fish.

Freshwater fishing is generally conducted on a small scale, and often in conjunction with other activities, such as caribou hunting. It is primarily for the purposes of subsistence. This type of fishing offers the opportunity for entire families to engage in outdoor activities for a day, overnight, or for entire weekends. Many women have spoken with nostalgia and fondness about fishing trips undertaken in recent years with husbands, children and extended family members. As with other subsistence activities, Arviarmiut highlight the value of these fishing trips as ways to relax, and spend time with family members away from the clamour of town life.

As the sea-ice continues to melt in spring, cracks appear and leads open, as the ice gets thinner with each passing day. Some families travel out to the leads closest to Arviat to jig for fish. While I have observed this phenomenon from time to time, it is a relatively rare occurrence. Many people consider the ice at this time of year to be too dangerous for travel, and while experienced hunters still travel to the thin floe edge to hunt adult seals and seal pups newly emerged from their dens, many parents are loathe to take their children onto the ever-thinning ice.

Once the ice has broken up, open water returns for three or four months in summer and early fall, and fishing activity on the sea increases dramatically. I will discuss most of this fishing activity in the next section. However, on many summer evenings, young and old alike can be observed standing along the community dock, fishing for sculpin and the occasional char with fishing rods. This type of fishing is most definitely for leisure purposes, with adults and their children doing this as a way to pass a few minutes or hours on a fine summer evening. Sculpin are usually caught and then released, and char rarely venture this far into the inlet in front of the town.

Both men and women in Arviat consider these fishing activities to be predominantly 'men's work.' While women accompany their husbands and other family members on some of these fishing trips, most often it is the men who decide where to go, and the type of fishing that will be undertaken. Despite the contribution women make, both in terms of the fish they catch, and in their role passing on fishing and other environmental skills to children and young people, both women and men perceive the female role in these activities to be more leisurely, and not as 'serious' as the male role. As with many hunting activities, women often participate more passively than men (Tyrrell 2005). Usually, women accompany men on beluga and seal hunting trips only when a husband cannot find a suitable male hunting partner. While women are more likely to participate in freshwater fishing and caribou hunting as part of larger family groups, there is one activity perceived by Arviat women to be an exception to this.

In the case of sea-run Arctic char, women say that they have traditionally been the predominant fishers, setting nets along the shoreline throughout summer, while men have more commonly fished for char by boats a short distance offshore. If this is the case, then these women who fish along the shoreline are actively engaged in much more than the mere act of landing fish. These fishers consult the physical and social environment in order to determine when, where and how to fish, and choose female fishing partners with whom to share the work and the harvest.

Sea-Run Arctic Char

By far the single most important fishery in Arviat is that of sea-run Arctic char. Over the five-year period of the Nunavut Wildlife Harvest Study (NWMB 2002), a total of 16,678 sea-run char were harvested, compared to only 79 land-locked char during the same period. Char (*salvelinus alpinus*) are anadromous fish, moving from the winter and spring fresh water of lakes and rivers to summer and fall salt water of the sea. They are to be found in abundance close to shore and at the beginning of summer and fall near river mouths as they migrate into and back out of the sea. Two types of net fishing take place on the open water. The first takes place by boat and the second from the shoreline. The former is acknowledged by all to be a male form of fishing, while women say that the latter is a female form of fishing.

Char Fishing by Boat

Using boats to fish for char is usually done in conjunction with other subsistence activities, most particularly beluga whale hunting. The seasons for beluga and char almost completely overlap, with char arriving and departing perhaps a week or two earlier than belugas. Boating is predominantly a male-dominated activity, and women tend to accompany their husbands when no other male hunting partner is available on a given day. While men boat regularly, women boat only occasionally. A few Arviat women, who are regular and active caribou hunters, have told me they have never been to sea by boat, have never seen a live beluga whale, and have no desire to ever do so, saying that the sea is too dangerous.

Arviat is situated on a shallow inlet of Hudson Bay, and thus safe and convenient access to the bay by boat is only possible at or near high tide. As a result, hunters and fishers often spend long hours at sea, departing Arviat on one high tide, and returning twelve hours later on the next high tide. Small 4-6 metre aluminium skiffs, or larger 8 metre wooden canoes, both equipped with outboard motors, are kitted out on any given day during summer and early fall with equipment required for beluga hunting, char fishing, and possibly some seal hunting too (although this is the least important season for hunting seal).

Sitting in a small boat for many hours awaiting beluga whales is a time consuming activity, and many hunters take the opportunity to set fishing nets. This is done in one of two ways. Landing on a reef or island, a man will set his nets along the shoreline in a manner similar to those set along the shoreline close to Arviat (detailed below). Alternatively, he will set them directly from his boat. This latter method takes place close to shore, in shallow water, where char are known to swim at this time of year. As there are usually two men (or occasionally a man and a woman) per boat, one member of the hunting party drives the boat slowly forward, while the second, having secured a rock, piece of iron, or other such weight to one end of the net, gradually feeds the net over the side of the boat and into the water. The net then hangs like a curtain in the water, entangling any passing char. One advantage of this form of fishing is that char can be removed from the net immediately, so they are fresher, and unlike shoreline char fishing, do not suffer the ravages of scavenging gulls when nets become stranded on emerging shoreline as the tide falls. Often, while in the midst of fishing for char, belugas are sighted, and nets are hurriedly retrieved from the water, so hunters can pursue the larger prize. If fishing does not come to such a sudden end, the net is cleaned of seaweed as it is hauled back into the boat, or the cleaning is postponed until the net can be spread out to dry outside the fisher's home, or on the outskirts of town, later in the day.

Char Fishing at the Shoreline

Even before the sea-ice has completely melted, net fishing from along the shoreline begins beyond the extreme limits of the diminishing ice, and continues throughout summer and early fall. Fishers travel the short distance by Honda to set nets around Nuvuk, the point of land to the east of the community. Most shoreline fishing takes place along the length of Nuvuk's sandy shores. At the extreme eastern end of Nuvuk a rocky and sandy reef, Huluraq, extends far out into the sea and this is also a prime location for shoreline fishing. At low tide, nylon nets measuring 8 to 15 metres are set along the beach running perpendicular to the shoreline and out into the sea. As the tide rises, char swim with the current, travelling close to shore and become trapped in the awaiting nets which hang like curtains in the water. Returning at the next low tide, the lucky and successful fisher cleans and guts any fish caught. This must be done quickly, as hungry gulls and polar bears are forever on the lookout for free food. Once the fish have been taken care of, nets must be cleaned. The fisher removes any seaweed that has become enmeshed in the net and sets the net out again in preparation for the next high tide. Depending on the number of fish caught and the amount of seaweed entangled in the net, this low tide job can take anywhere from half an hour to a couple of hours.

Two to three weeks after the break-up of the ice, the catch in nets along the shore can be substantial (McEachern 1978). More than any other important marine resource, char numbers fluctuate dramatically from year to year. Arviarmiut talk about good and bad years for char, and statistics support this. In 2000, 5,377 char were landed, compared to only 545 in 1999 (NWMB 2002:5). Some men told me that during summer 2002 they frequently landed 20-30 fish in their nets at a time, but in 2003, those same men said they considered themselves lucky to land four or five.

During the short summer months, while char are living that portion of their lives in the sea, activity along Nuvuk and Huluraq continues apace throughout the entire 24 hours of each day. Fishers make the most of the tides and if the optimum time for tending shoreline nets happens to fall in the middle of the night, then activity will continue throughout the night. At Arviat's relatively low latitude, summertime brings with it an average of three hours of twilight and, during these times when low tide occurs after the sun has briefly set, there is a constant hum of Hondas travelling between the community and Nuvuk and the headlights of machines are seen throughout the brief night. However, not everyone is willing or able to continue fishing when low tide falls at night and some give up fishing during those periods, resuming again once the tides have rotated around to more hospitable times of the day. But the season for sea-run char is a short one, and many Arviarmiut make the most of the brief eight to ten-week season of char fishing, by tending their nets at whatever time of day or night low tide occurs.

Fishing along the shore can be a time of great social activity. News often quickly spreads that one location along the shoreline is bringing in particularly large returns, leading to more fishers trying their luck there, thus increasing the number of nets set in the water and the number of people tending nets at each low tide. During these low tides, the sharing of conversation, gossip, information and fish frequently takes place. One person might share part of his or her catch with a less lucky fisher at a neighbouring net and the sharing of conversation, information, cigarettes, or chewing tobacco is ongoing. Fishing from the shoreline is rarely done alone. At this time of year when polar bears are on the prowl, there is always greater safety in numbers and women in particular have told me they prefer to set their nets in the company of others to provide more protection against bears.

Throughout the winter and spring of 2002 and 2003, in the course of participant observation, informal conversation, and more formal interviews, women often told me about summer char fishing from the shoreline at Nuvuk and Huluraq. Many women said that each summer they set their nets along the shore, and during winter and spring I began to look forward to engaging in this activity with women once summer 2003 arrived. However, my experience that summer turned out to be quite different. I shared a net that summer with a *qablunaq* woman friend, and for one month we regularly tended our net and harvested our char, at all times of day and in all types of weather. That summer I also regularly travelled around Nuvuk by Honda, either on my own, or in the company of Arviarmiut. I also spent time in boats, engaged in char fishing and beluga hunting with various Arviat hunters, close enough to shore to see the people who were working at their fishing nets along the shoreline. During that entire time, I never encountered any women setting their nets along the shoreline. I knew of four Inuit women and three *qablunaat* women who told me they set nets, but I did not ever meet these or other women while I was engaged in participant observation along the coastline. Instead, I regularly encountered men of all ages working at their nets along the shoreline. Why did this inconsistency exist between what women said was their role as shoreline char fishers, and the reality where very few women actually participated in this type of fishing?

Who are Arviat's Char Fishers?

According to Arviat men, fishing for char along the shoreline is predominantly men's work. According to Arviat women, it is predominantly women's work. During the course of my research, women said that men rarely ever fish, that only women fished in the past, but that now men are gradually beginning to participate in fishing. My research also revealed a perception amongst Arviat men that they had always fished, that fishing is 'man's work' and that, as with other marine subsistence activities, women only occasionally help their men out when nobody else is available.

Why do men and women have such different perceptions of their roles as fishers? Had men really not fished in the past, or was fishing activity so different in the past that it bore no resemblance to contemporary fishing practices? Jens Dahl (2000:112) has written, with reference to west Greenland, that "the division of labour during fishing and fish processing is different now to the old days. Women used to fish, but now they exclusively process." While fishing in west Greenland has become more of a commercial venture, could the same be true for subsistence practices in Arviat? The preliminary report of the Nunavut Wildlife Harvest Study makes no distinction between male and female hunters and, therefore, it is impossible to determine whether those reporting char catches are male or female. Nunavut-wide, women were included in the study (Jim Noble, personal communication) but it is not known whether any women were involved in the local Arviat study. This lack of information points to a need to collect wildlife harvest data disaggregated by gender, particularly with relation to fish, where there are varied perceptions with regard to which gender is the predominant harvester, and in relation to whether harvesting is for commercial or subsistence purposes. Nuttall's (1998) call for a greater understanding of gendered environmental knowledge and gendered impacts on the environment supports this.

By examining Arviat's recent past, we can better understand the involvement of both women and men in the sea-run char fishery. In their economic survey of the Keewatin in the early 1960s, Brack and McIntosh (1963:71) reveal, that compared to today, very little fishing took place in the vicinity of Arviat. They wrote that fishing near the community was 'casual and spasmodic' and that there were no more than five nets set along the shore on any given day. Given the increase in population in the forty years since their report was published (from 330 in 1962 to 2,060 in 2006), and the almost universal availability of Honda transport technology, nets are now set regularly all along the coastline around Nuvuk and Huluraq each summer. At the time of their study, a few families lived 100 kilometres north of Arviat at Sandy Point. They reported that these families were involved in intensive fishing activity.

In 1976 and 1977, in an attempt to bring alternative employment opportunities to the growing town, the Arviat Hunters and Trappers Organization was involved in a boat-building project with some local male fishers, the aim of which was to develop a commercial fishery. A *qablunaq* boat builder was contracted from southern Canada to teach local men how to build boats. At the time, shoreline fishing was seen as an impediment to developing a commercial fishery, and it was believed that if more fishing took place by boat, then a commercial fishery could establish itself. Seven boats were built and commercial fishing took place on both the Ferguson and Copperneedle rivers during the summers of 1977 and 1978. The char harvest both years was sold to a commercial cannery in Rankin Inlet (McEachern 1978). The boats were built by Arviat men, and the development of the fishing project was aimed solely at men (William Angalik, personal communication). However, following the first two years of the project, neither the boat-building nor the commercial fishing venture met with much success. Today, the only reminder of this fishing project is one small blue skiff that lies weathering in the sun, wind and snow above the high water mark to the west of Arviat. One man told me that Arviarmiut had not been interested in the commercial fishing venture, and this type of organized venture gradually faded away.

In the late 1980s, Kakivik Fisher Foods, a fish processing business, was set up by a local man in his late 20s. He bought fish from local fishermen, smoked it, and flew it out to Winnipeg and Toronto, from where it was sold to markets in the United States and Canada. Now an elementary school teacher, I asked this former businessman why he was no longer in the fish processing business. He ran through a list of those men who had been his main suppliers of fish, and noted that they were all now dead or too old to fish any more. "The fish are still there," he said, "but there's no-one to bring them in." The cost of transporting his product to southern markets was also prohibitively high, and in the end he quit the business and turned to a career in teaching. Keith *et al.* (1987) note that there exists little sustained support in northern communities for initiatives such as these, and this also appears to be true for Arviat.

In the present day, minimal commercial fishing takes place in Arviat. The retired manager of the Northern Store (originally from Scotland) smokes relatively large quantities of char he harvests, which he sells to the Northern Store, to a local hotel, and to businesses in Churchill. Some other Arviarmiut also sell char on a smaller scale to either the Northern Store, or to a small processing plant in Churchill. A principal deterrent to commercial fishing is a lack of preservation facilities in Arviat. Those transporting fish to Churchill run the risk of their fish rotting en route. However, those few who do sell their fish in this manner are exclusively men who conduct their fishing activity from boats. Occasionally, women call the local FM radio or announce via CB radio that they have one or two char for sale, ranging in price from $10-15 per fish. Compared to the low level of commercial fishing engaged in by men, this piecemeal sale of one or two char by women is indeed small-scale.

What is striking is the fact that all commercial fishing, no matter how small in scale, is conducted by men. This male involvement in commercial fishing dates back at least to the mid-1970s and perhaps beyond. Dahl (2000:130) argues that commercial fishing tends to be a male venture, while subsistence fishing, which carries low or no social status, is conducted by both male and female fishers. He writes that the transition to commercial fishing in Greenland has "rendered women's labour in the domestic-oriented process of production...superfluous." While the scale of commercial fishing in Arviat cannot be compared to villages in west Greenland, in both regions it is an exclusively male pursuit. Yet, despite this, women claim that they continue to be Arviat's predominant fishers, and that men, if they fish at all, have only adopted the practice in recent years.

My participation in and observation of the sea-run char fishery in 2003 and 2006 reveal a striking imbalance in the numbers of men and women involved. Of the very small number of women who told me they fished for themselves along the coastline, many lived in households without an adult male, and they fished with sisters, adult daughters or other female family members. They travelled a short distance from Arviat to Nuvuk and Huluraq to set nets at low tide. The number of women who told me they set nets (but none of whom I actually observed) was small in comparison to the number of men I observed setting nets. Throughout each day, as I worked at my own net, travelled around the coastline by Honda, or engaged in hunting and fishing activity by boat, I observed and interacted with many men coming from and going to their nets, extracting and gutting char, or laying their nets out to dry.

Therefore, a discrepancy existed between women's perceptions of male and female fishing roles and actual practices.This discrepancy is by no means Nunavut-wide. Anthropologist Peter Bates told me of his experiences setting nets with women fishers in Cambridge Bay (personal communication). An environmental explanation for this difference in practice could be related to regional tidal differences. In Arviat, where tides are central to fishing practice, nets are tended at each low tide. In Cambridge Bay, where tides are minimal, fishing can take place at any time of day or night, and fishers are less restricted by tidal cycles. This importance of tides to Arviat char fishers has the most impact on people engaged in formal employment. As the next section will demonstrate, female waged employment is on the increase in Arviat, and this may be related to the decrease in the numbers of women setting nets along the shore.

Female Waged Employment and the Decrease in Female Char Fishers

Full-time employment has risen in Arviat over the past few years, and in the 2006 census, 43.3% of the population over age 15 classed itself as 'employed.' The percentage is the same when broken down by gender (43.4% of men, 43.5% of women). Previously, the main employers were the stores, hamlet and schools. The three stores—the Northern Store, Padlei Co-op and Eskimo Point Lumber Supply—employ men and women full- and part-time as shop assistants, clerks, managers, etc. Seasonal employment is also available at these stores each summer when ships arrive with supplies. The hamlet employs women and men in clerical jobs and men are employed also to drive water, garbage and sewage trucks. The elementary school employs mostly local Inuit teachers, while *qablunaat* teachers are in the majority at the middle and high schools. Until recently, there were only a few other sources of full-time employment in Arviat. Secretarial and administrative positions were (and continue to be) held by women at the offices of Arviat Hunters and Trappers Organisation, the employment office, etc. Arviat holds one seat in the Nunavut Legislative Assembly, and in 2006, was also the home-town of Nunavut's Member of Parliament in Ottawa. Both of these elected officials employ female administrative assistants in their Arviat offices.

However, recent years have seen the arrival of a new source of full-time employment in Arviat, partially leading to the aforementioned population increase. The creation and decentralization of the Nunavut Government have led to the establishment of offices of various government departments in the larger communities throughout the territory. Arviat is home to the Departments of Education and Housing, as well as government-related offices such as the Nunavut Planning Commission and the Kivalliq Inuit Association. Nunavut Arctic College has its administrative headquarters in Arviat, while the Departments of Environment and Renewable Resources also have large Arviat offices.

The arrival of these government institutions has led to a marked increase in the population of the community, with newcomers arriving from across Nunavut and from southern Canada, as well as return migration of Arviarmiut who had previously moved away. However, these government departments are also important employers of local Arviarmiut, and in particular, local women. In recent years, there has been a dramatic increase in the number of women gaining clerical and other qualifications and taking up jobs in these offices. According to the 2006 census (Statistics Canada), 50 Arviat women are employed in 'business, finance and administration occupations,' compared to 25 men, and 85 women have 'occupations in social science, education, government service and religion' compared to 30 men. In comparison, men continue to work in more traditional fields of formal employment such as trades and construction work.

Therefore, despite the rapidly growing population and a still heavy reliance on social welfare payments, full-time formal employment in Arviat is on the increase, due both directly (government institutions) and indirectly (construction and services) to the creation of Nunavut. Wenzel (1989:17) argues that there are now "two strata of Inuit emerging: those who are employed and have direct access to money but little available time; and those who hunt but are critically short on cash support." It is this first group of Inuit who are of interest in the context of the sea-run char fishery. As more women advance further in formal education and employment, their opportunities for involvement in the shoreline char fishery appear to have decreased.

Challenges for Women Fishers

As outlined at the start of this chapter, fishing in Arviat is important both as a subsistence and as a leisure activity. This is true for both sea-run char and all other types of fishing. Setting nets is explained by both men and women as providing an opportunity to relax, get away from home and community for short periods of time, and to spend time with family and friends in a natural setting. At the same time, sea-run char make an important economic and nutritional contribution to Arviat households.

Some women in Arviat told me they were actively engaged in char fishing in the past, but no longer are. All of these women gave work-related reasons for ceasing to fish. Women in full-time employment said they no longer have the time to go to Nuvuk to set and check their nets, and so have not fished since taking up full-time formal employment at various government departments, or as clerical workers at other government-related institutions. These women all expressed regret at not being able to fish in summer, but said work came first, and the need for a cash income to support their families was greater than the need to fish. One woman said, "Low tide is the best time to fish, but if you have a job or something to do, you just can't do it."

Men are not constrained by work commitments in the same way women appear to be. During those few weeks of summer, many men work all day in formal employment, and then tend their nets in the evenings, some remaining overnight at their nets as tides dictate, returning to Arviat each morning in order to go to work. Working women are much more constrained by their duties within the home, having to care for children and other family members. Indeed, remaining within the home and town after work and at weekends actually facilitates male fishing practices. As mentioned above, fishing, despite its high economic value, is perceived by many Arviarmiut as a mere leisure pursuit. Framed in this way, women have chores to attend to in the home once the working day ends, leaving them with less time to pursue 'leisure' activities.

For these women, access to fish is not the central issue as they have access to fish through informal kin-based sharing networks, acquiring char from their husbands, sons, brothers, and other male family members. What is of greater importance is that these women say they miss the actual practice of fishing. One woman, who now works for the Department of Education, said fishing gave her the opportunity to get away from her family and from the bustle of town, and to chat with her fishing partner in the peace and quiet of the shoreline. Another woman said she enjoyed the solitude of fishing alone early in the morning: "It's a great feeling to be out there on a summer morning at 6 a.m. Just you and the sea. It's so quiet and peaceful. You are all alone and it is the most peaceful feeling in the world." These women conceded that, as a result of being in full-time employment, they were more confined to town during evenings and weekends, and did not have the same opportunities to get out of town they once enjoyed.

Do men in full-time formal employment experience the same confinement in town? From my experience and observations, they do not. Many men in full-time employment still take the time to tend nets and participate in other subsistence/leisure activities. These men are not restrained by household chores or child-minding duties to the same extent as women.

For both women and men, the experience of peace and solitude is one of the most important aspects of summer char fishing. Many find the confines of town and life lived in overcrowded houses difficult, and look forward with great anticipation to the chance to get away from it all for a few hours or days. Women, who once had these opportunities, are now more confined to the community due to their work and family commitments. Traditionally, entire families moved along the coastline in summer, going to Sandy Point or elsewhere. As indicated by Brack and McIntosh (1963), in the past, entire families camped at fishing sites throughout each summer. It is now a much rarer occurrence for families to go fishing for sea-run char. Fishing at Sandy Point has now become an exclusively male activity. At weekends throughout summer, groups of men regularly make the 100 kilometre boat journey north to Sandy Point, setting out on Friday afternoons, once work has ended. On some weekends, up to sixty men might depart for Sandy Point in twenty to thirty boats, setting their nets along the shore and engaging in beluga and seal hunting as opportunities arise. What was once a place of fishing activity for entire families of both sexes has now become an activity in which only men (and predominantly adult men) engage. Women and children have become excluded from these fishing opportunities. While male-only social and physical spaces have expanded, female spaces continue to become more restricted and confined to the built environment of the town.

Knowledge, Skill and Place

As with all other social and subsistence activities, fishing involves the sharing, production, reproduction and maintenance of information, knowledge, skill, techniques and materials. While engaged in fishing activities during summer 2003 and 2006, I have observed men regularly engaged in conversation with their fishing partners or with men at other nets. Information regarding good fishing locations or innovative techniques are shared during the course of conversation. Nuvuk and Huluraq, where most shoreline net fishing takes place, are places of intense social activity, where fish are netted within rich webs of social relationships. Fishers meet as they travel to and fro along the sandy trails of Nuvuk, and on the diurnal beaches on Huluraq. During the course of these meetings, social relationships are created and strengthened as people discuss their fishing experiences, share cigarettes, chewing tobacco, and char, and help each other out in various fishing activities. This informal social engagement leads to the production and reproduction of social and environmental knowledge—knowledge about fish, birds, mammals; knowledge about tides, winds, seaweed; and knowledge about fishing practice. As women are more and more excluded from fishing due to work commitments and the increasingly male-dominated fishery, they are also excluded from this social interaction and the reproduction of social and environmental knowledge. Younger women—daughters, younger female siblings—who might previously have accompanied women on their trips to the shoreline no longer have the opportunity to do so. Boys, while still accompanying their fathers on such trips, no longer accompany their mothers, and are thus denied access to female forms of knowledge, communication and sociality.

Knowledge is created and exists in its practice and in its accompanying embodied skills (Lave 1993; Palsson 1994; Ingold 2000). The increasing exclusion of women from fishing due to the greater demands made by full-time employment results in fewer opportunities for fishing skills as well as fewer opportunities for the body of environmental knowledge within which these fishing skills are embedded, to be practised or passed on to younger generations of women. As women spend more time at the office and less time at the shoreline, they become further excluded from these social interactions. While they may still have access to fishing information through the informal communication networks that exist within the community, they no longer have the opportunities to participate in or include younger generations of women in fishing. As Nuttall (1998) points out, fishing helps maintain a cultural identity that allows for the continuity of local livelihoods. It also embodies notions of human-environment and human-animal relationships. One can only speculate that if these trends away from the shoreline continue, then future generations of Arviat women will fish less and less, if at all. Will anyone feel the loss of environmental knowledge and fishing skills attendant on this move away from the shoreline? While we may never quite grasp this loss of knowledge and skill, what is clear is that this move away from the shoreline is already having an impact on notions of identity and belonging among Arviat women. This may explain why women still identify themselves as Arviat's predominant shoreline fishers, when in reality, they no longer are.

Conclusion

In Arviat, women are involved, to varying degrees, in different forms of fishing throughout the year. The fresh- and salt-water derbies in May attract entire families, with women, children and elders predominating in the salt-water event. Lake and river fishing in spring is often a family event, although fishing in these same places in fall is usually a more exclusively male activity. Despite the high economic and social value of all these events, they are very often perceived by Arviarmiut to be leisure pursuits, and are not attributed the same high economic or social status as other subsistence activities such as caribou, seal or beluga whale hunting.

The Arctic char fishery is unique in the consciousness of Arviarmiut, as women perceive themselves to have long been the predominant shoreline char fishers, while my observations suggest that male fishers predominate. I have no conclusive evidence regarding the historical involvement of women in this type of fishing activity. McEachern (1978), conducting research in the late 1970s, found that the amount of fishing taking place around Arviat had contracted since the 1960s, but like the Nunavut Wildlife Harvest Study almost 25 years later, he failed to distinguish between male and female shoreline fishing taking place close to the community.

Since 2001, many women in Arviat have told me that they fished along the shoreline each summer before taking up full-time employment, and many said they regretted no longer having the time to do so. Their continuing family and household commitments had to be attended to at the end of the working day, and this prevented them from tending fishing nets. Of the men engaged in shoreline fishing during summer 2003 and 2006, many were the husbands or partners of women in full-time employment. The employment status of these men (unemployed, semi-employed, employed) seemed to have little or no bearing on participation in fishing activities along the shoreline (although lack of access to boats prevented some men from participating in char fishing and beluga hunting by boat).

The impacts on knowledge, skill and fishing practice are potentially far-reaching for these women who are constrained through full-time employment and lack of time. Daughters and other female family members are effectively also excluded from the places and social spaces where fishing takes place. The social relationships that women create through fishing along the shore, and the knowledge and skills they gain through fishing alone or in the company of other women, are no longer taking place.

Men gave no indication as to whether their fishing activity had increased or decreased from the past. Fishing from boats has always been a male activity, but whether men were more recent converts to shoreline fishing remains unclear. While women in full-time employment had been removed from the sea-run char fishery, full-time employment did not place the same constraints on men. In the past, when there was less or no waged employment, entire families travelled together to fishing sites, camping for extended periods of times before moving elsewhere. These once multi-gendered and multi-generational spaces, have become transformed into male-only spaces of weekend activity. Socially, therefore, the char fishery has become a male domain, although it seems that commercially, it may have been so for a much longer time.

The greatest regret expressed by women was the loss of the social time they once shared along the shoreline. Few opportunities exist for women to get out of town in the company of other women, to relax and share stories and learn together. At the same time, an essential contribution to the subsistence economy is also being lost. As a result, women also expressed regret that their own daughters are not now acquiring fishing skills and environmental knowledge, and are not learning to negotiate the social relationships attendant on fishing practice in the company of mothers, siblings and other female family members. Therefore, with each passing year, shoreline fishing falls farther and farther from the world of women, as it becomes a more central feature of the summer world of men.

Acknowledgements: I wish to thank the many Arviarmiut who have contributed to and supported this research. The knowledge, skills and advice of William Angalik, Angie Eetak, Judy Issakiark, Jenny Kalluak, Rhoda Karetak, Suzie Muckpah, Theresa and Arden Nibgoarsi, Frank Nutarasungnik, and Melanie Tabvahtah, were particularly valuable with regard to understanding fishing practices in Arviat. I thank my char-fishing partner Crystal Burgess for her great company, and Martha and Gord Main for their valuable insights. I wish to thank Tim Ingold, Nancy Wachowich and Peter Bates at University of Aberdeen for their helpful comments and advice.

References

Bodenhorn, B. (1990). "I'm Not the Great Hunter, My Wife Is": Iñupiat and anthropological models of gender. *Études/Inuit/Studies* 14(1-2): 55-74.

Brack, D.M. and D. McIntosh (1963). *Keewatin mainland area economic survey and regional appraisal report.* Ottawa: Industrial Division, Department of Northern Affairs and National Resources.

Dahl, J. (2000). *Saqqaq: An Inuit hunting community in the modern world.* Toronto: University of Toronto Press.

Dahlberg, F., ed. (1981). *Woman the gatherer.* New Haven and London: Yale University Press.

DiLeonardo, M. (1991). *Gender at the crossroads of knowledge: Feminist anthropology in the post-modern era.* Berkeley: University of California Press.

Freeman, M.M.R. (2005). '"Just one more time before I die": Securing the relationship between Inuit and whales in the Arctic regions,' pp. 59-76 in Kishigami, N. and J. Savelle, eds. *Indigenous use and management of marine resources.* Senri Ethnological Studies 67. Osaka: National Museum of Ethnology.

Ingold, T. (2000). *The perception of the environment: Essays in livelihood, dwelling and skill.* London: Routledge.

Jolles, C. Z. and Kaningok (1991). Qayuutat and Angyapiget: Gender relations and subsistence activities in Sivuqaq (Gambell, St. Lawrence Island, Alaska). *Études/Inuit/Studies* 15(2): 23-53.

Keith, R.F., T. Fenge, P. Jacobs and S. J. Woods. (1987). Arctic fisheries: New approaches for troubled waters. Northern perspectives 15(4): 1-5.

Kuhnlein H.V., L.H.M. Chan, O. Receveur and G.M. Egeland (2003). 'Arctic Indigenous Peoples, traditional food systems and POPs,' pp. 22-40 in T. Fenge and D. Downey, eds. *Northern lights against POPs: Combating Toxic Threats at the Top of the World.* Montréal: McGill-Queen's University Press.

Lave, J. (1993). 'The practice of learning,' pp. 3-32 in S. Chaiklin and J. Lave, eds. *Understanding practice.* New York: Cambridge University Press.

McEachern, J. (1978). *A survey of resource harvesting, Eskimo Point, N.W.T.* Delta, B.C.: Quest Socio-Economic Consultants Inc.

Moore, H. (1989). *Feminism and anthropology.* Minneapolis: University of Minnesota Press.

NWMB (2002). *The Nunavut Wildlife Harvest Study: Five year preliminary report for Arviat.* Iqaluit: NWMB.

Nuttall, M. (1998). 'Gender, indigenous knowledge and development in the Arctic,' pp. 67-73 in T. Greiffenberg, ed. *Development in the Arctic. Proceedings from a Symposium in Slettestrand, Denmark.* Copenhagen: Dansk Polar Centre.

Ortner, S. (1974). 'Is female to male as nature is to culture?,' pp. 67-88 in M. Rosaldo, and L. Lamphere, eds. *Women, Culture and Socieety.* Stanford: Stanford University Press.

Ortner, S. and H. Whitehead, eds. (1981). *Sexual meanings: The cultural construction of gender and sexuality.* Cambridge: Cambridge University Press.

Palsson, G. (1994). Enskilment at sea. *Man* (Journal of the Royal Anthropological Institute) 29(4): 901-928.

Rousselot, J.L., W.W. Fitzhugh and A. Crowell (1988). 'Maritime economies of the North Pacific Rim,' pp. 151-172 in W.W. Fitzhugh and A. Crowell, eds. *Crossroads of continents: Cultures of Siberia and Alaska.* Washington, D.C.: The Smithsonian Institute.

Statistics Canada (2006). *Community Profiles: Arviat.* Electronic document: http://www12.statcan.ca/english/census06/data/profiles/community

Stern, P. (2000). Subsistence: Work and leisure. *Études/Inuit/Studies* 24 (1): 9-24.

Tyrrell, M. 2005. *Inuit perception, knowledge and use of the sea in Arviat, Nunavut.* Ph.D. thesis. Department of Anthropology. University of Aberdeen. Unpublished.

Wenzel, G. W. (1989). Sealing at Clyde River, N.W.T.: A discussion of Inuit economy. *Études/Inuit/Studies* 13 (1): 3-22.

Wenzel, G. W. (1991). *Animal rights–human rights: Ecology, economy and ideology in the Canadian Arctic.* Toronto: University of Toronto Press.

CHAPTER FOUR

Are Living Fish Better Than Dead Fillets? The Invisibility and Power of Icelandic Women in Aquaculture and the Fishery Economy

Anna Karlsdóttir

Abstract: This chapter examines women's status and participation in the Icelandic fisheries and in aquaculture. The first section focuses on women's involvement in economic activities in both the traditional and industrial fishery sectors. The second presents the results of studies on Icelandic women's participation in aquaculture and provides women's views on fisheries with an emphasis on gender and regional development. Women's contributions to the fishery economy have historically and currently not been emphasized. Women are active in different occupations in aquaculture and state that work in aquaculture is more rewarding than in the fish processing industry. The tendency for women to migrate from communities that are based on fisheries and aquaculture prevails. As more Icelandic women attain further education, their opportunities for jobs in managerial and research-based positions will improve.

Introduction

Iceland is traditionally a rural agricultural society in which fisheries have been the mainstay of the economy and one of the drivers behind modernization and industrialization of the Icelandic economy in the late twentieth century. The participation of Icelandic women in the labour market has, during the twentieth and early twenty-first centuries, been higher than in other Western neighbouring countries in all economic sectors. Even so, women rarely attain leading positions in the industry, despite the fact that women in the last several decades have, to a large extent, received a higher education than men. Furthermore, economic development in Iceland during this time has led to uneven regional development between the capital and rural coastal areas. De-population of the coastal areas has occurred as a result, with women remaining a marginalized labour force with few economic alternatives. Therefore, it is important to gain insight into women's views on what is promoting and maintaining this marginalization.

Methodologies

This chapter presents findings from studies conducted by the author on the role and status of Icelandic women involved in fisheries and aquaculture. The empirical data referred to in the chapter was collected between 2004 and 2007 and is supplemented by the relevant social science literature. Both quantitative and qualitative methodologies were used including surveys distributed by post as well as a phone survey and personal interviews with 105 individuals. The first fieldwork phase surveyed the status of women in aquaculture across the whole sector from production to research and development. The goal of this study was to assess women's involvement in decision-making at various levels (Karlsdóttir 2005, 2006). In 2005, the author was involved in a consultative committee established by the Icelandic fishery minister and was responsible for compiling perceptual survey data from leaders of 23 of the largest fishery firms in Iceland with an annual revenue greater than 500 million ISK (equivalent to $7.5 million USD). In 2006, research was conducted on women's roles in the fishery in the Eastfjords of Iceland, a region undergoing significant occupational changes (Karlsdóttir 2006b). The fourth fieldwork phase in 2007 included research in the north east region of Iceland and along the coast, where sustainable rural lifestyles were examined among farmers and fishers with emphasis on women's household roles (Karlsdóttir 2008; Benediktsson *et al.* 2008).

Better to Leave than to Stay at Home

Women's demographic status in rural areas and along the coast in Iceland has rarely been examined. A Nordic Congress and a Nordic Regional Policy Committee on women's movement from rural areas were convened in 1990 at the initiative of the Nordic Council of Ministers. At the instigation of the Ministry of Social Affairs, the National Economic Institute in Iceland gathered statistical data especially on women's participation in the labour market in the late twentieth century. By 1990, the migration of Icelandic women from the rural and coastal areas was considered grave on a comparative Nordic basis (Félagsmálaráðuneytið 2000; þjóðhagsstofnun 2000). This out-migration was due to the fact that women rejected out-dated working conditions, isolation and paternalism because these social patterns favoured men. The response of women was to escape this situation. In the beginning of the 1990s, the socio-economic prospects for women in the rural and coastal areas were not particularly good. According to sociologist Kerstin Gulbrand, the core of the problem was that young women were primarily perceived as potential wives who should be in their place when needed. They were not perceived as worthy individuals in their own right and very seldom were accorded status in the transformation of Icelandic society (Árnadóttir 1990). In recent years, demographic patterns relating to gender have attracted greater attention by scholars (Bjarnason and Thorlindsson 2006; Corbett 2007; Milbourne 2007; Rasmussen 2007; Rauhut *et al.* 2008).

A study by The National Economic Institute of Iceland on the status of women illustrated that the most significant movement between communities occurred amongst women in their thirties and that there was dominant unemployment amongst this group during the winter. The study also revealed that while the full-time employment rate for women increased in the 1990s in the capital city of Reykjavik, the overall full-time employment rate of women in rural areas declined. Additionally, the average salary of rural women was 6% lower than for women in the capital area and approximately 53% of the average salary for their male counterparts (þjóðhagsstofnun 2000). Women along the coast and in rural areas also had fewer higher educational opportunities. A significantly higher rate of women outside the capital in almost all age categories only have a primary education. From a regional development standpoint, the situation of women in rural and coastal areas, when compared to their counterparts in Reykjavík, was unequal.

The Invisibility of Women in Icelandic Fishery History

The involvement of women's perspectives and the recognition of women's roles in fishery management, fishery communities and the fishery economy has been limited. This is despite the fact that historical records show that by the fourteenth (but more obviously by the eighteenth) century, women were active agents in many aspects of the fishery (Þorláksson 2003; Bragadóttir 2003). However, the invisibility of women's agency in the fishery in written data over time is surprising. Women took an active part in fisheries during the herring era (approximately 1940 to 1960), and they were important workers in the harbour. There is a local, well known song about Marta who salts, all the time calling for a new barrel and salt. This song celebrates the efforts and spirit of the hardworking fisherwomen.

Hún Hríseyjar-Marta með hárið sitt svarta var fræg fyrir kátínu forðum á síld. Og það hressti okkur alla að heyra' hana kalla: "Hæ, tunnu! Hæ, tunnu! Hæ, salt,meira salt!" (Author unknown)	"Black-haired Marta from Hrisey (island by North Iceland) was renowned for spreading joy during the herring season. It pleased us all to hear her shout: Hey barrel, hey barrel, more salt, more salt!"

In the 1980s, historian Þórunn Magnúsdóttir studied fishery registries from 1891 to 1981. During this time, women from all parts of the country fished and 3,394 women were registered as 'seawomen.' It is therefore peculiar that it is often claimed that fisheries are not women's work. Even in 2003, 680 women, of a total fisheries labour force of 4,480 people (or almost 15%) were registered as active fisheries employees. This represents only 10% and a steadily decreasing part of total active labour force in Iceland.

Table 1 shows the total workforce involved in fishing and includes women's participation. It illustrates that women in north east Iceland, Reykjanes peninsula and in the west fjords have been most active in the fisheries.

In the mid-1990s, two social scientists began conducting research on the gendered nature of the work of fishery processing workers. Sociologist Guðbjörg Linda Rafnsdóttir focused on the status of female fish processing workers, their perceptions of income, and their affiliation to women's labour organizations (Rafnsdóttir 1997, 1998), while anthropologist Unnur Dís Skaptadóttir concentrated on the social construction of gender and the gender-based division of labour within the fishery industry through a qualitative study of fish processing women (Skaptadóttir 1997a, b). These scientists determined that technological development within the fishing industry has had an impact on the involvement of women. In the wake of the introduction of the production line, and later, when information and computer technology were incorporated into production processes, women became almost solely involved in trimming and worm picking as well as the packing of fish. This meant doing the work at the pace of the flow line (under a bonus system) where little concern was devoted to training and experience. Their work was marginalized and it was believed that anyone possessed the skills to do this work (Skaptadóttir, 1997a,b, 1998, 2001). Skaptadóttir found that the gender-based division of labour became even more defined in fish processing as a result of increased technological development and, as a result, women's jobs were devalued.

Women and Fisheries in Iceland Following the Implementation of the Quota System

In Iceland, the individual transferable quota system was implemented gradually from 1980 (capelin) to 1984 (cod) and then was predominant as the resource management system for all species in 1990. This transferable quota system then applied to all commercial fisheries (Icelandic Parliament 1990: *Laws on the Regulation of Fisheries* nr.38/1990). Anthropologist Hulda Proppé (2002) studied women's views, identities and experiences of the quota system. According to Proppé, no women were publicly involved in the decision-making related to the implementation of the new fishery regulation system and no women were selected as members on publicly appointed committees.

An inevitable redistribution of ownership rights to the resource occurred in the wake of the implementation of the new resource management system, both regionally as well as on a company level. Some of the families of fishermen with a registered history of fishing inherited the quota. Some kept the quota and made use of it by practising fisheries, others rented out quotas for longer or shorter periods to vessels, and still others sold it to increasingly consolidated companies. Among those who inherited quotas were daughters

Table 1. Number of active fishers in Iceland by gender and region (1998-2003) based on information from tax bills. Source: Statistics Iceland 2004

Region	Year					
	1998	1999	2000	2001	2002	2003
Capital region	1270	1300	1150	1040	1110	1030
♀	100	100	100	90	130	130
West Iceland	590	590	610	580	610	560
♀	90	100	120	100	110	100
West fjords	680	650	530	560	480	440
♀	180	170	80	90	70	70
North-west Iceland	280	270	250	230	230	230
♀	40	40	30	40	40	30
North-east Iceland	1070	1060	970	900	990	960
♀	120	130	170	170	150	150
East Iceland	600	530	510	510	530	470
♀	70	50	50	50	50	50
South Iceland	720	720	730	690	670	630
♀	50	70	100	100	100	80
Reykjanes Peninsula	710	720	650	540	520	460
♀	120	170	130	80	70	70

and fishermen's wives. However, since ownership of access to the resource and practising fisheries are not always linked, these women are not necessarily directly involved in the fishery sector but are mainly capital holders of the access to the resource. A few, however, still remain active in running operations.

In general, women are not visible in public debates on resource management which have been prominent in the media since the late 1990s. A review of the media coverage on the quota system and its impacts identifies 1,677 articles for the period 2000-2005. Only one of those articles referred to women and, in particular, to a survey conducted on women and their reasons for choosing political candidates during the election period (Morgunblaðið 2003). It is fair to conclude that in almost all levels of society related to fisheries, women are rendered invisible. In the public discourse on the impacts of the individual transferable quota system, the focus has been almost solely related to impacts on the economic and financial spheres (Forsætisráðuneyti 2000; Hall *et al.* 2001). In more recent years, the focus has shifted to the impact of the quota system on regional settlement. Discussions on how the system affects the household income of fisher families including women's lives and opportunities have been minimal. Certain policies in fisheries management significantly impact on the lives of individuals and families and therefore, will affect the sustainability and development of communities and have other less evident consequences. Resource management systems will also affect how people define their identities because the system contributes to a certain order in which people are assigned positions and roles.

In an extensive study by Proppé (2002), as well as my own research on Icelandic women's representation in official committees, it could be concluded that the quota distribution in Iceland did not involve women other than peripherally. However, as agents within fishery-families, women were affected to a great extent. Proppé found that women's agency and behaviour reflected resistance against the bureaucratic system as well as compliance with it. People who decide to keep living in a fishing village are both complying with the situation in the fisheries and resisting it. They accept it by participat-

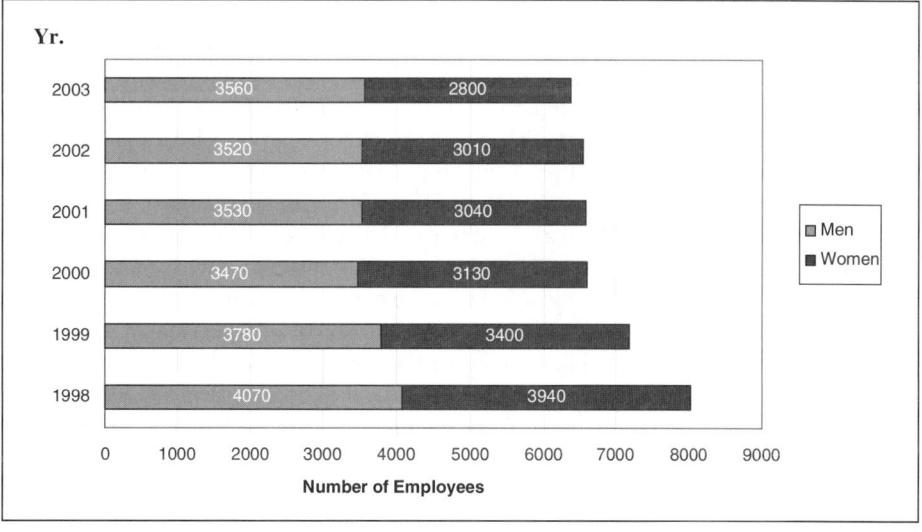

Figure 1. *Number of employees divided by gender in the fish processing industry (1998-2003). Source: Statistics Iceland 2004*

ing in the quota system and convince themselves and others that behaviour other than 'playing along' is unacceptable. They are, in fact, not content with the rules, but accept them believing they have no other options. Resistance includes maintaining a livelihood in a fishing village despite the fact that the quota system is rarely beneficial for their existence on a personal basis. Also, the quota system only grants fishing rights to some individuals and communities and excludes others. By deciding to continue life in an unstable economic environment, people oppose mainstream ideas on lifestyle and rational behaviour upon which the government's quota policy is based. People maintain and protect their lifestyle regardless of its challenges and this involves an opposition to the quota system. It is the resistance of those who do not hold power in the public sphere. However, this is, in some ways, a double-edged sword because it is the agent who experiences reduced rights rather than those against whom the resistance is directed. Women take part in this resistance through maintaining a collective spirit and collaboration in the community or through assuming stereotypical ideas about female behaviour by placing emphasis on their roles as wives and mothers. In this way, they strengthen their ties to men and oppose a societal pattern where a household requires two breadwinners (Proppé 2002).

Many stereotypical perceptions of women's abilities still exist in fisheries and this perception is not only limited to Iceland (*see* Thomas 2004). Among some women is a deeply ingrained belief that certain occupations within fisheries suit women better than men and vice versa. There are also other jobs where women and men are equally qualified (*see also* Karlsdóttir 2006a,b). Very rarely is this explained by anything other than natural attributes. Women in decision-making roles are often involved in small family businesses where a husband and wife work equally. The division of labour is not always clearly defined but, in more practical terms, women are often responsible for bookkeeping and tasks that, in more hierarchical settings, would be termed as managerial jobs. The paradox is that women often do not see themselves as being in charge and believe that they are the supporting rather than the decision-making arm of the business. This occurs even though their role is crucial in exercising power within a business. Women often express the view that they are not suited to rule and some even state that they need to mature before they can assume a leading role. Some of these opinions seem surprising from the viewpoint that they are not generally applied to men. Many female participants taking part in the study by Proppé (2002), considered themselves and the women they knew as suitable for managerial and decision-making roles in the community. This observation is also supported in other studies involving interviewing women both in fisheries and aquaculture (Karlsdóttir 2005, 2006b). At the same time, the tone was negative about strong and decisive women. One woman expressed the view that if a woman is too visible in a family fishery company, she is considered a 'she-devil' and would be nicknamed with a negative 'male-like' term by both women and men. It is worth considering why the few women who are visible in leading positions in fisheries are considered provocative and why they have to accept being the subjects of controversy in order to survive. Women who do not assume 'legitimate' roles and behaviour or who think first about their career and have no children, are exposed to scrutiny and criticism within society (Proppé 2002). These factors might very well keep women from stepping forth as leaders within the fishery sector, even though they possess the same qualities and qualifications as men.

Gendered Division of the Labour Market

In almost all societies, a certain gendered division of labour prevails and the caretaking of children and the sick, education and household tasks rests with the women (Korabik *et al.* 2003; Poelmans *et al.* 2003). According to public perception, women's tasks are usually related to domestic work, whereas physically demanding jobs (not to claim that domestic work is not physically demanding) are often relegated to men (Anker 1997). Men's tasks are often more technical or scientific in nature and demand more vocational training and education. Women who do not fit within that frame are deemed controversial (Anker 1997). These perceptions can prove dangerous over the long-term for the social adaptation of societies in transition where the need for dual-income families is well-established (Korabik *et al.* 2003). Along with these perceptions, the gendered division of work tasks is maintained and has negative effects for women's status in society. It shapes the relationship of men to women and the identity of women themselves. By that token, it is easier to conclude that women are less qualified labour, and that it is natural that they are responsible for less valued tasks. Their income is kept low and their careers are different from men (Anker 1998; Melkas and Anker 1998).

According to Richard Anker (1997), occupational segregation by sex has often been explained according to the neo-classical human capital model. This model assumes that workers and employers are rational and that labour markets function efficiently. According to this theory, workers seek out the best paying jobs after taking into consideration their own personal assets, constraints, and preferences.Therefore, the upbringing and care-taking of children and the primary responsibility for household tasks are constraints that influence women's preferences. Because of competition and efficient labour markets, employers pay workers their marginal product, which in many cases leads to women's lower income. This constraint means that women's opportunities for a career outside the home are disadvantaged compared to men. Obviously, women's responsibilities within the household affects their choices, and women might be more attracted to jobs offering more flexibility in working hours. Independent of whether scholars agree with the human capital model, studies have shown that women's work patterns are likely to be affected by workplace and residential location.The constraints of these workplace patterns will vary for different groups of women. Thus, geography has implications for the existence of occupational segregation (Burnell 1997). In spite of substantial feminist critique about the neo-classical models in the feminist literature (Pratt and Hanson 1990; Ferber 1995; Moe 2003) the ideas derived from neo-classical human capital theories in the 1970s and 1980s seem still to prevail indirectly to some extent, especially in Iceland's rural scene (Karlsdóttir 2007). No one would openly suggest that women are not seeking further opportunities or access to decision-making in the labour market because family and children are their main priority. However, though this approach and the ingrained norms and perceptions bear little relation to the daily lives of many women, it does not necessarily detract from their influence on people's behaviour and their contribution to gender-based discrimination against women. Some research on gender and competence indicates that the scientific and 'gender neutral' technique of measuring competence used by employers conceals values of masculine identities around which the organizational realities are structured (Rees and Garnsey 2003).

Since 1920, Icelandic women's rate of working outside the home has been significantly higher than that of Nordic neighbouring countries (Einarsdóttir 2004). According to Table 2, almost 80% of Icelandic women are working in the formal labour market. The majority of them or 63% work full time and 37% work part-time, while 90% of men work full time and only 10% work part-time. According to this statistical overview, Icelandic men use 49 hours on the average in fulfilling a full-time work week, while women use 43 hours on average. In 2001, the gender equality gap in research and development functions in the labour market was decreasing, indicating that women's educational level necessary for fulfilling positions in that strata of the labour market is being met. The unemployment rate in Iceland is low when compared globally. However, this overview does not show the connection between regional variances in women's participation in the labour market.

Table 2. Participation in labour market divided by gender (2003).
Source: Statistics Iceland 2004

	%	
	Women	Men
Participation in labour market 16-74 years	78	86
Ratio working full time (35+ hours /week)	63	90
Ratio working half time (<35 hours /week)	37	10
Average work hours full time	43	49
Unemployment rate	3	4

Women currently make up the majority of university students in Iceland (Háskóli Íslands 2005). However, women are still less favoured than men when it comes to governance and access to decision-making. There is a substantial pool of well-educated female labour in Iceland, so household priorities cannot solely be blamed for the low percentage of women in management.

Women in Fisheries and Aquaculture: Access to Decision-Making

It is my experience that according to public perception, women are thought to be only involved in fish processing. However, several hundred women are registered as fishermen as illustrated in Table 1. Women are also to be found in a wide array of jobs connected to the fishery sector. However, women's involvement in management and access to decision-making is low. Only one woman holds the chair of a board in a fishery company and only one woman has been a member of the stakeholder organization of boat-owners (Hauksson 2004; Karlsdóttir 2005). Both women are wives of entrepreneurs in the field. Family relations seem to provide access to power in the sector more so than personal, educational or other formal qualifications.

A study conducted on women's representation in aquaculture involved an examination of registered aquaculture companies (in the official firma register, 'firmaskrá', as well as members of the coalition of Icelandic Association of fish processors) and gathering data on staff representation and the gendered division of gender. The total participation of all relevant parties in this study was possible given the size of the Icelandic population and the small number of companies involved. Only 36 women are involved in

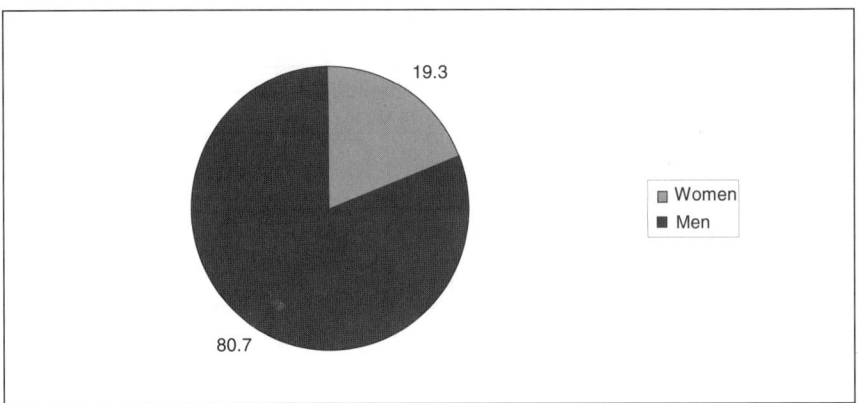

Figure 2. *Percentage of employees in aquaculture divided by gender (2004.)*

fish farming compared to 200 individuals overall. In aquaculture, only three women sit as board members in all 32 fish farming companies.

No woman has ever served on the board of The Iceland Aquaculture Association (TIAA) that is part of the coalition of the Icelandic Association of Fish Processors. Two women have served as board members of this latter organization since its inception. The Ministry of Fisheries has, for the last several years, emphasized the working conditions and other aspects of female labour in fish processing. In three different committees devoted to fish processing, the gender division of committee members has been even. Almost all other committees related to fisheries in Iceland demonstrate a more uneven gender division as shown in Table 3. In general, the situation since 1971 has improved regarding gender representation in public committees in Iceland. It is interesting to note that the two ministries in charge of aquaculture have the least number of women involved in committees or public working groups. Women's representation has improved slightly- increasing from 12-13% to 20% since 2003 when a study was conducted (Hlöðversdóttir 2003). This confirms a picture of male-dominated stakeholders.

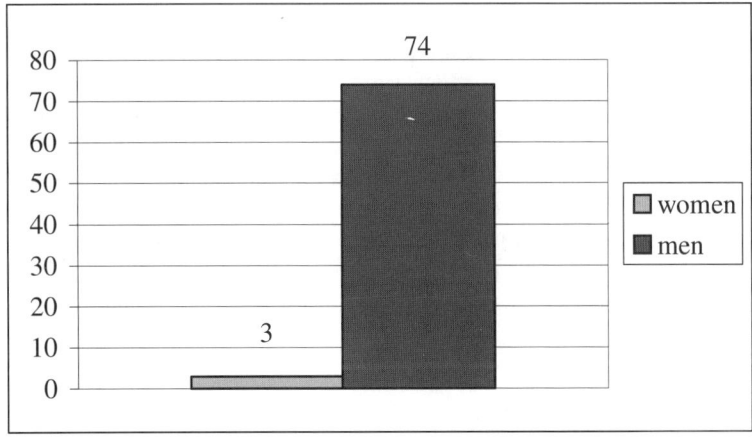

Figure 3. *Chairs of boards in aquaculture operations divided by gender (2004).*
Phone survey 2004.

Table 3. Gender distribution of committees related to aquaculture

Consultative Committees		Total membership	Female Members	% Female Members	Female Deputy	% Female Deputy
Ministry of Agriculture	4	23	4	23.5%	1	16.6%
Ministry of Fisheries	4	22	5	22.5%	0	-
Ministry of Environment	6	48	14	37%	6	60%

Table 4. State -initiated boards, committees and councils (1971-2003).
Statistics Iceland 2004.

Year	Percentage	
	Women	Men
1971	3	97
1980	7	93
1983	9	91
1987	11	89
1990*	17	83
1994*	21	79
2000	26	73
2001–2003	29	71

Several public institutes are charged with various monitoring, licensing and research tasks within fisheries and aquaculture. The Ministry of Fisheries is responsible for ocean-based species and the Marine Research Institute and the Icelandic Fisheries Laboratories are primarily involved in project and advisory roles in connection with operation permits, research and development collaboration on the farming of cod, haddock, halibut, turbot and leopard fish. The Directorate of Fisheries and The Institute of Freshwater Fisheries are the primary public institutes operating under the auspices of the Ministry of Agriculture. Besides being in charge of the administration of freshwater fisheries promoting the sustainable use of salmonid fisheries resources, these organizations also issue operation permits for fish farming production. Table 5 illustrates the number of women and men involved in the operation of public organizations and institutes that monitor and/or conduct research on aquaculture in Iceland. As illustrated, the relative number of female aquaculture specialists is striking, with the majority of these situated in urban centres such as Reykjavík or Akureyri.

Table 5. Gender representation in governmental institutions related to aquaculture and fisheries (2004)

Governmental bodies	Total number of female staff	Total number of male staff	Female aquaculture specialists	Male aquaculture specialists
Ministry of Fisheries	11	10	1	1
The Marine Research Institute	42	84	4	5
Icelandic Fisheries Laboratory	34	20	5	2
Ministry of Agriculture	14	10	0	1
Directorate of Fisheries	-	2-5	-	1
Institute of Freshwater Fisheries	5	13	2	7

Status of Fishery and Aquaculture Jobs

Thirty-one women involved in fish farming were interviewed in a study conducted between 2002 and 2004 on women's representation in aquaculture. The interviews were conducted during the summer of 2004 in four different locations in Iceland. The majority of these women were occupied in the slaughtering, foddering[1], fry and fingerling part of production or in fish processing. Compared to Norway, women in Icelandic aquaculture are involved in more diverse functions of this industry (Pettersen and Alsos 2004). In this study, seven women were owners or managers of fish farms and five women were involved in research and development. Many of the women interviewed had previous experience in conventional fish processing—the land-based plants for the processing of fish caught at sea. Some of the women lost their jobs following the implementation of the quota system and the Hazard Analysis and Critical Control Points (HACCP) as well as increased quality control requirements in the industry. Almost all of the women interviewed stated that their jobs in fish farming were perceived by society to hold more status than conventional jobs in fish processing even though job responsibilities were similar. Most of the women stated that their jobs in aquaculture were more multifaceted and flexible and not as monotonous as those in fish processing.

One limitation of this phase of the study is that we did not interview men and investigate their perceptions of women's participation in aquaculture but in a later work made for the Minister of Fisheries Committee (Sjávarútvegsráðuneyti 2007; Karlsdóttir

---

[1] Feeding the fish

2006a), executives, predominantly male persons, responded to a survey indicating their perceptions of women's roles in fisheries, fish farming, and fish processing. This could have enriched women's perspectives of working in fish farming. However, supplemental information from employers did paint a picture of women as comprising an effective and reliable workforce. Women themselves stated that working in aquaculture was livelier than working in fish processing and that it was more rewarding to deal with living fish than dead fillets. One of our informants with extensive experience in the field claimed that the mentality in the workplace had improved, whereas earlier, women in the more male-dominated workplace risked being harassed.

Even so, women's status in aquaculture in Iceland is not encouraging. A female research and development officer in aquaculture did perceive the situation in fisheries to be male-dominated because women do not put themselves forward. They look around and, as young women, realize that they do not have the same or equal opportunities as men. They do not gain power in fisheries and their voices are only heard to a limited extent. It is therefore logical that they look to other spheres, educate themselves further and leave this sector. There is not much space for their efforts and resources. There have been so many changes within the fishery industry and with technological development, a transformation has occurred. Fish processing is geographically moving to China and the jobs within the fish processing industry have become less attractive. Working life has changed and the workplace is not as lively. Instead of a collective break where there would be time for jokes and laughter, only three people are sent to break at a time, because the process is ongoing. The processing line is too broad for them to keep up conversations across the line while working. The jobs are less flexible and more monotonous. However, a female manager interviewed for this study stated that there had been much improvement in the work environment. For example, women are not standing in a cold and wet environment while also struggling with cystitis. She perceives aquaculture as a more systematic food industry that is dependent on different laws than the fishery sector. It is more in line with industrial manufacturing and she believes that women were as qualified as men to take on managerial or other senior positions.

Lack of Recognition?

Most of the women interviewed were modest when asked about their roles in aquaculture and downplayed the importance of their contributions. Many women living in rural areas stated that being involved in aquaculture secured a stable income throughout the year whereas other job opportunities were more seasonal. For example, jobs in tourism generally were only available during the summer, so year-round jobs in aquaculture were considered lucrative. The women involved in establishing fish farms were generally the wives or daughters of entrepreneurs in the field. Some stated that neighbours or other people in the village were, to some extent, suspicious and not very supportive toward the establishment of an aquaculture business. One woman stated that in the area she was living there was a tendency among inhabitants to hold people back and not let them "go too far with things." She explained that neighbours followed her and her family very closely and left them alone when the prospect of the business was not too good, but that the attitude changed when prospects improved and the company became profitable.

Only a minority of women participate in decision-making processes in aquaculture and these are owners or managers by virtue of family ties. Most women are never approached to take part in decision-making roles and the possibility of having influence was remote. Women who had attained decision-making roles felt lonely and isolated and stated that they felt more valuable in roles other than decision making. They felt they were not capable of mastering the communication skills necessary to deal effectively in power situations and that male colleagues treated them as a minority. Some stated that the domain of power relations was not appealing because it was a male-dominated world. Most of the informants were more positive toward involvement in governmental committees than the boards of the companies they worked for. Others would rather be in charge of the company they worked for. Most of the women felt more positively about the prospect of being in charge of the company they worked for, rather than being involved in governmental committee work. This may be because these women have first-hand experience of production processes and issues at the fish farm, while many believe that they have little awareness of political processes related to government committee work.

Many of the women stated that there was a difference in the way men and women carried out their jobs. Some of the oldest women claimed they had observed these differences for generations and that these varying attitudes likely influenced their career options. Women stated that women's approaches were characterized by responsibility and loyalty while men's approaches placed greater emphasis on impressing others and assuming more dominant roles. One of the researchers stated that it has been her experience that women are extremely conscientious and that they complete tasks for which they have been made responsible. She believes that a man will make sure his efforts are visible while a woman may not always be diligent about this. Being a sole woman among men in executive meetings also invites attention. One of the executives indicated that when she meets with foreign clients and partners, men seem surprised by her presence at the meetings. They will almost always ask her what her profession is when she enters the meeting room while her male colleagues are never asked. This bothers her as it is her presence, rather than that of her male counterparts, that is being questioned. In another study where executives of the 23 largest fishery firms were asked about different perceptual assumptions on women's work performance compared to male staff, over half of respondents agreed that women were more dutiful than their male colleges (Karlsdóttir 2006a).

One woman, who co-owned a fish farm with her husband, often feels that customers and others devalue her managerial position. Although she has worked alongside him for many years, customers will always ask to speak with her husband when they want to buy fish. She states that her husband does not have the overall awareness of certain business details. For example, her husband must always ask her if the fish is ready or not. In this way, she is bypassed by customers as a responsible agent within the company. Some of the women claim that it is more difficult for women to live in the coastal communities than in Reykjavik and that there were several examples of women leaving the area when their kids left to attend high school or college. There were not too many choices for women and they would apply for whatever job was available. One of the informants explained that part of the failure to acknowledge women's contributions to the rural economic sector in fisheries was due to a strong tendency to social uniformity—making it harder for women to move beyond traditional roles. Two women described how children who were not doing homework properly would be frightened by tales that they would end up in the fisheries, thus reflecting societal values not beneficial for the development and status of primary industries.

Conclusion

Will societal perceptions of men's and women's roles and abilities hinder the long-term sustainability of coastal and rural communities in Iceland? Can gender balance ever be attained in these areas? Are these contemporary values and attitudes determining the de-population of non-urban communities or are they signals of broader social change? If so, how will rural and coastal identities be reshaped and what impact will it have on women's opportunities to gain ground in occupations previously dominated by men? Many of the questions cannot be easily answered.

The most common response to decreasing job opportunities in the rural and coastal areas has been an increasing participation in higher education and further investigation of alternative employment especially in the growing tourism sector or handicraft production, small industrial activity, or construction. Results from recent studies of Icelandic women in aquaculture indicate that many women create job opportunities for themselves and their husbands by establishing businesses related to aquaculture. Many female employees in aquaculture perceive their options to be improved if they pursue further education or vocational training related to the field.

It can be concluded that the Icelandic government, labour unions, stakeholder organizations and employers within fisheries and aquaculture have not acknowledged women's participation in and contributions to this sector. These organizations recognize women's presence but there exists limited awareness that women represent important links in the connection between the choice to live in rural and/or coastal settlements and available economic opportunities. Women who are influential, either through owning a business, acting in management or supervisory roles or as researchers, are usually highly educated or possess family ties that assist them in tapping into decision-making processes.

Public authorities have not developed any strategies to encourage women to pursue non-traditional occupational choices nor to remain living in the rural or coastal areas in the way that Norway has done[2] (Jóhannesson 2001). During the 1990s, the number of women from rural and coastal areas seeking further education increased more rapidly than among women from urban areas. This is partly because of increased educational opportunities by virtue of more universities and higher education institutions located around the country and also because of improved opportunities for distance education. There is, therefore, some hope that the future survival of coastal and rural areas will be based on more equal terms between women and men. To make that possible, there needs to be greater emphasis placed on empowering women.

Various female fisheries grassroots organizations exist in other nations including Norway, the Netherlands, Spain and France. Their members are fisherwomen, employers in fish processing, wives of fishermen or aquaculture producers and employees in aquaculture. No organizations like this exist in Iceland. Through these organizations, it is possible to become influential so that women's contributions cannot be ignored. In some

---

[2] Since the 1950s, the central Norwegian authorities have practised a regional development policy with emphasis on sparsely populated areas in the northern regions and in areas with population density less than one person per 12.5 square kilometres. Various projects and strategies were pursued in the last decades of the twentieth century to slow down the migration of women from those areas and projects have been established to encourage income opportunities.

European women's fisheries and aquaculture organizations, members have influence in the most significant levels of government, stakeholder organizations and on regional and supranational boards for fishery management.

As indicated in discussion on social uniformity among women in small coastal and rural villages, there seems to be a contradictory sense of well-being among the women interviewed for my study. On one hand, the possibilities of women advancing in decision-making positions in fisheries despite greater opportunities in higher education, are not overly promising, so women are kept in place in conventional female positions if they enter primary industries at all. At the same time, their roles are not valued so that the attraction of living in the coastal villages turns into aversion and children are reminded that a livelihood in this region is not sustainable or beneficial for them in the future. This situation impacts on inhabitants and on women's perception and aspirations in a negative way. By that token, women 'vote with their feet' by weakening by degrees the future foundation of coastal and rural settlements while securing an education and a future for their offspring. It is crucial for the continued future of coastal villages and rural areas to initiate social change and to secure jobs and opportunities for all women. However, many women do not perceive fisheries or aquaculture as a viable career. It is hoped that when more women enter the management of fisheries and aquaculture, greater numbers of women will follow.

Aquaculture is an industry with future economic potential for women and men in Iceland as it becomes clear that fishery resources are being fully exploited and that there exists an extensive knowledge base on marine and freshwater fish biota within the nation. Icelanders have a long tradition of knowledge development within fisheries and freshwater resources, and this has, to some extent, been transferable to the arena of aquaculture (Gunnarsson 2004). New technological knowledge, greater precision and more stability have been accomplished with some negative effects, but also with better management, more research and an increasingly well-educated workforce. It is likely that small-scale char farming will increase and that there will be a greater number of sea-based species farmed. It is also possible that fish farming will implement and apply the use of biotechnical knowledge to a greater extent than is currently practised. To make it possible for aquaculture to thrive, the industry needs innovative spirit, a renewal of educated labour, a competent staff in all arenas, training and an acknowledgement of women's roles to an equal extent as men. Education plays a key role, not only to strengthen self-esteem but also to invigorate the industry. Women living rurally or in the coastal villages are a significant part of whatever changes will occur in the future of Iceland.

References

Anker, R. (1997).Theories of occupational segregation by sex: An overview. *International Labour Review* 136(3): 315-39.

Anker, R. (1998). *Gender and Jobs: Sex Segregation of Occupations in the World*. Geneva, International Labour Office.

Árnadóttir, B. (1990). "Því flykkjast þær suður": Interview with Gulbrand. *VERA, Tímarit um konur og kvenfrelsi* 9(6): 15-18.

Bjarnason, T and T. Thorlindsson (2006). "Should I stay or should I go?" Migration expectations among youth in Icelandic fishing and farming communities. *Journal of Rural Studies* 22(3): 290-300.

Benediktsson, K., M. Júlíusdóttir, and A. Karlsdóttir (2008). Litróf landbúnaðarsamfélagsins. In *Fræðaþing Landbúnaðarins*. Bændasamtök Íslands.

Bragadóttir, R. (2003). *Róðu betur, kær minn karl—Af sjókonum á 18. öld. Fyrirlestur ámálþinginu Hvar er minn sess?* Af 18. aldar konum. RIKK og Félag um 18. aldar fræði, 15 febrúar 2003, Háskóli Íslands.

Bryman, A. (2004). *Social Research Methods*. 2nd.Edition. Oxford: Oxford University Press.

Burnell, B. S. (1997). Some reflections on the spatial dimensions of occupational segregation. *Feminist Economics* 3(3): 69-86.

Corbett, M. (2007). All kinds of potential: Women and out-migration in an Atlantic Canadian coastal community. *Journal of Rural Studies* 23(4): 430-442.

Einarsdóttir, Þ. (2004). *Gender-Segregated Labor in Iceland: Characteristics and Causes*. Presented at the conference Women Power and The Law, Center for Gender Studies at the University of Iceland, August 27, 2004.

Félagsmálaráðuneytið 2000. *Athugun á stöðu kvenna á landsbyggðinni*, júní 2000: 4.

Ferber, M.E. (1995). The Study of Economics: A Feminist critique. *The American Economic Review* 85(2), 357-361.

Firmaskrá Íslands (1998 – 2005). *Firmaskrá Islands*. Electronic document: http://www.firmaskra.is/ Last accessed: May 1, 2008.

Forsætisráðuneytið (2000) *Auðlindanefnd – Álitsgerð, Reykjavík*. 1-192.

Gunnarsson,V.I. (2004). *Staða og framtíðaráform í íslensku fiskeldi*. Landbúnaðar- og sjávarútvegsráðuneyti, Reykjavík.

Hagstofa Íslands (2004). Konur og karlar 2004. *Hagtíðindi* 137: 1-72.

Hall, A., Á. Jónsson, S. Agnarsson, and T.Þ. Herbertsson (2001). Staðleysur og staðreyndir um íslenska kvótakerfið. *Morgunblaðið*, June 9.

Hauksson, J.G. (2004). Konur í viðskiptalífinu-áhrifamestu konurnar 11-70. In Frjáls Verslun— Sérrit um viðskipta-, efnahags- og atvinnumál. *Heimur hf. Reykjavík* 66(10): 64-74.

Háskóli Íslands (2005). *Jafnréttismál–Jafnrétti í tölum*. Electronic document: http://www.hi.is/page/jafnrettistolur Last accessed: July 10, 2005.

Hlöðversdóttir, B.Í. (2003) *Völd, tengsl og eðli nefnda, stjórna og ráða hjá hinu opinbera og fyrirtækja sem skráð eru hjá Kauphöll Íslands–þátttaka kvenna*, BSc. thesis. University of Iceland.

Icelandic Parliament (1990). *Laws on the Regulation of fisheries* nr. 38.

Jóhannesson, B. (2001). *Atvinnuþróun og stoðkerfi atvinnulífs á landsbyggðinni*. Reykjavík: Byggðastofnun.

Karlsdóttir, A. (2005). *Þáttaka kvenna í ákvarðanatöku í fiskeldi. Rannsóknarstofa í kvenna- og kynjafræðum, Jarð- og landfræðiskor*. Reykjavík: University of Iceland.

Karlsdóttir, A. (2006a). 'Women and fisheries follow-up survey,' pp.71-78 in L. Sloan et. al, eds. *Women and Natural Resource Management in the Rural North: Arctic Council Sustainable Development Working Group 2004– 2006*. Norway: Forlaget Nord.

Karlsdóttir, A. (2006b). 'Women´s role and situation in the fishery sector in the Eastfjords of Iceland,' pp.79-96 in L. Sloan et. al, eds. *Women and Natural Resource Management in the Rural North: Arctic Council Sustainable Development Working Group 2004-2006*. Norway: Forlaget Nord.

Karlsdóttir, A. (2007). 'Kvinders deltagelse i beslutningsprocesser i fiskeopdræt og fiskeri – kon sekvenser for regional udvikling,' pp.182-195 in G.L. Rafnsdóttir, ed. *Arbete och Välfärd: Arbete och Demokrati*. Reykjavík: Nordisk Ministerråd / Vestnorden / Háskólaútgáfan.

Karlsdóttir, A. (2008). 'Not Sure about the Shore! Transformation effects of ITQs on Iceland's Fishing Economy and Communities,' in M.E. Lowe and C. Carothers, eds. *Community Impacts of Fisheries Privatization*. American Fisheries Society Press.

Korabik, K. *et al.* (2003). A multi-level approach to Cross Cultural work-family research—A Micro and Macro Perspective. *International Journal of Cross Cultural Management: CCM* 3(3): 289-303

Magnúsdóttir, Þ. (1988). *Sjókonur á Íslandi 1891-1981.* Reykjavík: Ritsafn Sagnfræð-istofnunar.

Melkas, H. and R.Anker, eds. (1998). *Gender equality and occupational segregation in Nordic labour markets.* Geneva: International Labour Office.

Milbourne, P. (2007). Re-populating rural studies: Migration, movements and mobilities. *Journal of of rural studies.* 23(3): 381-386.

Moe, K.S., ed. (2003). *Women, Family and Work: Writings on the Economics of Gender.* Malden, MA:Blackwell Publishing

Morgunblaðið (2003). Kvótakerfið á landsbyggðinni og velferðarmál meðal kvenna. *Morgunblaðið* April 15. Reykjavík:Árvakur.

Pettersen, L.T. and G.A. Alsos (2004). 'The Role of Women in Norwegian Fish Farming,' in *AKTEA International Conference proceedings:Women in fisheries and aquaculture: lessons from the past, current actions and dreams for the future.* November 10-13, Santiago De Compostela: Université Bretagne.

Poelmans, S.AY, N. Chinchilla, and P. Cardona (2003). The adoption of family friendly HRM policies: Competing for scarce resources in the labour market. *International Journal of Manpower* 24(2): 128-47.

Pratt, G. and S.Hanson (1988). On the links between home and work: Family—household strategies in a buoyant labour market. *International Journal of Urban and Regional Research* 14: 55-74.

Proppé, R.H. (2002). Konurnar og kvótinn: Kynhugmyndir og upplifun kvenna af orðræðu og auðlindastefnu í sjávarútvegi, óútgefin MA ritgerð í mannfræði við Háskóla Íslands, nr. 1839.

Proppé. R.H. (2003). 'Ég sé kvótakerfið fyrir mér sem lopapeysu—kynhugmyndir og upplifun kvenna af orðræðu og auðlindastefnu í sjávarútvegi,' pp. 425-33 in F. Jónsson, ed., *Rannsóknir í Félagsvísindum.* Reykjavík: Félagsvísindastofnun Háskóla Íslands / Háskólaútgáfan.

Rauhut, D., *et al.* (2008). The demographic challenge to the Nordic countries, *Nordregio working Paper* 2008: 1.

Rees, B. and E. Garnsey (2003). Analysing Competence: Gender and identity at Work. *Gender, Work and Organisation* 10(5): 551-78.

Rannsóknarstofnun fiskiðnaðarins (2004). *Ársskýrsla 2003.* Reykjavík: RF.

Rafnsdóttir, G.L. (1997). 'Valkyrjur eða ambáttir? Sjálfsbjargarviðleitni íslenskra kvenna,' pp.130-135 in H. Kress and R. Traustadóttir, eds. *Íslenskar kvennarannsóknir.* Reykjavík: Háskóli Íslands og Rannsóknarstofa í kvennafræðum.

Rafnsdóttir, G.L. (1998). 'Kynferði og sjávarbyggðir—Sjónarhorn Félagsfræði,' pp. 233-44 in F. Jónsson, ed. *Rannsóknir í Félagsvísindum II.* Reykjavík: Félagsvísindastofnun H.Í. og Hag fræðistofnun H.Í., Háskólaútgáfan.

Rasmussen, R.O. (2007). 'Gender and Generation Perspectives on Arctic Communities in Transition,' pp 1-10 in P. Kankaanpää *et al.,* eds. *Knowledge and Power in the Arctic. Proceedings at a conference in Rovaniemi, Finland, April 16-18. 2007.* Arctic Centre Report No. 48. Rovaniemi: University of Finland and Arctic Centre.

Samtök fiskvinnslustöðva (2004a). *Inn- og útflutningur sjávarafurða.* Electronic document: http://www.sf.is/innutflutningur.asp. Last accessed: February 26, 2009.

Samtök fiskvinnslustöðva (2004b). *Landssamband fiskeldisstöðva.* Electronic document: http://www.lfh.is/ Last accessed: February 26, 2009.

Sjávarútvegsráðuneyti (2007). *Skýrsla nefndar um störf kvenna í stærstusjávarútvegsfyrirtækjunum á Íslandi.* Reykjavík.

Sloan, L. *et.al* (2004). *Women's Participation in Decision-making Processes in Arctic Fisheries Resource Management.* Nordfold: Arctic Council.

Skaptadóttir, U.D. (1997a). 'Breytingar í sjávarbyggðum: Sýn kvenna,' pp. 173-87 in G. Pálsson, H, Ólafsson & S.D. Kristmundsdóttir, eds. *Við og hinir. Rannsóknir í mannfræði*. Reykjavík: Mannfræðistofnun Háskóla Íslands / Háskólaútgáfan.

Skaptadóttir, U.D. (1997b). 'Birting margbreytileikans í íslenskri sjávarbyggð,' pp.121-29 in Kress, H and Traustadóttir, R, eds. *Íslenskar kvennarannsóknir*. Reykjavík: Háskóli Íslands.

Skaptadóttir, U.D. (1998). 'Verkaskipting kynja og vinnuskipulag í fiskiðnaði,' in F. Jónsson, ed. *Rannsóknir í Félagsvísindum II*. Reykjavík: Félagsvísindastofnun H.Í. og Hagfræðistofnun H.Í., Háskólaútgáfan.

Skaptadóttir, U.D. (2001). 'Kyngerving vinnunnar,' in F. Jónsson and I. Hannibalsson, eds. *Rannsóknir í Félagsvísindum III*, Reykjavík: Félagsvísindastofnun Háskóla Íslands, Háskólaútgáfan.

Thomas, M. (2004). "Get yourself a proper job girlie!": Recruitment, retention and women seafarers. *Maritime Policy and Management* 31(4): 309-318.

Þorláksson, H. (2003). 'Fiskur og höfðingar á Vestfjörðum: atvinnuvegir og höfðingar á Vestfjörðumfyrir 1500,' in E.Vestfirðir, *Ársrit sögufélags Ísfirðinga*, aflstöð íslenskrar sögu 43.

Þjóðhagsstofnun (2000). *Athugun á stöðu kvenna á landsbyggðinni*. Reykjavík: Flagsmálaráðuneytið.

CHAPTER FIVE

Everyone Goes Fishing: Gender and Procurement in the Canadian Arctic[1]

Kerrie-Ann Shannon

Abstract: This examination of fishing derbies in the Canadian Arctic provides insight into Inuit procurement and flexibility in gender roles. Through a focus on fishing derbies, this significant aspect of Inuit life is recognized. Many ethnographies and land use studies have previously concentrated on hunting and considered the activity as indicative of Inuit land use. The fishing derby provides an alternative ethnographic example of procurement; it is an activity in which women, children, elders and men participate. Women's roles in the Arctic have often been discussed in terms of gender division of labour or in terms of their complementarity to men's roles. From this example we may begin to understand how women are active participants in procurement without necessarily evaluating of their activity as a complementary aspect to men's hunting. A significant aspect of the fishing derby is that it demonstrates occasions when activities are not necessarily divided along gender lines and thereby reveals a degree of flexibility in gender roles.

Introduction

Hunting remains vital to Inuit life, identity and the economy and has been the dominant focus of much Arctic ethnography and land use studies (Balikci 1970; Birket-Smith 1936, 1945; Brody 1987, 2000; Caulfield 1997; Dahl 2000; Damas 1984; Freeman 1976; Freuchen 1961; Graburn 1969; Gubser 1965; McDonald *et al.* 1997; Nelson 1969; Nuttall 1992; Rasmussen 1908, 1929). In consequence, activities of procurement, such as fishing, have tended to be less emphasized. Within the context of gender, culture and northern fisheries, I suggest it is important to examine procurement beyond a gendered division of hunting and gathering. Fishing and fishing derbies are procurement activities in which women, children, and men participate.[2] An examination of fishing derbies leads to a more complete understanding of Inuit procurement and of gender relations in Inuit culture.

The fishing derby as an example of procurement is significant for three reasons. First, this important aspect of Inuit livelihood is recognized. Second, the fishing derby provides

---

[1] An earlier version of this paper was published in *Études Inuit Studies*, 2007: 30(1):9-29.
[2] A fishing derby is a contest to catch either the largest or the most fish, depending upon the rules of the particular contest.

balance to the previous gender bias and concentration on hunting. Because of this focused view on hunting in the Arctic, women's activities, especially those of procurement, have remained on the periphery. Women's activities have largely been examined and evaluated with respect to their contribution to hunting or as complementary to men's work. Circumpolar ethnographies and land use projects devote little attention to the activities of women outside of their contribution to hunting (Dahl 2000; Freeman 1976; McDonald *et al.* 1997; Brody 1987, 2000; Caulfield 1997; Nuttall 1992 and others). Even work which affords a broader definition of hunting to include women's work maintains a focus on hunting and has not specifically examined the significance of womens' activities outside the sphere of hunting (Bodenhorn 1990; Fienup-Riordan 1990a,b; Sharp 1981, 1994). It is this focus on hunting which I question. By examining fishing as procurement, we gain additional insight into Inuit skill and opportunity. Third, a more complete understanding of gender roles is attained through an examination of fishing and fishing derbies. Men and women have different roles in preparing for the fishing derby, which is consistent with an understanding of gender roles as complementary in Inuit society. However, a significant aspect of fishing is that it provides a counter example to gender division of labour in terms of hunting and gathering or complementary gender relations. I use the fishing derby example to argue that procurement has been too focused on hunting and suggest that by exploring fishing we can begin to understand women's involvement in procurement.

In this paper I will briefly describe the setting and methods. I will explore the implications of fishing as procurement within the hunter gatherer literature. I will discuss gender roles in the fishing derby specifically and in hunting and gathering societies more generally. The ethnographic example of the fishing derby provides insight to skill in Inuit procurement and people's relationship with the world around them as well as insight into gender roles in procurement as both men and women can be 'real fishermen.'

Setting

The inhabitants of Southampton Island have had a long history of contact with *Qallunaat* (the Inuktitut term for white people). The indigenous Sadlermiut of Southampton Island died out around 1902, as a result of disease (Bird 1953; Dunning 1962; Freeman 1969/70; Mathiassen 1945; Moyer 1970; Sutton 1932, 1934; VanStone 1959). Inuit were brought to the island through initiatives and relationships with *Qallunaat*, mainly with whalers and the Hudson's Bay Company.[3] The current population of Inuit on Southampton Island comprises primarily Avilimmiut and Uqummiut.

Fieldwork was conducted on Southampton Island in the predominantly Inuit community of Coral Harbour.[4] Coral Harbour has divisions based on kinship and religion. Community-wide contests, such as the fishing derbies, reflect a sense of community that can transcend these internal divisions. The residents of Coral Harbour travel in order to hunt, fish, or live for short periods of time in other locations, and they procure resources from most of the island. The concept of 'community,' therefore, not only refers to the

---

[3] In contrast with settlements such as Grise Fiord (Marcus 1992; Freeman 1969), the relocation was not forced.

[4] Community Permissions were attained and a Research License issued by the Nunavut Research Licensing Board.

hamlet but also incorporates all of Southampton Island, while the word 'town' is used to refer to the geographic area of the hamlet.

Coral Harbour is situated on the southern coast of Southampton Island, Nunavut, Canada. As the island's only community, Coral Harbour is located at 83 degrees longitude and 64 degrees latitude, well above the treeline and is characterized by an Arctic climate. The 'Community List,' updated in September 1999, numbers 737 people. The permanent settlement of Coral Harbour, established in the 1950s and 1960s, brought together Avilingmiut and Uqqumiut that had been living in dispersed camps on Southampton Island. These two groups of Inuit were brought to the island through their interaction with whalers and the Hudson's Bay Company.[5] The term 'community' is used throughout this paper for Coral Harbour, to refer to a broader area beyond the geographical confines of the town. The residents of Coral Harbour travel in order to hunt, fish, or live for short periods of time in other locations, and they procure resources from most of the island. The notion of 'community,' therefore, not only refers to the hamlet (or town) but also incorporates all other parts of Southampton Island where such activities take place. Despite divisions by place of origin or kinship, the majority of community residents participated in the fishing derby. I will briefly discuss my research methods, as I am in agreement with Emerson *et al.* (1995:11) that "what the ethnographer finds out is inherently connected with how she finds it out." Research was conducted employing a combination of methodological approaches from participant observation, formal and informal interviews to a public method of interviewing. As a female ethnographer, I spent much time with families and followed the daily routines of the households in which I lived. In addition I was able to attend the fishing derby in two consecutive years, 1999 and 2000. The participation aspect also involved apprenticeship learning, as I did not initially have the necessary skills to participate. I found it necessary first to spend time doing before asking more direct questions.

One of the special characteristics of participant observation is its very personal aspect.[6] Okely draws attention to the dangers of separating out the person collecting data from the theoreticians who interpret it. According to Okely, the field experience is not separable from theory, and fieldwork methods cannot be reduced to a set of 'laboratory procedures' (1992:3). In carrying out this project I not only relied upon suggested methods but also utilized an existing local system by conducting a radio call-in show (*see below*). Sunderland (1999), in her article *Fieldwork and the Phone*, describes her success in utilizing a technique which, like the radio call-in show, was not found in research methods books. I believe it is important to be open, flexible, and creative in choosing the most appropriate methods for research.

Past personal experience has been influential in data collection. Previous fieldwork in the same community was valuable for gaining a greater general understanding of Inuit culture as well as familiarity with local sensibilities and circumstances. This familiarity and understanding aided the design of research questions, as well as providing insight into the strategies that might prove most effective in accomplishing my research goals.

To initiate my research, a 'local radio call-in show' as a public form of interviewing was very useful (Shannon 2003).[7] It also proved influential in continuing the research

---

[5] The indigenous population of the Sadlermiut are thought to have died out around 1902.
[6] Bernard 1988; Crick 1982; Ellen 1984; Hastrup 1992; Okely and Callaway 1992.
[7] Local radio is an important part of daily life in Arctic communities of Canada (Briggs 2000; Creery 1994; Kishigami 2000). The local radio is an established practice used in

process and created an interview list, as well as a format for informal interviews. This combination of methodological approaches and strategies allowed me to collect a wide range of information initially and then, later, to seek answers to more specific questions.

Regarding my own learning processes, I was given some pointers on how to do things, but for the most part my instruction came from watching. This is how Inuit learn to do things. Instructions and step-by-step directions are not given. Often the way someone will show another how to do something is simply by beginning a project without saying a word. In his discussion of learning the blacksmith trade in East Africa, Coy states that "while there was little in the way of formal instruction, observation and trial and error predominated in the learning process" (1989:120). This is similar to what has been suggested more generally concerning the acquisition of traditional, local, or indigenous knowledge.

For Coy, apprenticeship learning is about engagement with others. In a similar vein, Palsson (1994a) discusses his own 'enskilment' in both doing ethnography and learning about fishing. Palsson states that the learning process is not merely about internalising knowledge but also about being "actively engaged with an environment" (Palsson 1994a: 901). Likewise, as a participant or apprentice, I was learning not only about the lives of Inuit, but also about how to conduct anthropological research. This learning process, begun in the field, still continues. As Palsson states, "we never actually leave the field as long as we take part in the ethnographic enterprise" (1994a:921-922).

Hunting-Gathering and Fishing

Recognizing the Importance of Fishing as Procurement

As a consequence of previous foci on hunting, other activities of procurement have tended to be less thoroughly addressed in the relevant literature. Some land-use studies and ethnographies do include procurement activities apart from hunting (Department of Education 1996; Freeman 2000; Ikuutaq 1984; McDonald *et al.* 1997; Riewe 1992; Usher 1975); however, this area of inquiry needs further elaboration. It is important to also explore these other, non-hunting procurement activities, which are often under-represented in Canadian Arctic ethnography. Similarly, Stewart (2005) argues that fishing is a crucial element of Inuit subsistence which has been under-reported in Canadian Arctic ethnography because of the major emphasis placed on hunting. In this paper, I focus on fishing as procurement which may be considered separate from hunting and may offer additional insight into Inuit procurement. As will be discussed in greater detail, fishing serves multiple functions which go beyond the subsistence importance of catching fish.

Fishing has often been difficult to classify. In differentiating between hunters and gatherers, Ingold suggests the terms refer "to what people do rather than what they eat" (Ingold 1986:80). He argues that the distinctions are based on the activities of pursuit and

---

communicating and soliciting information and therefore, it was a useful tool for my research for a variety of reasons. First, the radio was an important way to gather information. Second, the shows generated a list of potential interviewers. Third, the radio provided a context for informal interviews through communicating information about my research to the community. Furthermore, this tool provided an immediate dissemination of information.

collection, yet, when one examines fishing, it is sometimes an activity of pursuit and at other times one of collection (Ingold 1986:80).

I have specifically chosen the word 'procurement', as it does not specify a method of how a resource is obtained nor have implications with respect to gendered division of labour that the words 'hunting' and 'gathering' could. I define 'procurement' as any method of obtaining resources, whether it is by hunting, trapping, gathering, or fishing.[8] Procurement can include activities which are separate from hunting and therefore incorporates activities which also include participation of women. For example, a 'procured' resource can be fish, mussels, eggs, plants, or other animals. In this paper, I use 'procurement' to mean those resources obtained from the land,[9] but recognize the usefulness of the term given the complexities of goods exchanged which can include purchased material goods. The notion of procurement refers to more than just the action of obtaining a resource; it also indicates the way in which hunter-gatherers approach the world. Hence the activity of fishing is more than just the processes of getting fish but it is reflects how Inuit approach the world around them.

As Bird-David (1992:40) argues, the word procurement is "accurate enough for describing modern hunter-gatherers who apply care, sophistication and knowledge to their resource-getting activities". Ingold supports Bird-David's choice of this word:

> The notion of procurement nicely brings out what I have been most concerned to stress: that the activities we conventionally call hunting and gathering are forms of skilled, attentive 'coping' in the world, intentionally carried out by persons in an environment replete with other agentive powers of one kind and another (1996:149).

Thus, the notion of procurement helps us to move beyond a reductionist concept of hunting and gathering, or foraging, as an interaction in nature as well as providing a term which does not invoke gendered divisions of labour. In the next section, I will focus on fishing as procurement.

I suggest, that in order to understand how Inuit engage in procurement, it is also necessary to incorporate an understanding of how they 'fish.' Fishing by methods such as jigging or setting nets has been important to the economy and social life in the Arctic. Numerous ethnographic accounts have mentioned fishing in many parts of the Arctic. Boas (1964) identifies fishing as an aspect of documenting the traditional culture among the Baffin Island Inuit. More recent accounts, such as Dahl's (2000) depiction of a Greenlandic community, discuss fishing as part of a mixed economy. Other authors describe fishing as part of contemporary community life in the Canadian Arctic (*see* Dorais 1997). Fishing, of course, has varying importance in different locations, but is mentioned as a subsistence activity across the Arctic (Balikci 1970; Barker and Barker 1993; Briggs 1970; Brody 1987; Burch 1988; Graburn 1969; Gubser 1965; Hensel 1996; Honigmann and Honigmann 1965; Mauss and Beuchat 1979; Nuttall 1998c; Riches 1982; Wenzel

---

[8] It is important to note Inuit would probably not use one overarching term to refer to all their activities of obtaining a resource.

[9] The phrase "on the land," which is used in English by Inuit, refers to activities which are quite literally on terra firma, and sometimes to activities on lake-ice or sea-ice.

1991, amongst others). However, fishing itself and fishing derbies have not been the focus of detailed ethnographic inquiry in the Canadian Arctic.

The fishing derby is an event in which all participate, therefore, fishing may be more representative of procurement than hunting. In the same way, Palsson (1988, 1991) suggests that fishing in Iceland is as indicative of people's perception of, and engagement with, their environment as hunting. Although hunting has often been held to be representative of Inuit life, I suggest that fishing is also an important procurement activity. I am using the fishing derby as an example of how Inuit procure and by doing so we can move away from a focus on hunting as the main form of procurement. By switching the focus to fishing women's actions and activities can be viewed as an integral part of procurement rather than how their actions contribute to hunting or are complementary to hunting. As everyone participates in the fishing derby, I suggest it could be viewed as an epitome of procurement more so than hunting, which has limited participation.

Fishing contests and derbies are significant ways in which people utilize the land and interact with each other. Within the community, competitive games take place as part of everyday life at many levels and frequently involve some kind of hunting, gathering or fishing. The annual large fishing derbies as a large community sponsored event with significant prizes has a relatively recent history and has been organized in this manner for approximately fifteen years. Competition and games, however, are not a contemporary phenomenon.[10] Fishing contests are often a part of celebrating a holiday; for example, derbies are organized during Christmas, Easter and on Nunavut Day. Often, times to fish are limited with specific start and finish times; a contest may be as short as a few hours. The local contests are easy to participate in because people do not have to prepare as when they camp. Local contests, sometimes announced spontaneously, are met with an immediate response as people seize the opportunity to participate. In essence, people are always ready to fish.

Inuit are often involved in procurement activities through games and contests. I argue that it is often difficult to distinguish between a contest, such as the fishing derby, and a subsistence activity and perhaps this division is unnecessary. Working in the Inuvialuit Territory, Stern (2000:10) asserts that "traditionally, there was no need to distinguish between work and leisure activities." She argues that as Inuit have become accustomed to work schedules, they are more inclined to separate subsistence and leisure. For young adults, she states, "hunting is no longer a form of work. Rather, it is a leisure activity" (Stern 2000:10). Contrary to Stern, I argue that one should not separate the notions of work and leisure with respect to subsistence activities. Perhaps there is a separation between what is considered wage work and what is considered an activity of livelihood or subsistence work. A subsistence activity such as fishing or egg picking may be enjoyable but that does not make it recreation. In the same way, Hensel (1996) shows how "subsistence practices also connect areas of people's lives that are often separate in Euro-American culture. This is particularly true of what might be termed 'recreation'" (Hensel 1996:67). For example, if Yup'ik are asked what they do for fun and what they do for

---

[10] Games have been mentioned in various reports and ethnographies concerning the Arctic (Ager 1977; Attikutsiak 1999; Balikci 1970; Bennett *et al.* 1994; Boas 1964; Bodenhorn 1995; Briggs 1970; Burch 1988; Chance 1990; Graburn 1969; Glassford 1976; Gubser 1976, 1965; Heine 1998; Jansen 1979; Morrison and Germain 1995; Nelson 1969; Spencer 1959; Sprott 1997; Wilkinson 1955).

work, the answer is the same: hunting and fishing (Hensel 1996:67). In fact, competition is often used as a motivator for people and is involved in many activities of livelihood without classifying the activity as recreation.

Fishing derbies, like other activities of procurement, are not simply for catching fish. Many anthropologists comment on the cultural importance of subsistence activities. Writing of the Yup'ik relationship with their environment, Fienup-Riordan (1990a:47) observes that:

> it is possible but altogether inappropriate to reduce subsistence activities to mere survival techniques and their significance to the conquest of calories. Their pursuit is not simply a means to an end but an end in itself.

Fishing is an important cultural activity for the entire family and is a popular activity in many parts of the Arctic. In reference to fishing, Balikci (1970:35) suggests, "This was probably the happiest season of the Netsilik year—food was plentiful, the weather was warm, and there was no immediate cause for anxiety". People genuinely enjoy fishing. Oakes and Riewe mention that "almost everyone enjoys ice fishing..." (1995: 85). In other areas of the Arctic, such as the northern Bering Sea, "fishing on the spring ice was often a pleasant social occasion" (Rousselot et al. 1988:155). Without doing further field research, it is difficult to know whether similar contests take place throughout the Arctic, or whether they are unique to the Keewatin and Baffin regions of Nunavut.

Fishing and fishing contests become the focus of town life in the spring, dominating discussions and occupying people's time with the preparation and fundraising for such events. Much excitement is generated over the derbies. Indeed, fishing is one activity in which most community residents participate. Elders, children, men and women all fish during the derby and babies are taken along in the *amautik* (parka to carry a baby). Through my own participation in two annual fishing derbies in successive years, as well as participation in several other fishing contests, I provide a detailed ethnographic account of fishing elsewhere (Shannon 2003) and a summary here.

The preparation for the derby begins months before the derby itself. Male and female volunteers form a committee and plan the derby as well as raise funds for prizes. Prizes range from money ($5,000 CDN dollars) to a round-trip plane ticket worth over one thousand dollars to more modest prizes. Some resources are secured outside of the community but a majority of the funds are raised locally by donations to and participation in fundraisers. Penny sales, a popular fundraiser, are a mixture between a rummage sale, raffle, and bake sale. Various people or organizations donate goods to the penny sale, individuals then purchase tickets for a chance to win the item in a draw.[11] There may be several penny sales leading up to a fishing derby.

Gender and the Fishing Derby

Men and women are active in community preparation for the fishing derby and also for their own family preparations. Gender does influence the type of work that men and women perform for their own family. Family preparations are primarily divided along

---

[11] These sales are an important way of distributing goods in the community (*see* Shannon 2003 for more detail).

gender lines as women sew and prepare the clothing and food, while men focus on transportation by building the sleds and repairing snowmobiles. As work space is divided along gender lines, this influences social space. As will be discussed in more detail in the following section, the division of labour among Inuit is often considered complementary (Balikci 1970; Briggs 1974; Gubser 1965; Guemple 1986, 1988). Men and women typically undertake different tasks but the work is valued by the other gender and creates a complete household.

The gender division is flexible and as necessary tasks are carried out, men and women may perform the role of the opposite sex. For example, men usually drill the holes for fishing with a large ice auger. Although this is generally viewed as the task of men, on one occasion, there were an insufficient number of holes for all the women fishing. Although the women had never used an ice auger before, they drilled another hole. Similarly, Guemple discusses how the roles each gender performs does not necessarily relate to skill but rather to convention. This was similar to how tasks were divided in preparation for the fishing derby. Moreover, the way to learn something is to watch it. Men and women watch the work of the other gender and therefore, are ready to perform the task of the opposite gender if need be. When the women begin to use the ice auger, I asked them if they knew how to use it and they simply answered that they had watched it in use. This way of learning through observation coupled with a strong sense of autonomy contributes to flexibility in the performance of gendered tasks. For example, if a woman wants to hunt, she does and if a man wants to sew, he does. Not only is there flexibility in gender roles but there are times when activities are not necessarily divided by gender. The preparations for the derby are divided along gender lines, yet complementarity is not always part of the activity of fishing or part of the skill in fishing.[12] Significantly, the activity of fishing is not necessarily divided by gender.

Gender Roles in Hunting-Gathering and Fishing

Fishing and fishing derbies demonstrate that procurement is not necessarily divided by gender. As fishing is an activity in which women are active participants, we gain insight about women's activities in procurement which have been underrepresented. An examination of fishing also leads to a more complete understanding of gender roles in Inuit society. Gender in hunter-gatherer literature has previously focused on discussions concerning divisions of labour and the role of women in subsistence. Of key significance are the divided roles of men and women in resource attainment. Beginning with the 1966 *Man the Hunter* Conference (Lee and Devore 1968), research on hunter-gatherer societies in other regions outside of the Arctic has recognized the importance of women's activities because of their contributions to the food supply through gathering activities. The importance of women's contribution was elevated as the caloric significance of their contribution was recognized.

Despite this recognition of the importance of women, it took some time before the focus of research became more balanced. Slocum argued the bias was within the discipline itself and that culture is examined "almost entirely from a male point of view…"

---

[12] It is often difficult to separate the preparation for the derby and the activity of fishing. The preparation in town may be divided by gender but when men and women are fishing on the ice, it is important to recognize that this is an activity not necessarily divided by gender..

(1975:49). To offset the bias in anthropology, research began to focus on women, highlighting the significance of their contribution in many parts of the world (Dahlberg 1981). Women's participation through gathering was emphasized, and their overall input to food supply recognized as important. This focus has led to a more inclusive understanding of hunter-gatherers' lives and brought attention to women's roles in subsistence where they they had previously been overlooked.[13] In anthropology, significant contributions have been made which concentrate on women's roles.[14] Within the context of the hunter-gatherer literature, there is a limitation to shifting the focus and emphasizing one activity over the other. Gathering was considered the more important of the two activities but the same division of labour was still emphasized.

Because of the reliance of Arctic peoples upon hunting, the Arctic has been viewed as an exception to the conclusions of the *Man the Hunter* Conference. Gathering and other activities of procurement were considered unimportant because they offered little caloric contribution in the Arctic. Therefore, attention has remained primarily on hunting while other activities of procurement have been less emphasized. As activities of procurement can be valued for social importance as well as their contribution to the food supply, I suggest it is essential to consider procurement activities apart from hunting. It is a goal of my research to concentrate on this less addressed component of Inuit procurement and thereby achieve a broader understanding of procurement, where the active roles of women and children in procurement are also recognized.

If the Arctic has been viewed as an exception to the prevalent model of the division of labour-with men hunting and women gathering, then how have women's roles been discussed in the relevant literature? As previously mentioned, a division of labour among Inuit is often viewed as complementary. Briggs (1974) draws attention to the important role of women in her article *Eskimo Women, Makers of Men*. Briggs demonstrates the important position of women in an Inuit cultural context by highlighting the value of their complementary roles. In doing so, Briggs also makes a contribution to anthropological gendered studies by providing empirical evidence that runs counter to Ortner's (1974) argument regarding the universal subordination of women.[15] Although Ortner formulates universal claims that women are subordinate to men, Briggs demonstrates the valued role of women in Inuit society and thereby illustrates ethnographic evidence of egalitarian societies. Empirical work highlighting the varied positions of women, including women in egalitarian societies,[16] further weakened Ortner's argument.

---

[13] For example, Meehan (1982) focused her work on the collection of sea molluscs amongst the Anbarra, an Australian Aboriginal society...

[14] Previously, gender studies have stressed the inclusion of women in ethnographic research. This focus was intended to balance the previous male bias in anthropology. Many works have led to an increased awareness and consideration of women in various societies (Bodenhorn 1990, 1993; Briggs 1970, 1974; Dalhberg 1981; Fedorova 2000; Giffen 1930; Jolles 1994; Landes 1969; Leacock 1978, 1981; Leacock *et al.* 1986; Matthiasson 1974; Mearns 1994; Reiter 1975; Rival 1993; Rosaldo and Lamphere 1974; Sharp 1981, 1994; Strathern 1972, 1987, 1988; and, Weiner 1976, among others).

[15] Ortner's (1974) renowned article "Is Female to Male As Nature Is to Culture?", concentrated on the theme of the universal subordination of women.

[16] For a collection of the various roles of women from different societies, *see* Matthiasson (1974).

Guemple emphasizes the importance of examining gender within the terms of "natives' own conception"[17] and discusses gender in terms of what men and women do. Guemple (1986) describes the complementary roles of men and women in a way that is consistently reported in the literature on the Canadian Arctic; men were traditionally identified with the hunting of large game while women were traditionally identified with the home. Briggs describes similar divisions of labour between men and women and also recognizes that "the sexual division is not rigid" (Briggs 1974:270). Condon and Stern (1993) discuss the flexibility of gender roles in Holman and report on recent changes that have influenced understandings of gender identity. They express similar concerns and question the dichotomy in Guemple's model, as he does not explicitly address flexibility in gender roles. Recognizing that gender roles can be shifting is an important aspect of understanding these roles. The activity of fishing was a procurement activity not necessarily divided by gender. Furthermore, there was evidence of flexibility in gender roles during the fishing derby. While fishing, women sometimes performed men's work when necessary, as in the example of women drilling the fishing hole with the ice auger.

Gender roles in the Arctic are not always viewed as complementary. Other Arctic and Subarctic accounts have evaluated women's work in terms of how women contribute to hunting (Bodenhorn 1990; Fienup-Riordan 1990a,b; Sharp 1981, 1994). For instance, in writing of the Subarctic Chipewyan, Sharp explores the insignificant role of women in attaining food. He suggests women, "represent a null case with which our growing recognition of the significance of women's roles in subsistence must deal" (Sharp 1981:221). From this starting point, Sharp explores the actions of women in securing the food supply. In her work entitled "I'm Not the Great Hunter, My Wife Is," Bodenhorn (1990) explores women's contributions to subsistence economies through their hunting-related activities. According to Bodenhorn, hunting also incorporates many additional activities performed by men as well as women. This definition of hunting is valuable, for it illustrates that for Iñupiat, the division of labour by gender does not necessarily mean that there is a lack of involvement in hunting. The focus, however, remains on hunting and a gendered division of labour. In another study, exploring women's roles in subsistence among the Subarctic Chipewyan, Brumbach and Jarvenpa (1997:18) question the dichotomy between man the hunter and woman the gatherer:

> If these rigid assumptions have merit, then what of the role of women in circumpolar arctic and subarctic societies where plant foods contribute very little to the diet in terms of calories? Do women play any role in the food quest in these environments?

These are intriguing and promising questions to ask. Although there is merit in exploring women's activities in hunting, my research focuses on fishing as a procurement activity. I suggest that without also exploring fishing and other activities of procurement, hunting is viewed as being of sole importance and women's roles are evaluated in how they

---

[17] In this respect, the discourse of gender studies, similar to kinship studies, has come under close scrutiny and suffered analogous problems. Gender, like kinship, is viewed as being useful only when explored in terms of how people themselves make use of the concept. Schweitzer (2000) explains it is not important 'what kinship is but rather what kinship does." Similarly, it is important to ask what gender does..

contribute to hunting. It is this emphasis on hunting which I have questioned and has led me to examine fishing as a way to understand Inuit procurement. I suggest that without also examining fishing, as well as other activities of procurement, hunting dominates as being representative of Inuit procurement.

The Fishing Derby

The large fishing derby entails jigging through lake ice for Arctic char (*Salvelinusalpinus* and in Inuktitut *iqaluk*). Fishing is intense and people fish for long periods of time taking advantage of the extended hours of daylight. Some people fish late in the evening since there is a twilight glow until after midnight, while others fish early since dawn comes quickly. As people have been returning to the same lakes for several years, many camping spots are already determined. Although 'ownership' of land is not the norm, there is a certain sense of usufruct rights over certain areas. In contrast to a tent, a cabin has permanence to its structure and therefore creates a more or less exclusive claim to the use of the surrounding area. The increasing emergence of cabins also indicates how people are combining life in permanent settlements with a commitment to spending time out on the land. The cabins and tents are arranged around kinship which is different from town where living space is assigned.[18] Although people travel and camp with their relatives, they also socialize with non-kin during the derby and, at times, kin groups merge out on the ice.

Since the fishing derby involves participation from almost everyone in the community, the derby provides a social opportunity for people to interact in a way that may be different from the way they relate within the physical parameters of town. For example, people who might not visit one another in town may stand talking while fishing. In town, social interaction is divided both by kinship relations and gender. As previously mentioned, men and women have divided work and social space. Women socialize with relatives of the same sex, visiting in homes or in women's cabins. Visiting may entail helping in large tasks, such as cleaning a polar bear hide. Likewise, men spend time with male relatives frequently working in a garage. Work and socialization are often combined and relatives help one another, as labour is also part of a larger reciprocal sharing network. Relatives spend a lot of time together and, essentially, friends are relatives. Men and women may have friends who are not relatives but they are usually of the same sex. Gender divisions exist even within the kinship groups. For example, there is very little (if any) social interaction between women and their brother-in-laws. The separation of genders begins in childhood, as children imitate and follow the roles of adults of the same gender. The separation of genders among children is not rigid as children's autonomy is an important aspect of childrearing. Children may at times choose to imitate the roles of the opposite gender. As teens become young adults, this separation of genders is more apparent. The fishing derby, however, creates an occasion where both genders and kin and non-kin are closely interacting.

During the fishing derby, a gendered division of space is not always apparent. As the fishing derby lasts for several days, the social dynamics change. Some individuals may

---

[18] The settlement pattern in town is determined by a housing association. People may not necessarily live near kin. However, when people camp and set up cabins, they do so in close proximity to kin. In this way, spending time on the land is similar to how people lived in kin groups before the change to permanent settlements.

develop friendships because of a shared love of fishing, in this way, fishing may allow for a freedom of interaction outside of kin relations (but these are individuals usually of the same gender). A noticeable alteration in social interaction occurs when the fishing is intense and people occupy holes in close concentration to non-kin or individuals of the opposite gender. While fishing in this close proximity, Inuit may chat and socialize. Both men and women will fish intensely for periods of time. This interaction may also include joking and teasing, which may not necessarily occur in town, with the exception of large community-wide games. The fishing derby thereby creates an occasion for social interaction that may cross kinship and gender boundaries.

Opportunism and Skill

Fishing involves social interaction between people and is a significant component of the social relationship between people and the world around them. In the following section, I will explore how fishing can lead to an understanding of procurement in terms of opportunity and skill.

Procurement in an Arctic climate is greatly dependent upon dexterity and alacrity; people must be able to respond to opportunities with an eager readiness. As Brody describes, "readiness to move to ensure successful hunting can hardly be exaggerated" (1987:95). This sense of mobility was also apparent in the way Inuit engaged in fishing. The movement to different fishing holes is a typical example of how Inuit seize opportunities. It is not random movement but movement tied into previous experience and a sense of awareness. Through this knowledge and awareness, people are able to adjust their positions to increase their chances in fishing. Although the holes themselves provide a static indication of what is occurring as people's movements are restricted by the ice, this does not accurately reflect the amount of movement that takes place during the fishing derby. People are often moving in search of fish. Understanding the movement of fish is crucial to seizing an opportunity.

When someone began catching a lot of fish, people would rush over to the area with their ice augers to make new holes as close as possible to that person. As soon as a big fish was caught, it was quickly removed from the hook in order to try to catch another fish. A sequence of movements—catching the fish, removing it, and catching a new one—could happen repeatedly. I watched one fisherman catch about twenty good-sized char, one after the other, in less than fifteen minutes. The catching of this many fish got people excited, and they would competitively race to the holes surrounding the person in order to get a hole as close by as possible. Although fishing is a competition, this does not mean that competition and cooperation are necessarily mutually exclusive (*see* Ager 1977; Balikci 1970; Gubser 1965). The competition was based on the opportunity to catch the fish and was not necessarily directed against individuals.

There was not only a great deal of movement from one fishing hole[19] to another; there were also significant shifts in areas of the lake where people fished. Distances between groups could be a kilometre or greater. People would try fishing in another part of the lake, depending on their success in a previous location.

---

[19] Fishing holes are not owned by anyone, although it was expected someone in a group may bring an ice auger to drill holes.

People's movements are related to keen observations. Inuit learn from a young age to pay attention to their surroundings and the movement of others. As previously mentioned, observing is a key aspect of learning. In fact, watching and attending to the movement of terrestrial animals, fish or other people is reinforced by a caregiver's questions. Children are often asked things like "*Nowk* birdie?" ("Where is the bird?"). As the question is being asked, the person will take the baby to the window to look at the snow bunting. The way in which both young boys and girls are encouraged to 'watch' is not necessarily divided by gender. When I asked a woman how she knew the land, the response was that she was always taught it was important to watch. In this way, watching the land is inseparable from learning or knowing the land. Although boys often have more experience out on the land as they are taken on hunting trips, young girls are also encouraged to watch the land. For example, on the way home from the fishing derby, a three-year-old girl was praised for knowing the significance of passing sea-ice in relation to home while travelling.

As children grow older, there is a more apparent separation of their activities based on gender. Young boys may follow their fathers or uncles on hunting trips as young girls may watch as their mothers prepare a skin. Children learn the complementary roles of each gender by example but children have the autonomy to try activities that are typically done by the other gender.

Despite the apparent separation of tasks by gender, it is important to recognize that not all procurement is divided by gender. A significant aspect of fishing as procurement is that it is an activity not necessarily divided by gender. Fishing illustrates that women and girls know about the land in ways that are equivalent to male knowledge. I am not suggesting that knowledge of the land or environment is gendered but argue that without including women's activities in procurement, there exists a gendered bias in the literature which equates hunting and thereby, men's activities, as being representative of Inuit knowledge of the land. How people know the land separate from hunting is often overlooked and knowledge of the land is simply equated with knowledge gained through experience in hunting, as Brody (1987:71) comments:

> Hunters and trappers know about animals. Their knowledge is detailed and intimate. The details and intimacy are a personal science, a system of understanding that reveals and secures the people's absolute dependence on the land.

Brody's statement illustrates the general assumption that people's involvement with hunting can be equated with people's relationship to the land. What has been evident from the fishing derby is that both men and women, boys and girls, have knowledge and awareness about the land and their surroundings. By understanding procurement apart from hunting, we can value women's activities in procurement and appreciate that they too have knowledge about their environment in a manner that is similar to Brody's comment about hunters. Gender does not necessarily determine how people watch their surroundings or how they fish.

Awareness and knowledge are crucial to being able to act upon an opportunity. People are aware of the movements or locations of fish, not only through their own encounters but also through sharing this information with others as this is an important aspect of social discourse (Bodenhorn 2000; Hensel 1996). Awareness not only applies to fishing but also supports how other anthropologists have described people's actions in the Arctic and Subarctic. Ridington (1990) suggests that the hunter's acquisition of knowledge is a form of empowerment. Fienup-Riordan likewise highlights awareness, which allows people "a sense of control over their destiny" (1990a:168). Although Ridington and Fienup-Riordan both draw attention to the way knowledge empowers the hunter, awareness is not sufficient in itself. Fishing is not only dependent upon knowledge or awareness, it must also be coupled with opportunity and with the skill or dexterity to act. This is not to suggest that skill in seizing an opportunity and skill in awareness can be easily separated; both, however, must be present.

Awareness is often discussed in terms of traditional knowledge or traditional ecological knowledge but skill may be more applicable in order to describe Inuit procurement. The fishing derby provides an ethnographic example of Ingold's (2001) discussion of skill and exemplifies his argument regarding knowledge as skill. Skill involves knowledge learning by doing and with constant adjustment to the environment. During the fishing derby, people are not only proficient in their awareness but importantly, their ability to seize an opportunity. Skill is the successful coming together of both awareness and action.

Previous experience is influential in one's awareness as well as in one's actions in responding to an opportunity. The ability to know how and when to do something is not learned through prescriptive instructions but rather in the actual practice of doing it. Ingold and Kurttila show how skill is "a property of the whole organism-person, having emerged through a history of involvement in an environment" (2000:1). With this definition of skill in mind, both men and women have equal potential for skill in fishing.

An important aspect of focusing on skill is that it brings forth the importance of opportunity or good fortune. In fact, the two are intimately connected, and part of being skilled is getting the timing right to seize an opportunity. Skill can be looked at as a coin, with one side of the coin as awareness and the other side as opportunity. How is this opportunity expressed and how is success in fishing explained for men and women? I will investigate how skill and opportunism are consistent with ideas about how fish may come to certain individuals.

Getting Fish: Do They Come to Certain People?

Does everyone have the same chance to catch fish? Men and women could and did have equal success in fishing. Success in fishing was not dependent upon gender or age. During the spring, women often express a love of fishing and may fish more frequently than men thereby increasing their chances of success. During the fishing derby, however, both men and women fished approximately equivalent amounts. Fishing requires mobility; if one does not get a fish, one switches holes; an impatience in fishing is compensated by mobility. Did the fish come to certain people or did the people seek them out? Various anthropological accounts discuss how animals come to certain people (*see* Bodenhorn 1990; Feit 1994; Fienup-Riordan 1983, 1990a,b, 2000; Hallowell 1960; Hensel 1996; Ingold *et al.* 1988; Ingold 1980, 1986, 1994, 2000a; Nuttall 1992; Paine 1971; Ridington 1990; Tanner 1979; and Wenzel 1991 for anthropological accounts discussing the relationship between people and animals). I wondered if this was also the case with fish. In the context of Icelandic fishing, Palsson (1994b) shows that in the past the ability that some people had to obtain fish was described as a quality of 'fishiness.' The relationship between fish and humans was thought to have been controlled by supernatural forces. Palsson explains that the quality of 'fishiness' was something an individual either did or did not possess, but one could not control the amount of 'fishiness.' Inuit similarly express the notion that fish come to certain people; however, the role of the fisherman is not conceived to be entirely passive. Moreover, it was not the only explanation given for fishing success. Inuit in Coral Harbour would mention that a person might be 'lucky' or be a 'real fisherman.' The term 'fisherman' is used to describe both men and women and affords the same opportunities to men and women. There is an equality of opportunity assured through the rules of the fishing derby and men and women have the same opportunities in fishing. Rules to maintain good social relationships apply to both men and women. For example, one should not boast about the number of fish caught.

Being A Real Fisherman

Certain people have luck in fishing, and if fish do come to a person, it is partially due to luck. I was informed about which people had luck in fishing in the past and in the present. Some people told me that this luck could shift and that there was no assurance that if fish came to you one year they would come to you again. When I asked for further explanation about this notion of luck, it was made clear that the people who have good luck are deemed 'real fishermen.' A 'real' person is also someone who maintains the proper social relations with humans and animals (Wenzel 1991). Wenzel discusses the qualities of a 'real person' as generous, patient, obedient and co-operative and he emphasizes that they learn these qualities from hunting (Wenzel 1991). I suggest that these qualities are also learned from fishing and are the qualities of a 'real fisherman.'

A real fisherman is a person who really loves fishing and is the one who has luck or is successful.[20] There is no Inuktitut word for luck; rather, there is a notion of success. The people who catch many fish are actively seeking the fish, or they really love fishing, or they are real fishermen. This is expressed by the ending or suffix—*sujuk,* a term used to describe those people who really love fishing. These people are ready to seize an opportunity, since they put a lot of time and focus on the activity of fishing. Luck, therefore, is not independent of one's actions. At the 2000 fishing derby, one woman caught almost fifty fish. When I asked how she did it, she shrugged her shoulders and said, "Luck, I guess." By attributing the situation to luck, she also humbled herself; thus, luck can work as a social levelling mechanism. Luck can still play a role; she was lucky the fish came to her, or luck is involved in her readiness to catch the fish. Although there was no prize for catching the most fish, everyone knew who caught many fish; this was well regarded by others and comments were made about that person being a good or 'real' fisherman.

I suggest that procurement is a combination of the fish coming to certain people, along with good fortune and an ability or skill to seize an opportunity when presented. Procurement is possible only through a combination of knowledge, experience, skill and luck or fortune.

Luck may have different meanings, and, according to Gubser (1965), understanding luck involves effort: "The Nunamiut do not think in terms of pure luck as many white men do... *Pilyautaktuni* means 'to have good luck,' but it implies effort or involvement rather

---

[20] Anderson discusses a similar concept of the successful fisherman being the one who loves fish (2000:128). As in Anderson's account, the person who loves fishing is also the one who the fish may come to.

than success merely by chance" (Gubser 1965:226). Luck, or people being real fishermen may not be pure chance, as Gubser suggests. Writing of the Sub-Arctic Cree, Feit also considers luck differently from pure chance. Feit discusses the Waswanipi meaning of luck as being about "expressions of the cycles of power" rather than a notion of 'luck' as unexplainable chance occurrence (1994:436).

A sense of luck, chance, or fortune can often be intertwined with beliefs in other powers. People's actions and behaviour might be able to bring about luck. For the Inuit, treating fish and animals with respect might influence their luck. This is different from the explanation which Palsson provides in describing the Icelandic notion of 'fishiness', in which people are seen as passive recipients of fish. In the Inuit context, there are seemingly inconsistent views that reflect both control and lack of control over one's own success. On the one hand, a person's own actions are thought to influence whether or not the animal presents itself (Balikci 1970; Fienup-Riordan 1990a; Gubser 1965; Spencer 1959). The generosity of a person, for example, is thought to influence future human-animal interactions. On the other hand, one cannot control whether or not a fish or any animal comes to someone. Skill in fishing as in other kinds of procurement, is about the ability to perceive and respond to an opportunity. The fish may come to certain people, but in no way is this expressed in a passive sense. Despite the notion that fish may come to certain people, people remain mobile to increase their chances of finding them.

Fortune or opportunity can be a part of the unpredictability of resource procurement without surrendering passively to chance. An experienced and knowledgeable person may return from a fishing trip and explain that she had no luck. Luck or fortune refers to the lack of control over whether or not an opportunity presents itself. However, this lack of control may be attributed to the inherent wilfulness of the animal/fish. Thus, an appeal to luck does not rule out the idea that fish are empowered with properties of sentience. This notion of opportunism is not always highlighted with respect to Inuit hunting. As hunting has often been taken to be representative of the Inuit relationship with the land or their environment, an examination of fishing thereby contributes to a greater understanding of Inuit procurement and engagement with the world around them. In examining fishing, the importance of opportunity and skill come to the fore.

Conclusion

An exploration of fishing as a procurement activity, as demonstrated with the example of the fishing derby, has beneficial consequences for understanding gender roles. First, because the activity of fishing itself is not divided by gender, it demonstrates the variability in these roles. I do not dismiss the complementary nature of men's and women's roles, but rather highlight the variability existing within this dichotomy. Many aspects of Inuit daily life in Coral Harbour reflect the complementary gendered division of labour. However, fishing at the fishing derby is not divided by gender. The model of men hunting and women gathering, or of women being active in roles complementary to men's hunting, does not always apply. Gender may influence specific tasks one performs; yet, the activity of fishing itself was not defined by gender. In fact, understanding when gender is not a defining characteristic informs a broader understanding of Inuit gender relations and flexibility in gender roles.

Second, an exploration of women's involvement in the fishing derbys and their active participation in fishing extends our understanding of women's activities beyond an investigation of how women contribute to hunting or the complementary nature of women's work. It would be misleading to discuss fishing as solely women's work. Yet, it is through paying attention to women's activities of livelihood that we can have a broader understanding of procurement and a balance to the previous attention to hunting. Nuttall states, "few researchers anywhere in the Arctic have documented in detail the daily routines of women and the vital contributions they make to the social and economic vitality of their communities..." (1998a:70). Procurement, apart from hunting, is one area in which women are actively involved. By exploring procurement through an examination of fishing, I address this lacuna in the literature and, in part, respond to this imbalance in research. This does not mean that it is necessary to concentrate solely on women's activities. I suggest that we must examine whole families' engagement in fishing and, through this, gain insight into women's procurement activities. In fishing, women's contributions to Inuit livelihood are not restricted to their complementary roles or how they contribute to hunting.

Fishing, as exemplified by the fishing derby, could be equally representative of Inuit procurement as hunting, because everyone—men, women, children and elders—participate. A focus on activities in which women are participants adds balance to an understanding of how Inuit engage in procurement. Focusing mainly on hunting as a way to grasp people's relationship to their environment limits our understanding, and, as Nuttall points out, "human-environmental relations are often based on men's practical knowledge" (1998b:25). I maintain that it is difficult to discuss perceptions of environment or relations to the environment without also exploring other ways in which northern peoples participate in procurement.

I suggest that the fishing derby provides an ethnographic look at procurement where skill comes to the fore. This may have implications for thinking of hunter-gatherer procurement as opportunistic where awareness is coupled with the skill to seize opportunities. Fishing demonstrates that skill in procurement for men and women need not be divided by gender. Inuit are not only skilled in awareness but also in seizing opportunities. Skill and success in fishing is not divided by gender and both men and women have the potential to be 'real fishermen.'

References

Ager, L. (1977). 'The Reflection of Cultural Values in Eskimo Children's Games,' pp. 92-8. in D Lancy, and A Tindall, eds. *The Study of Play: Problems and Prospects.* West Point, New York: Leisure Press.

Anderson, D. (2000). *Identity and Ecology in Arctic Siberia: The Number One Reindeer Brigade.* Oxford: Oxford University Press.

Attagutsiak, K. (1999). *Poems and Stories.* Iqaluit: Nunavut Arctic College.

Balikci, A. (1970). *The Netsilik Eskimo.* Garden City, New York: The Natural History Press.

Barker, J. and R. Barker (1993). *Always Getting Ready, Upterrlainarluta: Yup'ik Eskimo Subsistence in Southwest Alaska.* Seattle: University of Washington Press.

Bennett, K., P. Nelson, and J. Baker (1994). 'Meaning in Mud: Yup'ik Eskimo Girls at Play,' pp. 179-209 in J. Roopnarine, J. Johnson and F. Hooper, eds. *Children's Play in Diverse Cultures.* Albany: State University of New York Press.

Bernard, R. (1988). *Research Methods in Cultural Anthropology.* Newbury Park: Sage.

Bird, J.B. (1953). *Southampton Island. Vol. Memoir 1. Geographic Branch.* Canada Department of Mines and Technical Surveys. Ottawa: Edmond Cloutier, Queen's Printer and Controller of Stationery.

Bird-David, N. (1992). Beyond "The Hunting and Gathering Mode of Subsistence" Observations on the Nayaka and other Modern Hunters. *Man* 27: 19-44.

Birket-Smith, K. (1936). *The Eskimos.* London: Methuen and Co.

Birket-Smith, K. (1945). *Ethnographic Collections From the Northwest Passage. Report of the Fifth Thule Expedition 1921-1924. VI (2).* Copenhagen: Gyldendalske Boghandel, Nordisk Forlag.

Boas, F. (1964). *The Central Eskimo.* Lincoln: University of Nebraska Press.

Bodenhorn, B. (1990). "I'm Not the Great Hunter; My Wife Is," *Iñupiat and Anthropological Models of Gender* 14: 55-74.

Bodenhorn, B. (1993). 'Gendered Spaces, Public Places: Public and Private Revisited on the North Slope of Alaska,' pp. 169-203 in B. Bender, ed. *Landscape Politics and Perspectives.* Oxford: Berg.

Bodenhorn, B. (1995). 'Christmas Present: Christmas Public,' pp. 193-216 in D. Miller, ed. *Unwrapping Christmas.* Clarendon Press: Oxford.

Bodenhorn, B. (2000). '"It's Good to Know Who Your Relatives Are But We Were Taught to Share with Everybody": Shares and Sharing among Iñupiaq Households,' pp. 27-60 in G. Wenzel, G. Hovelsrud-Broda, and N. Kishigami, eds. *The Social Economy of Sharing: Resource Allocation and Modern Hunter–Gatherers.* Senri Ethnological Series no. 53. Osaka: National Museum of Ethnology.

Briggs, J. (1970). *Never in Anger: Portrait of an Eskimo Family.* Cambridge, Massachusetts: Harvard University Press.

Briggs, J. (1974). 'Eskimo Women: Makers of Men,' pp. 261-304 in C Matthiasson, ed. *Many Sisters: Women in CrossCultural Perspective.* New York: The Free Press.

Briggs, J. (2000). 'Conflict Management in a Modern Inuit Community,' pp. 110-24 in P. Schweitzer, M. Biesele, R. Hitchcock, eds. *Hunters and Gatherers in The Modern World.* New York: Berghahn Books.

Brody, H. (1987). *Living Arctic: Hunters of the Canadian North.* Vancouver and Seattle: Douglas and McIntyre and University of Washington Press.

Brody, H. (2000). *The Other Side of Eden: Hunters, Farmers, and the Shaping of the World.* New York: North Point Press.

Brumbach H.J. and R. Jarvenpa (1997). 'Woman the Hunter: Ethnoarchaeological Lessons from Chipewyan Life-Cycle Dynamics,' pp. 17-32 in C. Claassen, R Joyce, eds. *Women in Prehistory: North America and Mesoamerica.* Philadelphia: University of Pennsylvania Press.

Burch, E. (1988). *The Eskimos.* Norman and London: University of Oklahoma Press.

Caulfield, R. (1997). *Greenlanders, Whales, and Whaling: Sustainability and Self-Determination in the Arctic.* Hanover: University Press of New England.

Chance, N. (1990). *The Iñupiat and Arctic Alaska: An Ethnography of Development.* Case Studies in Cultural Anthropology. Orlando: Holt, Rienhart, and Winston.

Condon, R. and P. Stern. (1993). Gender-Role Preference, Gender Identity, and Gender Socialization Among Contemporary Inuit Youth. *Ethos* 21: 384-416.

Coy, M. (1989). 'Being What We Pretend to Be: The Usefulness of Apprenticeship as a Field Method,' pp. 115-135 in M. Coy, ed. *Apprenticeship: From Theory to Method and Back Again.* Albany: State University of New York Press.

Creery, I. (1994). 'The Inuit (Eskimo) of Canada,' pp. 105-46 in MR Group, ed. *Polar Peoples: Self-Determination and Development.* London: Minority Rights Publications.

Crick, M. (1982). 'Anthropological Field Research, Meaning Creation and Knowledge Constructtion,' pp. 15-37 in D. Parkin, ed. *Semantic Anthropology*. ASA Monograph 22. London: Academic Press.

Dahl, J. (2000). *Saqqaq: An Inuit Hunting Community in the Modern World*. Toronto: University of Toronto Press.

Dahlberg, F., ed. (1981). *Woman the Gatherer*. New Haven: Yale University Press.

Damas, D., ed. (1984). *Handbook of the North American Indians. vol. 5: The Arctic.* Washington, D.C.: Smithsonian Institution.

Department of Education, Culture and Employment, Northwest Territories (1996). *Inuuqatigiit: The Curriculum From the Inuit Perspective*. Yellowknife: Education, Culture and Employment, Northwest Territories.

Dorais, L.-J. (1997). *Quaqtaq: Modernity and Identity in an Inuit Community*. Toronto: University of Toronto Press.

Dunning, R. (1962). A Note on Adoption Among the Southampton Island Eskimo. *Man* LXII: 163-166.

Ellen, R., ed. (1984). *Ethnographic Research: A Guide to General Conduct*. ASA Research Methods in Social Anthropology. London: Academic Press.

Emerson, R., R. Fretz, and L. Shaw (1995). *Writing Ethnographic Fieldnotes*. Chicago Guides to Writing, Editing, and Publishing. Chicago: The University of Chicago Press.

Fedorova, E. (2000). 'The Role of Women in Mansi Society,' pp. 391-398 in P. Schweitzer, M. Biesele, and R. Hitchcock, eds. *Hunters and Gatherers in the Modern World*. New York: Berghahn Books.

Feit, H. (1994). 'The Enduring Pursuit: Land, Time, and Social Relationships in Anthropological Models of Hunter-Gatherers and in Subarctic Hunter's Images,' pp. 421-39 in E. Burch, and L. Ellanna, eds. *Key Issues in Hunter-Gatherer Research*. Oxford: Berg.

Fienup-Riordan, A. (1983). *The Nelson Island Eskimo: Social Structure and Ritual Distribution*. Anchorage: Alaska Pacific University Press.

Fienup-Riordan, A. (1990a). *Eskimo Essays: Yup'ik Lives and How We See Them*. New Brunswick, NJ: Rutgers University Press.

Fienup-Riordan, A. (1990b). Introduction. *Études/Inuit Studies* 14: 7-22.

Fienup-Riordan, A. (2000). *Hunting Tradition in a Changing World: Yup'ik Lives in Alaska Today*. New Brunswick, NJ: Rutgers University Press.

Freeman, M.M.R., ed. (2000). *Endangered Peoples of the Arctic: Struggles to Survive and Thrive*. Westport, Connecticut: Greenwood Press.

Freeman, M.M.R. (1976). *Inuit Land Use and Occupancy Project*. Ottawa: Department of Indian and Northern Affairs.

Freeman, M.M.R. (1969). Adaptive Innovation Among Recent Eskimo Immigrants in the Eastern Canadian Arctic. *Polar Record* 14: 769-781.

Freeman, M. (1969/1970). Studies in Maritime Hunting I. *Folk* 11-12: 155-171.

Freuchen, P. (1961). *Book of the Eskimos*. Cleveland: The World Publishing Company.

Giffen, M.N. (1930). *The Roles of Men and Women in Eskimo Culture*. The University of Chicago Publications in Anthropology: Ethnology Series. Chicago: The University of Chicago Press.

Glassford, R. (1976). *Application of a Theory of Games to the Transitional Eskimo Culture*. Studies in Play and Games. New York: Arno Press.

Graburn, N. (1969). *Eskimos Without Igloos*. Boston: Little Brown and Company.

Gubser, N. (1965). *The Nunamiut Eskimos*. New Haven: Yale University Press.

Guemple, L. (1986). Men and Women, Husbands and Wives: The Role of Gender in Traditional Inuit Society. *Études/Inuit Studies* 10: 9-24.

Guemple, L. (1988). 'Teaching Social Relations to Inuit Children,' pp. 131-49 in T. Ingold, D. Riches, J. Woodburn, eds. *Hunter and Gatherers 2: Property, Power and Ideology.* Oxford: Berg.

Hallowell, I. (1960). 'Ojibwa Ontology, Behavior and Worldview,' pp. 19-52 in S. Diamond, ed. *Culture in History: Essays in Honor of Paul Radin.* New York: Columbia University Press.

Hastrup, K. (1992). 'Writing Ethnography: State of the Art,' pp. 116-133 in J. Okely and H. Callaway, eds. *Anthropology and Autobiography.* London: Routledge.

Heine, M. (1998). *Arctic Sports: A Training and Resource Manual.* Yellowknife: Arctic Sports Association and MACA (GNWT).

Hensel, C. (1996). *Telling Ourselves: Ethnicity and Discourse in Southwestern Alaska.* Oxford: Oxford University Press.

Honigmann J. and I. Honigmann (1965). *Eskimo Townsmen.* Ottawa: Canadian Research Centre for Anthropology, University of Ottawa.

Ikuutaq, S. (1984). *Inuit Places Out on The Land.* Rankin Inlet: Keewatin Resource Centre.

Ingold T. and T. Kurttila (2000). Perceiving the Environment in Finnish Lapland. Special issue, 'Bodies and Nature.' P. MacNaughten and J.Urry, eds. *Body and Society* 6: 183-96.

Ingold, T., D. Riches, and J. Woodburn, eds. (1988). *Hunters and Gatherers 2: Property, Power and Ideology.* Oxford: Berg.

Ingold, T. (1980). *Hunters, Pastoralists and Ranchers: Reindeer Economies and Their Transformations.* Cambridge: Cambridge University Press.

Ingold, T. (1986). *The Appropriation of Nature: Essays on Human Ecology and Social Relations.* Manchester: Manchester University Press.

Ingold, T. (1994.) 'From Trust to Domination: An Alternative History of Human-Relations,' pp. 22 in A. Manning, and J. Serpell, eds. *Animals and Human Society: Changing Perspectives.* London: Routledge.

Ingold, T. (1996). 'Hunting and Gathering as Ways of Perceiving the Environment,' pp. 117-55 in R. Ellen, and K. Fukui, eds. *Redefining Nature: Ecology, Culture, and Domestication.* Oxford: Berg.

Ingold, T. (2000a). *The Perception of the Environment: Essays in livelihood, dwelling and skill.* London: Routledge.

Ingold, T. (2000b). 'Evolving Skill,' pp. 225-46 in H. Rose and S. Rose, eds. *Alas Poor Darwin! Arguments Against Evolutionary Psychology.* New York: Random House.

Ingold, T. (2001). 'From the Transmission of Representation to the Education of Attention,' pp. 113-153 in H. Whitehouse, ed. *The Debated Mind: Evolutionary Psychology Versus Ethnography,* Oxford: Berg.

Jansen, W.H. (1979). *Eskimo Economics: An Aspect of Cultural Change at Rankin Inlet,* Canadian Ethnology Service No. 46. National Museum of Man Mercury Series. Ottawa: National Museum of Canada.

Jolles, C.Z. (1994). Cutting Meat, Sewing Skins, Telling Tales: Women's Stories in Gambell, Alaska. *Arctic Anthropology* 31: 86-102.

Kishigami, N. (2000). 'Contemporary Inuit Food Sharing and Hunter Support Program of Nunavik, Canada,' pp. 171-92 in G. Wenzel, G. Hovelsrud-Broda, N Kishigami, eds. *The Social Economy of Sharing: Resource Allocation and Modern Hunter–Gatherers.* Osaka: National Museum of Ethnology.

Landes, R. (1969). *Ojibwa Woman.* New York: AMS Press.

Leacock, E. (1978). Women's Status in Egalitarian Society: Implications for Social Evolution. *Current Anthropology* 19: 247-275.

Leacock, E. (1981). *Myth of Male Dominance: Collected Articles on Women Cross-Culturally.* New York: Monthly Review Press.

Leacock, E., H. Safa, *et al.*, eds. (1986). *Women's Work: Development and Division of Labour by Gender.* South Hadley, Massachusetts: Bergin and Garvey Publishers.

Lee, R, and I. DeVore, eds. (1968). *Man The Hunter.* Chicago: Aldine Publishing Company.

Marcus, A. (1992). *Out in the Cold: The Legacy of Canada's Inuit Relocation Experiment in the High Arctic.* IWGIA Document 71. Copenhagen: IWGIA.

Mathiassen, T. (1945). *Material Culture of the Iglulik Eskimos, Report of the Fifth Thule Expedition 1921-1924 Vol. IV (1).* Copenhagen: Gyldendal.

Matthiasson, C., ed. (1974). *Many Sisters: Women in Cross Cultural Perspective.* New York: The Free Press.

Mauss, M. and H. Beuchat (1979). *Seasonal Variations of the Eskimo.* London: Routledge and Kegan Paul.

McDonald, M., L. Arragutainaq, and Z Novalinga (1997). *Voices from the Bay: Traditional Ecological Knowledge of Inuit and Cree in the Hudson Bay Bioregion.* Ottawa: Canadian Arctic Resources Committee.

Mearns, L. (1994). 'To Continue the Dreaming: Aboriginal Women's Traditional Responsibilities in a Transformed World,' pp. 267-287 in E. Burch and L. Ellanna, eds. *Key Issues in Hunter-Gatherer Research.* Oxford: Berg.

Meehan, B. (1982). *Shellbed to Shell Midden.* Canberra: Australian Institute of Aboriginal Studies.

Morrison, D. and G.-H. Germain (1995). *Inuit: Glimpses of an Arctic Past.* Hull, Quebec: Canadian Museum of Civilization.

Moyer, D. (1970). *The Dimensions of Conflict in an Eskimo Community.* Saint John's: Atlantic Association of Sociologists and Anthropologists.

Nelson, R. (1969). *Hunters of the Northern Ice.* Chicago: University of Chicago Press.

Nuttall, M. (1992). *Arctic Homeland: Kinship, Community, and Development in Northwest Greenland.* Toronto: University of Toronto Press.

Nuttall, M. (1998a). 'Gender, Indigenous Knowledge and Development in the Arctic,' pp. 67-7 in T. Greiffenberg, ed. *Development in the Arctic.* Copenhagen: Danish Polar Centre.

Nuttall, M. (1998b). 'Critical Reflections on Knowledge Gathering in the Arctic,' pp. 21-35 in L-J Dorais, M Nagy, L Muller-Wille, eds. *Aboriginal Environmental Knowledge in the North.* Quebec: GETIC.

Nuttall, M. (1998c). *Protecting the Arctic.* Amsterdam: Harwood.

Oakes, J. and R. Riewe (1995). *Our Boots: An Inuit Women's Art.* Vancouver/Toronto: Douglas and McIntyre.

Okely, J. (1992). 'Anthropology and Autobiography: Participatory Experience and Embodied Knowledge,' pp. 1-28 in J. Okely and H. Callaway, eds. *Anthropology and Autobiography.* London: Routledge.

Okely, J. and H. Callaway, eds. (1992). *Anthropology and Autobiography.* ASA Monographs 29. London: Routledge.

Ortner, S. (1974). 'Is female to male as nature is to culture?,' pp. 67-87 in M. Rosaldo, L Lamphere, eds. *Women, Culture, and Society.* Stanford: Stanford University Press.

Paine, R. (1971). Animals as Capital: Comparisons Among Northern Nomadic Herders and Hunters. *Anthropological Quarterly* 44: 157-72.

Palsson, G. (1988). 'Hunters and Gatherers of the Sea,' pp. 189-204 in T. Ingold, D.Riches, and J. Woodburn, eds. *Hunters and Gatherers 1: History, Evolution and Social Change.* Oxford: Berg.

Palsson, G. (1991). *Coastal Economies, Cultural Accounts: Human Ecology and Icelandic Discourse.* Manchester: Manchester University Press.

Palsson, G. (1994a). Enskilment at Sea. *Man* 29: 901-927.

Palsson, G. (1994b). 'The Idea of Fish: Land and Sea in the Icelandic World-View,' pp. 119-133 in R. Willis, ed. *Signifying Animals: Human Meaning in the Natural World, One World Archaeology*. London: Routledge.

Rasmussen, K. (1908). *The People of the Polar North: A Record*. Philadelphia: J.B. Lippincott Company.

Rasmussen, K. (1929). *Intellectual Culture of the Iglulik Eskimos*. Copenhagen: Gyldendal.

Reiter, R. ed. (1975). *Toward an Anthropology of Women*. New York: Monthly Review Press.

Riches, D. (1982). *Northern Nomadic Hunter-Gatherers: A Humanistic Approach*. London: Academic Press.

Ridington R. (1990). *Little Bit Know Something: Stories in Language of Anthropology*. Iowa City: University of Iowa Press.

Riewe, R, ed. (1992). *Nunavut Atlas*. Edmonton: Canadian Circumpolar Institute and Tungavik Federation of Nunavut.

Rival, L. (1993). The Growth of Family Trees: Understanding Huaorani Perceptions of Forest. *Man* 25: 635-652.

Rosaldo, M. and L. Lamphere, eds. (1974). *Woman, Culture, and Society*. Stanford: Stanford University Press.

Rousselot J-L, W. Fitzhugh, and A. Crowell. (1988). 'Maritime Economies of the North Pacific Rim,' pp. 151-72 in W. Fitzhugh and A. Crowell, eds. *Crossroads of Continents: Cultures of Siberia and Alaska*. Washington, D.C.: Smithsonian Institution Press.

Sanner, A. (1979). *Bringing Home Animals: Religious Ideology and Mode of Production of the Mistassini Cree Hunters*. Saint Johns: Institute of Social and Economic Research, Memorial University.

Schweitzer, P., ed. (2000). *Dividends of Kinship: Meanings and Uses of Social Relatedness*. European Association of Social Anthropologists. London: Routledge.

Shannon, K.A. (2003). *Readiness and Skill in An Arctic Environment: Procurement, Distribution and Skill*. Unpublished PhD thesis. Department of Anthropology, University of Aberdeen.

Sharp, H. (1981). 'The Null Case: The Chipewyan,' pp. 221-244 in F. Dahlberg, ed. *Woman the Gatherer*. New Haven: Yale University Press.

Sharp, J. (1994). 'The Power of Weakness,' pp. 35-58 in E. Burch, L. Ellanna, eds. *Key Issues in Hunter-Gatherer Research*. Oxford: Berg.

Slocum, S. (1975). 'Woman The Gatherer: Male Bias in Anthropology,' pp. 36-50 in R. Reiter, ed. *Toward an Anthropology of Women*. New York: Monthly Review Press.

Spencer, R. (1959). *The North Alaskan Eskimo: A Study in Ecology and Society*. Washington D.C.: United States Government Printing Office.

Sprott, J. (1997). Christmas, Basketball, and Sled Dog Races: Common and Uncommon Themes in the New Seasonal Round in an Iñupiaq Village. *Arctic Anthropology* 34: 68-85.

Strathern, M. (1972). *Women in Between: Female Roles in a Male World: Mount Hagen, New Guinea*. London: Seminar Press.

Strathern, M., ed. (1987). *Dealing With Inequality: Analysing Gender Relations in Melanesia and Beyond*. Cambridge: Cambridge University Press.

Strathern, M. (1988). *The Gendered Gift: Problems With Women and Problems with Society in Melanesia*. Cambridge: Cambridge University Press.

Stern, P. (2000). Subsistence: Work and Leisure. *Études/Inuit Studies* 24: 9-24.

Stewart, H. (2005). The Fish Tale That is Never Told: The Importance of Fishing in Inuit Societies. *Senri Ethnological Studies* 67: 345-361.

Sunderland, P.L. (1999). Fieldwork and the phone. *Anthropology Quarterly* 72(3): 105-117.

Sutton, G. (1932). 'The Exploration of Southampton Island,' pp.1-78 in *Memoirs of the Carnegie Museum XII, (Part 1, Sections 1, 2, 3)*. Pittsburgh: Carnegie Museum.

Sutton, G. (1934). *Eskimo Year: A Naturalist's Adventure in the Far North.* New York: MacMillan Company.

Usher, P. (1975). *Historical Statistics Approximating Fur, Fish and Game Harvests within Inuit Lands of the N.W.T. and Yukon 1915-1974.* London, ON: University of Western Ontario.

VanStone, J. (1959). *The Economy and Population Shifts of the Eskimos of Southampton Island.* Ottawa: Northern Co-Ordination and Research Centre, Department of Northern Affairs and National Resources.

Weiner, A. (1976). *Women of Value, Men of Renown: New Perspectives in Trobriand Exchange.* Austin: University of Texas Press.

Wenzel, G. (1991). *Animal Rights, Human Rights: Ecology, Economy and Ideology in the Canadian Arctic.* Toronto: University of Toronto Press.

Wilkinson, D. (1955). *Land of the Long Day.* Toronto: Clarke, Irwin and Company.

CHAPTER SIX

Gender, Knowledge, and Environmental Change Related to Humpback Whitefish in Interior Alaska

Melissa Robinson,[1] Phyllis Morrow[2] and Darlene Northway[3]

Abstract: In the Upper Tanana Athabascan village of Northway, in Eastern Interior Alaska, humpback whitefish (*Coregonus pidschian*) are an important component of the local native culture and the primary species in the subsistence fishery. In 2002 a collaborative project with the Northway Village Council, the U.S. Fish and Wildlife Service, and the University of Alaska Fairbanks began to examine the relationship between local knowledge and fisheries science. Through three years of semi-directed interviews and participant observation it became clear that, partially due to gender-specific fishing roles, women and men differed in their knowledge about how changes they perceive in their environment affect humpback whitefish. Women were able to document patterns that men had not observed such as seasonal and annual variation in the prevalence of parasite-infected whitefish. Men and women observed increased sedimentation in area lakes and increased air and water temperatures over the past fifty years. The inclusion of both men's and women's diverse perspectives led to new questions about possible trends in parasites and the impacts of environmental change and weather patterns on fish behavior and health. We argue that because men and women often differ in their knowledge base regarding a resource, the combination of their observations results in a more holistic understanding of a social and ecological system than either can accomplish alone. This, in turn, can lead to a more sound basis for resource management and enhance the ability of a system to cope with expected and unexpected change.

---

[1] Regional Resilience and Adaptation Program, Institute of Arctic Biology, University of Alaska Fairbanks.
[2] Professor of Anthropology, University of Alaska Fairbanks.
[3] Elder, Northway Village, Alaska.

Introduction

Women's roles and knowledge related to their environment are receiving more attention than in the past, as are their contributions to ecological research (McDowell 1984; Satia and Wétohossou 1996; Siar 2003; Shanley 2006). However, the relevant literature still reflects a general bias toward men's roles and knowledge in relation to farming, hunting, and other resource uses, while largely ignoring the crucial and complex role women play in these activities (Slocum 1975; Davis and Nadel-Klein 1992; Estioko-Griffen and Bion Griffen 1993; Madge 1994). The kinds of information that men can provide have traditionally been viewed as more valuable. This is also the case in terms of women's fisheries-related understanding and experience. Women in many different cultures and communities hold knowledge that is different from men due to the division of labour and the often gendered patterns by which informal and formal knowledge passes from person to person (Van de Ploeg 1993; Nygren 1999; Siar 2003). Knowledge within and between communities is not homogeneous or static and varies depending on gender and other factors such as age, religion, occupation, and social class (Guyer 1991; Davis and Nadel-Klein 1992; Madge 1994; Huntington 2002; Jewitt 2002; Crona and Bodin 2006) which intersect to shape, direct, and limit the experiences of individuals. Taking gender into account provides new and different perspectives, expanding not only our knowledge base, but possibly our practices of science itself (Davis and Nadel-Klein 1992; Schiebinger 1997; Jewitt 2002). Where the literature emphasizes men's knowledge, it misses the information that is embedded in women's activities, such as their particularly focused observations of fish anatomy gained by processing large numbers of fish for consumption.

Women's perspectives may help researchers understand environmental changes because women make different observations than men and may voice new concerns or relate information or knowledge passed on to them by other women. The information for this chapter, which explores women's environmental knowledge and concerns in a subsistence-based community in Alaska, results from a project linking local knowledge and fisheries science about humpback whitefish (*Coregonus pidschian*) in the Alaska Native (Upper Tanana Athabascan) village of Northway, along the Canada-Alaska border (Fig. 1). Research on environmental change in the Arctic and the role that local or indigenous people play in this process continues to grow and receive attention from scientists of various disciplines (*see* Riedlinger and Berkes 2001; Cochran and Geller 2002; Krupnik and Jolly 2002). In the Arctic, local people are already aware of changes that also have a significant impact on the rest of the world (ACIA 2005). For this reason, it is a particularly appropriate site for the study of climate and environmental change. Environmental changes such as increasing temperatures, thawing permafrost, and changes in wildlife populations or health are often noticed by local people, especially those who are engaged in daily subsistence activities including fishing and fish processing (Huntington 2002; Jolly *et al.* 2002). Therefore, arctic residents, both men and women, are important players in this process and possess crucial insights and observations about environmental change.

In Northway, Alaska, humpback whitefish are the main subsistence fish and are central to people's lives (McKennan 1981; Case 1986; Marcotte 1991), yet certain aspects of their life history are unknown in the scientific community (Tallman and Reist 1997; Brown *et al.* 2002). Through semi-directed interviews (Fig. 2) and five months of participant observation, it became apparent that men and women vary in the breadth and depth of their knowledge related to whitefish and their environment. A growing awareness of the need to listen to both men and women was a significant part of this project and led to

Gender, Knowledge, and Environmental Change

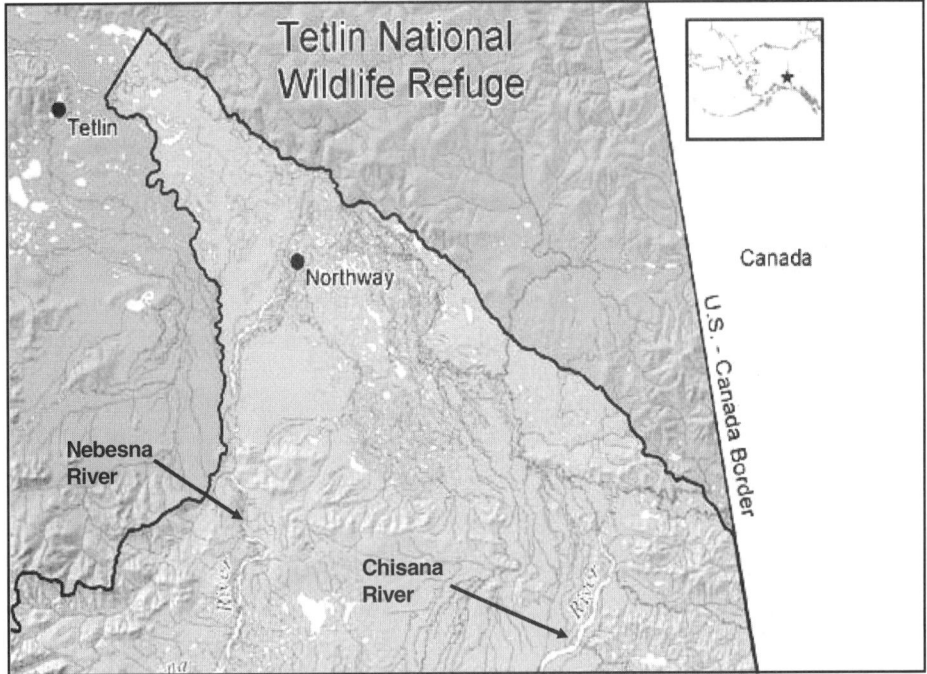

Figure 1. *Map depicting the localtion of Northway in relation to the Tetlin National Wildlife Refuge (TNWR). Map courtesy of the U.S. Fish and Wildlife Service.*

Figure 2. *Howard Fix of Northway shows Melissa Robinson seasonal areas he fishes for whitefish. Photo by G. Marunde Jr., July 2004, Northway Whitefish Project*

new and more in-depth questions and research directions, and a realization that women have an important role to play in understanding environmental change and human adaptation to change. Although women and men have been included in harvest assessments and traditional ecological knowledge studies about fish (Anderson and Fleener 2001; Georgette 2002; Simeone and Kari 2002; Brown *et al.* 2003), there are few studies highlighting how the differing roles that men and women play in these subsistence activities shape their knowledge base. Nor has there been sufficient research on the importance of these differing experiences, how local people link these experiences together, and how they may contribute to environmental studies in the Arctic. This chapter moves toward filling that gap by highlighting how working with women and men and understanding these gendered differences in knowledge leads to new insights concerning a fishery and related changes in an ecosystem. In particular, this chapter focuses on how men and women see and experience environmental change in the Upper Tanana region of Alaska in terms of changing water levels and siltation, parasites in whitefish, and shifting weather patterns.

Background

This study occurred from 2002 to 2004 in the Interior Alaskan community of Northway. Located 400 kilometers southeast of Fairbanks, Alaska, Northway is surrounded by the Tetlin National Wildlife Refuge (United States Fish and Wildlife Service). Approximately 75% of the 270 people living in Northway are Athabascan Indian (U.S. Census Bureau, Census 2000). The region's landscape is dominated by black spruce (*Picea mariana*), birch (*Betula neoalaskana*), and aspen (*Populus tremuloides*) along with small wetland areas. Common fish species in the drainage include humpback and round whitefish (*Coregonus pidschian, Prosopium cylindraceum*), pike (*Esox lucius*), grayling (*Thymallus arcticus*), burbot or lingcod (*Lota lota*), and long-nose sucker (*Catostomus catostomus*) (Case 1986; Brown 2006). The Nebesna and Chisana Rivers meet to form the Tanana River. Due to their glacier run-off pattern, these rivers maintain water flow throughout most of the summer, experience constant turbidity, and retain a high level of silt during open water periods (April-October). Depths in the Chisana and Nebesna Rivers range between 1 m-5 m (Brabets *et al.* 2000). The Nebesna, Chisana, and Tanana Rivers, along with their associated small lakes, creeks, and sloughs, comprise important habitat for humpback whitefish and are thus traditional fishing areas for local residents.

In Alaska and other parts of the Arctic and sub-Arctic, various whitefish species (subfamily: *Coregoninae*) are important food and cultural resources (Alt 1979; Anderson and Fleener 2001). In Northway, humpback whitefish comprise the largest percentage, by mass, of subsistence resources consumed (36%) with local residents consuming an estimated 220 kilograms of whitefish per capita per year (Case 1986; Marcotte 1991). As Northway resident Cherie Marunde said, "My mom's doing it forever, it's just a part of my life, or our lives" (Marunde and Marunde 2003).

In Northway as in other areas of the Arctic, residents expressed concern over perceived changes in whitefish abundance and health (Tallman and Reist 1997; Anderson and Fleener 2001; Brown 2006). Specifically, Northway residents reported that whitefish in their region were less abundant than in the past, as well as tasting different and being smaller in size (Brown *et al.* 2002). Prior to 1998 there was limited information available to fisheries biologists regarding these fish in the Upper Tanana drainage. In the fall of 2002, the Northway Whitefish Project, a collaborative project between the Northway

Village Council, the Regional Resilience and Adaptation Program at the University of Alaska Fairbanks, and the U.S. Fish and Wildlife Service, began. The overall goal of the project was to link the experience and knowledge of local experts with that of fisheries biologists to create a clearer picture of how the ecosystem is working and possibly changing. Similar studies linking local knowledge and Western science have been used not only to examine potential effects of climate change in the Arctic (Riedlinger and Berkes 2001), but also to research rare species in South America (Colding 1998) and to design protected areas or reserves in tropical regions (Johannes 1998) among other projects.

The cooperative nature of this project highlights critical challenges facing environmental researchers across the globe (Huntington 2002). This project employed creative methods to facilitate knowledge-sharing between scientific and local experts. A key element in this process is trust, which is formed through ongoing and effective communication between and among researchers and community members. This project created a local advisory board made up of local whitefish experts, hired a local high school research assistant, made project posters for the village, hosted community feasts and presentations and created interdisciplinary teams that worked together in the field, village, homes and offices. During the summers of 2003 and 2004, graduate student and researcher Melissa Robinson lived in the village and during this time, she and community members interviewed local experts identified by the village council using a semi-directed method. She also participated in all aspects of fish harvesting and processing. Twelve local experts were interviewed, sometimes individually and sometimes as a husband-wife or parent-child pair. Six women and five men were interviewed. Simply drinking tea or talking while washing clothes was invaluable in not only learning about whitefish and the ecosystem, but also gaining an understanding of the social structure and context into which fishing is woven. Without an understanding of this social context, the possible impacts of environmental change cannot be fully understood (Huntington 2002).

Gender Roles in Northway

Among the six common fish species in the area humpback whitefish are harvested in the largest quantity (McKennan 1981; Case 1986; Marcotte 1991). Although whitefish are harvested throughout the year, the majority of fishing is focused during the spring run of whitefish into lakes (late May–early June) and the fall run of whitefish out of lakes to spawning grounds in the main rivers (mid August–early October) (McKennan 1981; Case 1986). Prior to the 1970s, seasonal fish camps were located on small streams or lake outlets where people collectively harvested whitefish using weirs with large dip nets and occasionally cylindrical willow fish traps (McKennan 1981). Since the 1970s, most people have used gill nets where smaller creeks enter main rivers or lakes (Fig. 3). This is a very effective fishing method during peak fish runs where a family's catch of 100-200 humpback whitefish per day is not unusual (Marunde and Marunde interview 2003) (Fig. 4).

Whitefish are cut and dried for consumption by humans (*ba*) or dogs (*tsalkeey*), frozen whole, or fermented (*dzenaxł*) in buried birch bark baskets. Residents often eat whitefish eggs baked or fried, boil fish meat for soups, fry half-dried fish, or bake whitefish together with moose organs. Fried whitefish stomach (*ch'itsaan'*) is a delicacy among local residents. The oil rendered from boiling fish stomachs can be used to fry other foods such as biscuits or mixed with berries and eaten during holidays.

Figure 3. *Typical set gill net in a creek near Northway. Photo by J. Marunde, 2003, Northway Whitefish Project*

Figure 4. *Humpback whitefish caught in a gill net are usually placed in buckets and taken home in boats to be cut. Photo by J. Marunde, 2003, Northway Whitefish Project*

In Northway Village, fishing roles are generally divided based on gender. Women almost always cut fish while the men gather wood and drive the boats. It should be clear, however, that the division of labour is not rigid. In Northway, as in other areas, gender roles are dynamic and can change in response to socio-economic conditions (Guyer 1991; Madge 1994; Nygren 1999; White 1999; Jewitt 2002). There are times when men may cut fish and women drive the motor boats, depending on family or community needs and structure. For example, if the individual who usually does a task is ill or away from the community during whitefish harvesting (*e.g.,* for firefighting, a common summer wage earning activity), another adult with the requisite skills, whether male or female, may pick up the task. One resident commented that in her family, fishing roles also changed with age. She stated:

> They [men] rarely cut fish, but my uncle Kenny did. But it was later in years, you know, when he got older. My dad does too, but when we were younger most of us did it, us girls and my mom (Paul interview 2004).

Some fishing jobs in Northway are shared by men and women, while other jobs are distinctly male or female. In Northway, both men and women used to dip net for fish and now they paddle canoes to check and set gill nets in creeks and lakes. Thus, both women and men handle fish in and out of the water. However, beyond scaling and de-heading, fish processing is a woman's task. Women and young girls almost exclusively cut the fish, clean their stomachs, and prepare the fish for storage or eating (Case 1986). Understanding the importance of whitefish to Northway residents and the many and complex roles residents play in this process is required prior to understanding how men and women perceive environmental change.

Noting change

In Northway as in other areas, an individual's role in fishing influences one's experiences and knowledge about a fishery and an ecosystem (Siar 2003; Crona 2006). This section focuses on how men and women experience changing water levels and siltation, parasites in fish, and changing weather patterns.

Changing Water Levels and Siltation

In the past thirty years, Fish Lake, which was the location of two main fishing camps, has changed considerably (Sam and Sam interview 2004). Increased water levels and land/bank erosion in the lake have sunk 'fish frames' (fish drying racks), damaged a cemetery, and washed out fishing weirs (Fig. 5). Along with the changing water levels, increased siltation to the area is contributing to changes in fishing locations and methods from dip netting to setting gill nets as the silt obscures dip netting visibility. According to local residents, siltation in particular, is contributing to the rapidly changing channels and newly emerging sandbars. Although there is a considerable amount of fisheries literature examining the affects of sedimentation on aquatic communities (Murphy *et al*. 1981; Grant *et al*. 1986; Welsh and Ollivier 1998), little is known in the scientific community about this relationship with regard to whitefish in the Upper Tanana drainage. However, Northway residents believe that siltation is having a negative effect on whitefish and other wildlife species including ducks and muskrats. One resident explained:

Figure 5. *Traditional fish frame (drying rack) sinking into the water at Charlieskin Creek, near Northway. Photo by M. Robinson, July 2004, Northway Whitefish Project*

> It was clear yeah, people used that water for drinking and eating, cooking you know. It was good water. But you can't use it now. And it was full of good food for the ducks and the fish. That silt just killed off all that vegetation. It's just dead looking. You go back over there and it's just a dead, you don't see that vegetation that was in there. (Fix interview 2004)

This new siltation is due in part to the backing up of the Chisana River into creeks flowing into Fish Lake. It remains unclear what is causing the Chisana River to backup, although local people have their own theories including thawing permafrost and increasing temperatures. Also not surprising is the wide array of concerns and observations about increased sedimentation expressed by men and women based on their different forms of fishing participation.

Travelling on the water, both men and women observe changing sedimentation patterns in Fish Lake. Men commonly drive boats fitted with prop motors along the main rivers and lakes. As drivers, men are very aware of subtle differences in channels or sand bars as they wish to avoid hitting bottom, getting stuck, or breaking a propeller miles away from the village. When paddling a canoe, this awareness may not be as necessary since canoes draw less water, rarely get stuck, and do not have propellers that can break, leaving a person stranded (Fig. 6). Because men almost always drive the boats, they seem to have a keener grasp of how the channels in Fish Lake shift over time and the location and growth of new sandbars in the area. This is not to say that women do not have a good sense of channel location gained from years of canoeing or riding in the boats, but in general, men spoke in greater detail about the subtleties of channel shifts and new sandbars. For example, men were able to give dimensions and measurements for how far the channels move each year; "You'd go through it and the next year it would be over six feet or something" (Marunde interview 2004). Men were also able to describe the time scale and overall distance that sand has moved. One fisher stated:

> But to us it seems like the sand is almost made it all the way down, to almost near where we set our fish net. I mean within a mile. It didn't seem like that ten to fifteen years ago. (Marunde and Marunde interview 2003)

Although channels in Fish Lake have probably shifted continuously for years, knowledge about changes in this shift over the past twenty years is important since it appears to be happening at a faster pace. Similarly, identifying new locations where sand has 'migrated' is significant as well. Men's observations about sedimentation has also added context to this issue by illustrating how increased siltation impacts the community's ability to access fishing areas. Once again these are crucial contributions for natural resource managers and fisheries biologists since, even if increased siltation has little direct impact on whitefish health, it affects people's lives by altering their fishing locations and changing fishing methods. Thus, it can have an accumulated effect on stock health and structure since fish are exploited differently because of perceived increases in sediment loads.

Northway residents also believe siltation negatively affects whitefish and other species. Men and women discussed catching fewer whitefish in Fish Lake than twenty years ago. They also said that the fish they do catch are smaller and taste different than in the 1980s. Men and women pointed to the lack of muskrats, ducks, and certain vegetation that were once abundant in Fish Lake. Since women and children are often the ones who clean fish stomachs prior to being fried and eaten they also observe what fish feed on. Darlene Northway stated that she can tell if whitefish are eating 'right' by their stomach contents. She explained:

> When we get that fish stomach we can tell where they eat and we can tell if it eat those little, those little insect in creek or lake. We can tell like that they eat those.

Figure 6. *Northway residents using a canoe to check a gill net set in a creek. Both men and women frequently use this method to set and check gill nets. Photo by J. Marunde, 2003, Northway Whitefish Project*

And they eat nothing but sand too, sand going to be in their stomach. They eat, when they don't eat right they don't have those little bugs in there.

(Northway interview 2004)

These observations supplied by women have led some locals to believe that siltation covers crustaceans, a prime food resource for whitefish (Alt 1983), causing what some perceive as a decrease in the whitefish population. A few local residents also think that siltation covers food for ducks, muskrats and beavers, providing a reason for their absence (Fig.7). As Cherie Marunde said:

There used to be lots of muskrats in there and there isn't any cause there is no food for them I don't think. And then there was a beaver in there too, a few beaver. There isn't any now. I don't know if that affects them very much. And then the ducks. That used to be the biggest duck, you know I mean there were ducks everywhere when I was growing up, and now they're hardly anywhere.

(Marunde and Marunde interview 2003)

It is important to acknowledge that the effect siltation may have on whitefish is unknown to fisheries biologists and managers. However, hearing multiple hypotheses and personal experiences broadens the scientist's perspective by adding to or placing in question the variables being studied. This can help direct or focus future biological research on whitefish. For example, further science-based research may confirm local residents' perception that there is a direct cause and effect relationship between siltation and fewer ducks. If nothing else, these two examples in Fish Lake illustrate the ability of people to connect multiple experiences together in order to understand their world. The combined insights of men and women are integral to this process.

Parasites

Along with the changes at Fish Lake, Northway residents, women in particular, also have concerns about fish parasites. Women's concerns stem from their close contact with the internal anatomy based on their focused labour role of cutting and gutting fish. This, in turn, exposes them to internal parasites more often than men. There have been studies on parasites in broad whitefish (*Coregonus nasus*) in Canada (Choudhury and Dick 1997) and whitefish species in Russia (Bauer and Nilol'skaya 1961; Kogteva 1961), yet more complete understandings of what species exist within whitefish, what is a healthy parasite load, what factors contribute to differing parasite loads, and their impacts on whitefish, are still lacking. Although it is uncertain how parasites may impact human health, in Northway, heavily parasitized whitefish are not eaten by humans and are usually thrown to dogs (Marunde and Marunde interview 2003; Sam and Sam interview 2004). Therefore, regardless of their impact on fish or human health, the prevalence of parasites in whitefish can influence human consumption (and thus harvest) of the species if residents harvest more fish in order to meet their subsistence needs. Consequently, changes in whitefish parasites and their affects on local harvest are valuable pieces of information for fisheries biologists resource managers. Information about these parasites also provides a motivation to monitor and assess changes in the future to determine possible parasite-related impacts on whitefish health. As this next section highlights, women are crucial to raising and addressing these questions regarding parasites. Women indicated that whitefish parasites are not a new phenomenon. They mentioned hearing their mothers or grandmothers talk about parasites and recalled seeing parasites when they were young girls helping adults cut fish. One woman stated:

> Well, my mom always said it was worms or something. I don't know what she called it but I remember it was, yeah that it was there then too (Marunde and Marunde interview 2003).

This 'historical' information provided by women helps set a standard upon which to build knowledge about parasites and whitefish in this region. The process of passing information from older to younger women conditions girls to look for certain cues, such as particular abnormalities. As a result, women may notice changes that others not trained to see would miss, including new parasites or a new parasitic location. It is a combination of what people see, what they are taught to see through social relationships, and what they learn through their own experiences that can make a difference in their knowledge base. Parasites in whitefish are a good example of this selective observation.

Women experienced at cutting fish possess the historical knowledge and experience to know what is an unusual or a normal parasite load and are the judges of what constitutes an unacceptably large number of parasites for their family. While men also observed parasites and voiced concerns, women provided more detailed descriptions of seasonal and long-term differences in parasite prevalence. For example, one elder woman mentioned that she had never seen as many parasites as she did in 2003 (Sam and Sam interview 2004). Another woman noted that white cyst-type parasites are always worse in the spring (Marunde and Marunde interview 2003). Women gain these skills over time by handling whitefish in addition to the cumulative history and experiences of their older female relatives. The reasons for a 'bad' year or seasonal parasitic cycles in the Upper Tanana are unknown to fisheries biologists or local residents, although it may be due in part to seasonal and yearly food fluxes or parasite life cycles (Kogteva 1961; Robinson *et al.* 1998). However, linking women's observations of yearly and seasonal differences to each other can aid in identifying the causes of these temporal variations.

Women also spoke in much greater detail than men about parasite size and variety. In general, men noted colour and location of parasites in the fish, but beyond that, their knowledge was limited. When husband and wife pairs were interviewed together, women took over the conversation when it turned to parasites, describing in depth their location in whitefish, their size, colour, and type. Women in Northway described three different parasites: 1) a cyst-type parasite in the flesh which is almost 'woven into the meat,' 2) a long, thin, white worm located in or on the stomach, and 3) another white parasite located on the back part of the ribs. Without cutting the fish or otherwise paying close attention to the body cavity these discrete locations and associated parasites are likely to be overlooked.

Parasites provide a solid example of how men and women differ in their depth of knowledge based on social relationships and fishing roles. Not only can women provide detailed descriptions and historical information about parasites, but they also voiced an important concern. Parasites can play a significant role in regulating fish populations by directly affecting their growth and weight, and indirectly through increasing their risk of predation (Szalai and Dick 1991). Thus, women's observations of parasite loads and types may help identify poor-quality fish or changes in fish health. This information, along with scientific understandings about parasite-fish interactions, can help answer questions that local residents have about the variation in parasites loads and their impact

Gender, Culture, and Northern Fisheries – Chapter Six

on whitefish and human health. Overall, women's knowledge in this area can raise new questions about whitefish, suggest new research directions, and expand the knowledge base about about whitefish-parasite interactions.

Changing Weather Patterns

Arctic residents, including those in Northway, are noticing changing weather patterns, such as longer falls, shorter winters, and warmer summers (Greenland and Walker-Larsen 2001; Cochran and Geller 2002; Krupnik and Jolly 2002). Again, women provided a unique perspective based on their role cutting and preparing the fish. Acknowledging and then incorporating these differences and similarities into research creates a clearer picture of the ecological and social changes related to climate and weather patterns.

Knowing when to set and check a fish net depends partially on both air and water temperature. Since men and women check and set fish nets, both are in tune with changing water temperatures throughout a season and year. As water temperatures rise, fish nets need checking more often since fish caught in gill nets spoil faster (Georgette 2002). One man mentioned that during a hot July day fish can spoil overnight, but by August fish left in a net overnight remain firm (Sam and Sam interview 2004). However, some residents stated that summers are longer and the August month hotter than twenty years ago,

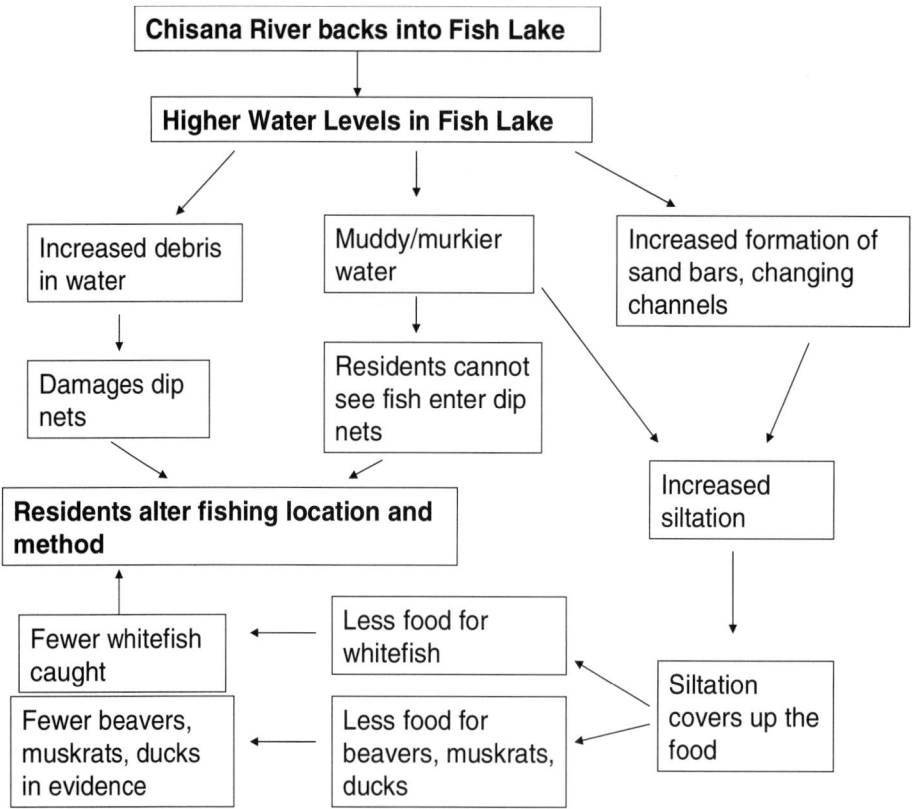

Figure 7. *Diagram illustrating the relationships specified by local residents among water level, siltation, and changing fishing locations and methods*

therefore, they check their nets more often than they used to. Men and women also talked about a daily movement of whitefish out of lakes into creeks during especially hot days when they believe lakes heat to a point whitefish cannot handle. Movements of humpback whitefish in response to daily temperature changes have not been studied by scientists, but local residents like Cherie Marunde and Howard Fix, consider it fact. They stated:

> If it's hot, I know [the whitefish] runs that night. I know it has to, I know temperature has to, you know, be a part of it when they run and when they don't run.
> (Marunde and Marunde interview 2003)

> For some reason the whitefish like cold water. I think that's why they come out of the lake in summer time, soon as the water starts getting warm and hot they head out of there and go to colder water in the river.
> (Fix interview 2004)

Some Northway residents voiced concern that this warming trend will negatively impact whitefish, based on their belief that whitefish need cold water and thus leave areas where the water temperature gets too warm. In this case, both men and women are linking observations of hot days and hot water to fish movements. While this only suggests a causal relationship, it points to an area for further investigation. Water temperature does play a role in whitefish egg and larval survival and development (Price 1940; Stalnaker and Grusswall 1974). As Magnuson (1991) pointed out, changes in temperatures due to global climate warming will alter thermal structure in lakes impacting the distribution and well being of fish. Thus, understanding the thermal limits of fish, such as humpback whitefish, is important to current and future management of the species.

Based on their experience processing fish, women also have unique observations concerning weather and temperatures related to the firmness of fish flesh. Fish firmness is related to how long fish lie dead in the net, the water temperature, how long they are out of the water, and the air temperature. Women use certain cues to establish the degree of firmness, making different processing decisions based on that determination. The firmness of a fish determines whether that fish should be cut for dogs or humans, and then whether it should be boiled, half dried, or fried whole. Women base this decision on how the flesh looks, how it smells, and how it feels to the touch and to the blade as it is being cut. One woman described that when the fish are too soft for traditional drying, the flesh feels mushy and crushes a certain way as it is cut (Paul interview 2004). Another woman said that when it gets too hot in July she can tell a fish will spoil as it is drying by the way it smells. She explained:

> In July when it's too hot then your fish spoils when you cut them so it's kinda a...you know hard to deal with. If I cut those fish then a lot of times it's too hot and they, they don't smell so good and so we give them to the dogs anyway
> Marunde and Marunde interview 2003).

These are experiences that most men do not have because they rarely cut or process fish. On the other hand, both men and women have the responsibility of turning the fish while they hang to dry in the smoke house (Fig. 8). On hot days fish need to be turned more often so that the air circulates and the fish dries without spoiling. Over the years, residents noticed trends in flesh firmness, fish smell, and the frequency at which fish need turning

Figure 8. *Humpback whitefish cut and drying for use as dog food (tsalkeey) dogs inside Northway elder Ada Gallen's smokehouse. Photo by M. Robinson, July 2004, Northway Whitefish Project.*

in relation to air temperature which can suggest climatic and weather pattern change over time. These observations come not just from men or from women, but through their combined and cumulative experiences.

Knowledge and experience gained from subsistence activities other than fishing, such as snow machine travel and trapping, are also important to one's knowledge about whitefish and the related changes that people see on the land. According to Northway residents lakes and rivers are freezing up later than they did thirty or forty years ago (Fix interview 2004; Sam and Sam interview 2004). While discussing harvesting whitefish under the ice one resident wondered about the effects of climate change on harvesting methods. He said:

> But now it's done warm up so the creek don't even freeze over until Christmas, so I don't know how that would work now. You know back then it froze over. Like the first of November we'd have no problem walking a trap line. So I don't know how that would work now, it's not even froze the first of November.
>
> (Fix interview 2004)

Besides just ice fishing, ice thickness is noted by the date that it is safe to snowmobile for recreation or trapping purposes (Marunde and Marunde interview 2003). In other contexts, some note an earlier thaw by the timing of roto-tilling their land for gardens. Others see the impact of thawing permafrost simply by the sinkholes in their yards (Fix interview 2004). Some of these activities may be more common to one or the other gender, so again, both men's and women's perspectives are essential to gain a full picture. Also, from a non-local perspective, observations related to gardening or trapping are unrelated to fishing, so are not necessarily an obvious area of inquiry. Along with gendered difference, these crucial experiences might be easily overlooked. Finally, it is the incorporation and combination of this variety of experiences that people use to make sense of their world.

The above examples illustrate how residents in Northway witness and observe changes in weather and temperature. It is not only their unique perceptions that add to the knowledge base, but also the similarities of observations and understandings of those observations that back one another's claims. The connections between weather, temperature, and whitefish are also better understood when one takes into account the combined observations from peoples' fishing and other subsistence activities.

Conclusion

In Northway, environmental change is affecting people's lives and influencing their fishing methods, locations, preparation, and harvest. This chapter has focused on how men and women experience change and how gender influences their knowledge base about whitefish and the ecosystem as a whole. Listening to both women and men is important because 1) they possess different information based on their roles related to subsistence activities such as fishing, 2) their concerns are based on what they see and hear in their respective roles, and 3) they have diverse historical knowledge bases passed down from other men and women. This historical knowledge is crucial when examining environmental change as scientists or researchers may have limited knowledge concerning their occurrence, relative intensity, and potential impacts to humans. It is not simply the varying perspectives people have that are important. Linking their observations is what creates a more complete understanding of a system over time. Crossing cultures and disciplines is a challenge, and recognizing the heterogeneity of communities and cultures is crucial in this process. Researchers must consider not only gender differences, but other factors such as age when examining local knowledge. For example, in Northway children usually clean the fish stomachs preparing them for frying and eating. Thus, children often have observations about current whitefish diet that adults may miss. This complexity should be embraced since it adds depth to the issue and in a sense gives it multiple dimensions.

The linkages and observations presented in this chapter have broadened the knowledge base about whitefish and the Northway area (Table 1), generated new hypotheses such as the potential connection between water temperature and fish behavior, focused or redirected fisheries research, and helped in understanding the social complexities that exist and how environmental shifts affect lives. Understanding a social or ecological system and the implications of changes to those systems is not possible without an awareness of the social context in which such processes occur. This requires a comprehension of the roles and perceptions of both men and women.

Table 1. Summary of men's and women's observations related to whitefish in the Upper Tanana drainage.

	Siltation	Parasites	Environmental Change
Women's Observations	Whitefish that feed in Fish Lake have stomachs full of sand	Detailed descriptions of parasite size, color, and location in whitefish	Flesh firmness due to their experience cutting and gutting fish
	General Knowledge of changes in channels and sand bars	Seasonal differences in parasite loads (type and amount of parasites)	Fish spoiling due to its smell
		Long-term changes in parasite loads (type and amount of parasites)	
Men's Observations	Finer grasp of changes in channel locations over time	General knowledge of parasites in whitefish	
	Detailed knowledge of locations of new sand bars		
Shared Observations	Bank erosion, murkier water than in the past		Changes in temperature, note frequency of checking nets and turning fish, thickness of ice, later freezing and earlier thawing

Acknowledgments: On behalf of the Northway Whitefish Project, we are grateful for the opportunity to work with the people of Northway Village and the many others who have contributed their time, knowledge, and skills to this project. This project is a collaboration of many experts from local fishermen to fisheries biologists who all added their unique perspectives in a co-operative nature. The authors want to thank the Northway Village Council, the Regional Resilience and Adaptation Program and the Long-Term Ecological Research Program at the University of Alaska Fairbanks (UAF), the Tetlin National Wildlife Refuge (U.S. Fish and Wildlife Service), the Fairbanks Fisheries Resource Office (U.S. Fish and Wildlife Service), and the Alaska Department of Fish and Game (ADFG). There are so many individuals in Northway Village who deserve recognition for their time, patience, and humour, in particular, the whole Marunde family, research assistant Glen Marunde Jr., Ada Gallen, Roy and Avis Sam, Julius and Martha Sam, Daisy Northway, Lorraine Titus, Verda Paul, Howard Fix, Cora Demit, Ida Wilson, and Oscar Albert. This project would not be possible without the assistance and encouragement of Randy Brown (U.S. Fish and Wildlife Service), Dr. Terry Chapin (UAF), Dr. Gary Kofinas (UAF), Caroline Brown (ADFG), Gary Holton (UAF), and William Schneider (UAF).

References

Arctic Climate Impact Assessement (2005). *Arctic Climate Impact Assessment.* Cambridge University Press. New York, NY, USA, 1042p.

Alt, K.T. (1983). Inventory and cataloging of sport fish and sport fish waters of western Alaska. Alaska Department of Fish and Game, Federal Aid in Fish Restoration Project F-9-15, Study G-1, Volume 24. Juneau: Alaska Department of Fish and Game.

Alt, K.T. (1979). Contributions to the life history of the humpback whitefish in Alaska. *Transactions of the American Fisheries Society* 108: 156-160.

Andersen, D.B., and C.L. Fleener. (2001). *Whitefish and beaver ecology of the Yukon Flats, Alaska.* Alaska Department of Fish and Game, Division of Subsistence, Final Report No. FIS00-06. Fairbanks: Alaska Department of Fish and Game.

Bauer, O.N. and N.P. Nikol'skaya (1961). 'Dynamics of the parsitofauna of the whitefish *Coregonus lavaretus* from Lake Lodoga and its epizootic importance,' pp. 224-238 in G.K. Petrushevskii, ed. J.I Lengy, I. Paperna *et al.*, trans. *Parasites and Diseases of Fishes: A Bulletin of the All Union Scientific Research Institute of Freshwater Fisheries, Leningrad 1957.* Jerusalem: Israel Program for Scientific Translations.

Brabets, T.P., B. Wang, and R.H. Meade (2000). *Environmental and Hydrological Overview of the Yukon River Basin, Alaska and Canada.* U.S. Geological Survey, Water Resources Investigations Report 99-4204. Anchorage: U.S. Geological Survey.

Brown, C., J. Burr, and M. Smith (2003). *Harvest monitoring of subsistence non-salmon fish in the lower Yukon River.* U.S. Fish and Wildlife Service, Office of Subsistence Management, Fisheries Resource Management Program Annual Report No. FIS 02-037-2, Fairbanks: U.S. Fish and Wildlife Service.

Brown, R.J. (2006). *Humpback whitefish (*Coregonus pidschian*) of the Upper Tanana River Drainage.* Alaska Fisheries Technical Report Number XX. Fairbanks: U.S. Fish and Wildlife Service.

Brown, R.J., C. Lunderstadt, and B. Schultz (2002). *Movement patterns of radio-tagged adult humpback whitefish in the Upper Tanana River Drainage.* U.S. Fish and Wildlife Service, Fisheries Resources, Alaska Fisheries Data Series Number 2002-1, Fairbanks: U.S. Fish and Wildlife Service.

Case, M.F. (1986). *Wildlife resource use in Northway, Alaska.* Alaska Department of Fish and Game, Division of Subsistence, Technical Paper No. 132. Fairbanks: Alaska Department of Fish and Game.

Choudhury, A. and T.A. Dick (1997). 'Parasites of the broad whitefish from the Mackenzie Delta, The Proceedings of the Broad Whitefish Workshop—The Biology, Traditional Knowledge and Scientific Management of Broad Whitefish (*Coregonus nasus* (Pallas)) in the Lower Mackenzie River,' in R.F. Tallman and J. D. Reist, eds. *Canadian Technical Report of Fisheries and Aquatic Sciences* 2193: 167-179.

Cochran, P. and A.L. Geller (2002). The Melting Ice Cellar: What Native Traditional Knowledge is Teaching Us About Global Warming and Environmental Change. *The American Journal of Public Health* 92: 1404-1409.

Colding, J. (1998). Analysis of hunting options by the use of general food taboos. *Ecological Modeling* 110: 5-17.

Crona, B.I. (2006). Supporting and enhancing development of heterogeneous ecological knowledge among resource users. *Ecology and Society* 11(1). Electionic document: http://www.ecologyandsociety.org/vol11/iss1/ Last accessed: February 28, 2009.

Crona, B. and O. Bodin. (2006). What you know is who you are? Communication patterns among resource users as a prerequisite for co-management. *Ecology and Society* 11(2): 7. Electronic document: http://www.ecologyandsociety.org/vol11/iss2/art7/ Last accessed: February 28, 2009.

Davis, D.L. and J. Nadel-Klein (1992). Gender, culture, and the sea: contemporary theoretical approaches. *Society and Natural Resources* 5: 135-147.

Estioko-Griffen, A. and P. Bion Griffen (1993). 'Woman the hunter: The Agta,' in C.B. Brettell and C. F. Sargent, eds. *Gender in a Crosscultural Perspective*. New Jersey: Prentice-Hall Inc.

Fix, H. (2004). *Personal interview by M. Robinson and G. Marunde Jr. July 2004*. Northway Whitefish Project, Northway, AK, USA.

Georgette, S. (2002). *Traditional knowledge of whitefish in Selawik, Alaska*. Progress Report to the U.S. Fish and Wildlife Service. Alaska Department of Fish and Game, Division of Subsistence. Kotzebue: Alaska Department of Fish and Game.

Greenland, B. and J. Walker-Larsen (2001). *Community concerns and knowledge about broad whitefish (Coregonus nasus) in the Gwich'in Settlement Area*. Gwich'in Renewable Resource Board Report 01-08. November 2001. Inuvik, NWT: Gwich'in Renewable Resource Board.

Grant, J.W.A., J. Englert, and B.F. Brietz (1986). Application of a method for assessing the impact of watershed practices: effects of logging on salmonid standing crop. *North American Journal of Fisheries Management* 6: 24-31.

Guyer, J. (1991). Female farming in anthropology and African history. Pp. 257-277 in *Gender at the Crossroads of Knowledge: Feminist Anthropology in the Post-Modern Era*. M. di Leonardo, ed. Berkeley: University of California Press.

Huntington, H.P. (2002). 'Human understanding and understanding humans in the Arctic system,' pp. xxi-xxvii in I. Krupnik and D. Jolly, eds. *The Earth is Faster Now: Indigenous Observations of Arctic Environmental Change*. Fairbanks, AK: Arctic Research Consortium of the United States.

Jewitt, S. (2002). *Environment, Knowledge, and Gender: Local Development in India's Jharkand*. Burlington, VT: Ashgate.

Johannes, R.E. (1998). The case of data-less marine resource management: examples from tropical nearshore finfisheries. *Trends in Ecology and Evolution* 13: 243–246.

Jolly, D., F. Berkes, J. Castelden, T. Nichols, and Sachs Harbor (2002). 'We can't predict the weather like we used to,' pp. 92-125 in I. Krupnik and D. Jolly, eds. *The Earth is Faster Now: Indigenous Observations of Arctic Environmental Change*. Fairbanks, AK: Arctic Research Consortium of the United States.

Kogtova, E.P. (1961). 'Fish Parasites from Pskov-Chud Water Reserve,' pp. 239-265 in G.K. Petrushevskii, ed. J.I Lengy, I. Paperna *et al.*, trans. *Parasites and Diseases of Fishes: A Bulletin of the All Union Scientific Research Institute of Freshwater Fisheries*, Leningrad 1957. Jerusalem: Israel Program for Scientific Translations.

Krupnik, I. and D. Jolly, eds. (2002). *The Earth is Faster Now: Indigenous Observations of Arctic Environmental Change*. Fairbanks, AK: Arctic Research Consortium of the United States.

Madge, C. (1994). Collected food and domestic knowledge in the Gambia, West Africa. *The Geographical Journal* 160: 280-294.

Magnuson, J.J. (1991). Fish and fisheries ecology. *Ecological Applications* 1: 13-26.

Marcotte, J.R. (1991). *Wild fish and game harvest and use by residents of five Upper Tanana communities, Alaska, 1987-88*. Alaska Department of Fish and Game, Division of Subsistence Technical Paper No. 168. Juneau, AK: Alaska Department of Fish and Game.

Marunde, G. and C. Marunde (2003). *Personal interview by M. Robinson and F.S. Chapin III. August 2003*. Northway Whitefish Project, Northway, AK, USA.

Marunde, Jr., G. (2004). *Personal interview by M. Robinson. July 2004*. Northway Whitefish Project, Northway, AK, USA.

McDowell, N. (1984). 'Complementarity: The relationship between female and male in the East Sepik village of Bun, Papua New Guinea,' pp. 32-52 in D. O'Brien and S. Tiffany, eds. *Rethinking Women's Roles*. Berkeley: University of California Press.

McKennan, R.A. (1981). 'Tanana,' p. 562-576 in J. Helm, ed. *Handbook of North American Indians, vol. 6: Subarctic*. Washington, D.C., Smithsonian Institution.

Murphy, M.C., C.P. Hawkins, and N.H. Anderson (1981). Effects of canopy modification and accumulated sediment on stream communities. *Transactions of the American Fisheries Society* 110: 469-478.

Northway, D. (2004). Personal interview by M. Robinson and G. Marunde. July 2004. Northway Whitefish Project, Northway, AK, USA.

Nygren, A. (1999). Local knowledge in the environment-development discourse. *Critique of Anthropology* 19: 267-288.

Paul, V. (2004). Personal interview by M. Robinson and C. Brown. October 2004, Northway White fish Project, Northway, AK, USA.

Price, J. (1939). Time and temperature relations in incubations of the whitefish (*Coregonus clupeaformis* Mitchill). *Journal of General Physiology* 23: 449-468.

Riedlinger, D. and F. Berkes (2001). Contributions of traditional knowledge to understanding climate change in the Canadian Arctic. *Polar Record* 37: 315-328.

Robinson, A.T., P.P. Hines, J.A. Sorenson, and S.D. Bryan (1998). Parasites and fish health in a desert stream, and management implications for two endangered fishes. *North American Journal of Fisheries Management* 18: 599-608.

Sam, R. and A. Sam (2004). *Personal interview by M. Robinson and G. Marunde Jr. July 2004*. Northway Whitefish Project, Northway, AK, USA.

Satia B.P. and C.Z. Wétohossou, eds. (1996). *Report of the working group on women's key role and issues related to gender in fishing communities*. Cotonou: Benin: Programme for the Integrated Development of Artisanal Fisheries in West Africa (IDAF).

Schiebinger, L. (1997). Creating sustainable science: women, gender, and science. *Osiris* 12: 201-216.

Shanley, P. (2006). Science for the poor: how one woman challenged researchers, ranchers, and loggers in Amazonia. *Ecology and Society* 11(2): 28. Electronic resource: http://ecologyandsociety.org/vol11/iss2/art28. Last accessed: February 28, 2009.

Siar, S.V. (2003). Knowledge, gender, and resources in small-scale fishing: the case on Honda Bay, Palawan, Philippines. *Environmental Management* 31: 569-580.

Simeone, B. and J. Kari (2003). *Ahtna traditional knowledge of resident species of fish in the Copper River Basin, east central Alaska*, Draft Final Report. U.S. Fish and Wildlife Service, Office of Subsistence Management, Fisheries Resources Monitoring Program Annual Report Project No. FIS01-110, Anchorage, AK, USA.

Slocum, S. (1975). 'Woman the Gatherer: Male Bias in Anthropologym,' pp. 36-50 in R.R. Reiter, ed. *Toward an Anthropology of Women*. New York: Monthly Review Press, New York, NY, USA.

Stalnaker, C.B. and R.E. Gusswell (1974). *Early life history and feeding of young mountain whitefish*. Ecological Research Series, Project 18050 DPL, Program Element 1B1021. Washington, D.C.:United States Environmental Protection Agency.

Szalai, A.J. and T.A. Dick (1991). Role of predations and parasitism in growth and mortality of yellow perch in Dauphin Lake, Manitoba. *Transactions of the American Fisheries Society* 120: 739-751.

Tallman R.F. and J.D. Reist, eds. (1997). The Proceedings of the Broad Whitefish Workshop—The Biology, Traditional Knowledge and Scientific Management of Broad Whitefish (*Coregonus nasus (Pallas)*) in the Lower Mackenzie River. *Canadian Technical Report on Fisheries and Aquatic Sciences* 2193: xi-219.

United States Census. (2000). *United States Census Bureau.*

Van de Ploeg, J. D. (1993). 'Potatoes and knowledge,' pp. 209-227 in M. Hobart, ed. *An Anthropolological Critique of Development.* London: Routledge, London.

Welsh Jr., H.W. and L.M. Ollivier (1998). Stream amphibians as indicators of ecosystem stress: a case study from California Redwoods. *Ecological Applications* 8: 1118-1132.

White, B. (1999). The woman who married a beaver: trade patterns and gender roles in the Ojibwa fur trade. Ethnohistory 46: 109-147.

SECTION TWO:

GOVERNANCE PRACTICES

CHAPTER SEVEN

"I Have Always Wanted to go Fishing": Challenging Gender and Gender Perceptions in the Quota-Oriented Small-Scale fishery of Finnmark, Norway

Siri Gerrard

Abstract: Since the quota system was implemented for the small-scale fishing fleet in Norway in 1990, the number of fishers has decreased substantially. While this is true for both male and female registered fishers, I argue that men and women are differentially impacted because of women's under-representation in the harvesting sector. Applying a feminist-inspired gender perspective with an emphasis on women, I link changes in the fishery policy, especially the quota system in Norway, to changes in women's and men's fishing practices and identities. I focus primarily on the challenges faced by female and male fishers in Finnmark, the northernmost province in Norway.

Introduction

In this chapter,[1] I investigate changes in women's, and to a lesser extent men's, practices, identities and perceptions about gender in the context of fisheries decline and policy change. In 1989, the head of the Norwegian Directorate of Fishery declared a moratorium on the cod fishery, and in 1990 launched the quota system. The pronounced purpose of the moratorium and the quota system was to protect the fish stocks, especially the cod, which is a very important stock in the Finnmark and Norwegian fishery economies. During the 1980s, the stock had declined drastically. The moratorium had substantial consequences for fishers and their families, particularly in the northern part of Norway. From the fishers' point of view, the moratorium and quota system represent changes from the outside. However, external and internal changes are not unusual either for the Norwegian or for the Sami fisher populations. Fishery people have always had to adapt to the imposition

---

[1] This chapter is developed from a paper delivered at the international conference AKTEA: "Women in fisheries and aquaculture: lessons from the past, current actions and dreams for the future," Santiago de Compostela, Galicia, Spain, November 10-13, 2004 (Gerrard 2005a). Some of the data and arguments can also be found in Gerrard (2005b). The article is based on data collected during 2005-2006.

of national policies, variations in fish stocks, fluctuations in the global market and the introduction of new technologies including fishing equipment, geographical information systems, cell-phones and computers. Because of such connections and relations, I argue that the local fisher population is connected to national as well as global processes.

Even though many fishers and their families in the three northernmost provinces of Norway (Nordland, Troms, and Finnmark) have Sami roots, it is only during the last few decades that attention has been paid to Sami fishery policy in the media or in the public arena in general. The Alta-Kautokeino demonstrations in 1979 and 1981 against the construction of a power plant in Alta River located in the heart of the Sami area of Finnmark, directed attention to the issue (Nilsen 2003). Since the creation of the Sami Parliament in 1989, members and staff of the Parliament and others have worked to put ethnicity on the Norwegian political agenda regarding fisheries. The coastal population seldom raises questions about ethnicity. In many coastal areas in Finnmark, to be a Sami is to be connected to the reindeer owners, herders and the population from Inner Finnmark, not to the coastal fisheries. Likewise, gender-related issues are seldom part of the fishery policy agenda or resource policy more generally. In the fall of 2004, however, the Sami Parliament placed considerable emphasis on women in fisheries in a report entitled 'The Sami Parliament's Report on Fishing as Industry and Culture in Coastal and Fjord Areas.' This report also addressed fishery households, communities and fishery culture.

In this chapter, I highlight relevant gender issues pertaining to the coastal fishery in Finnmark. In order to have a better understanding about women and men in the small-scale fishery, I start by situating it within the context of political change, with a focus on the quota policy. Thereafter, I introduce my perspective focusing particularly on gender. My concern for women and gender questions in fisheries, especially in Finnmark's coastal areas, is due to my interest in the small-scale fishery where men and women traditionally have had different tasks. At the local level, I focus on Åshild and Rolf Ove Pettersen, a fishing couple living in Skarsvåg in the municipality of Nordkapp, Finnmark. Their story helps me to elaborate on some of the changes that have taken place up to 2005. They have always been helpful and shared their stories about the pleasures and the challenges they encounter as small-scale fishers in a little fishing village in Finnmark. I have also discussed the findings and analyses with them. They have also read several drafts of the chapter and have expressed the importance of passing on information about their way of life to others interested in small-scale fisheries. I should also mention that I have had a close relationship with them and many of the inhabitants of Skarsvåg since I first started my work studying fishery cultures in the 1970s. I conclude the chapter by connecting these different elements and discussing some of the analytical challenges concerning gender-related issues in the fisheries. This discussion leads me to pose the question, "What does it means to be a fisher under today's quota regime?"

Quota Systems and Policy questions:
Increased Emphasis on Market Orientation

Quotas have played an important role since the system was launched in 1990. All over the world where quota systems have been applied, the implementation has had different consequences for women and men in fisheries and on femininity and masculinity (Porter, 1993; Gerrard 1995; Munk-Madsen 1996; Binkley 2000; Skaptadóttir 2000; Grzetic 2004; Power 2005). The quota system can be considered as a part of the restructuring of the fisheries and has lead to many changes in political and socio-economic aspects of life as well as in the environment (Neis *et al.* 2001).

I also contend that the quota system, in its various forms, has seemed to increase the market orientation in fishing. Today, it is therefore not only fish as a commodity in the market, but also, the right to fish organized in a quota system that is brought into the market system. Even though Norway continues to be a large exporter of fish, resource conflicts and how to resolve them have played an important role in the public debate. The quotas are therefore an important part of the Norwegian fisheries management system. All registered fishers in Norway have to pursue fishing within this system according to specific rules for each fishery. This must be done regardless of where each fisher lives and whether they identify themselves as Norwegian or Sami. However, during the first years of the quota system, the allowable catch of cod for part-time fishers who were entitled to a maximum sector quota for all the fishers that belonged to this group, was caught before many of the fishers from the Sami fjord districts had started the season in March or April. This season begins when the cod stock that has been spawning in Lofoten is on its way back to the Barents Sea.

Norway has a system of boat quotas for full-time registered fishers. They belong to group B, while the part-time fishers belong to group A. For the full-time fishers, the quotas are connected to the boat, not to an individual person, but there are requirements for people who want to obtain a boat with a quota. The owner of the boat can hire as many crew as he or, in a very few cases, she wants. However, since the quotas for most coastal boats are limited, many boat owners with a crew search for more boats with quotas. They say they need more quotas to be economically viable.

The formal quota system has gone through many changes over the years. From the very beginning, both Sami and non-Sami fishers who own a boat with historical rights in cod fishing in the northern part of the country were entitled to a quota commensurate with boat size.[2] In 2004, a boat of 10 metres and 46 centimetres could deliver 27 metric tonnes of gutted and headless cod in addition to certain quantities of haddock and saithe or coalfish. As of 2005, a boat of this size could deliver about 25 metric tonnes of cod;[3] with a reduction in the cod quota but an increase in the haddock and coalfish quotas. This does not necessarily represent an advantage for the Finnmark fishers since the cod fishery has been and still is one of the most important fisheries in that area.

In Norway, quotas can be rented and bought and sold indirectly.[4] During the first few years of the quota system, a fixed number of tonnes was reserved for the whole group of part-time boats (group A fishers), many of whom are located in Sami areas. In other words, individual boats in the small boat sector competed with each other for fish until the sector quota was filled. After a few years and protracted discussions, these boats also

---

[2] From 2004, it was also possible to transfer a quota from a larger boat to a smaller boat within the same size group.

[3] It is also possible to fish more coalfish (US: pollack) and haddock, but then one has to reduce the amount of cod. These days, cod is the most valuable fish with the highest price.

[4] The quotas are allocated to a boat and the amount of cod, coalfish or haddock that can be fished is dependent on the size of the boat and historical rights. If a boat from the southern part of Norway has been fishing in the northern part of Norway, it is allowed to continue to fish in the northern areas. It is not legal to buy or sell quotas, but one can buy and sell a boat with a quota. As of 2004, there is also a system for renting out a quota.

received a fixed quota. From the mid-1990s onward, all registered fishers could fish for cod when they wanted, normally in the spring season.

The Ethnic Dimensions of Fishery Policy

As mentioned, the ethnic dimensions of Norway's fishery policy is not well developed. In the 1980s, the Sami movement undertook initiatives to raise fishery-related questions with the Minister of Fisheries.[5] In the middle of the 1980s, two leaders of the Fishermen's Union in Porsanger Fjord highlighted the ethnic dimensions in order to claim better conservation of the fjord to protect the spawning area of the coastal cod. The conservation claim was an old one, but, for the first time in history, Sami heritage was brought forward through Samenes Landsforbund (the Sami Country Alliance), one of the Sami organizations in the area. The Fishermen's Union in Finnmark excluded them "on the flimsy pretext that they had both put forward a political claim relating to the fishing industry via an organization other than the Fishermen's Union" (Nilsen 2003:175). The Porsanger fishers went to court and won.

Svanhild Andersen (2001) states that fisheries authorities and the Fishermen's Union have, for many years, ignored the coastal Sami fishers' request for protection of the spawning grounds in the fjords. Even though some changes can be observed, the resistance from the Fishermen's Union and the Ministry of Fisheries continues even though most of the claims from Sami associations, committees and the Sami Parliament propose that all fishers should have the same rights and advantages independent of ethnic heritage. The resistance to Sami issues from those in power in the Norwegian Fishermen's Union and the Ministry of Fisheries can be better understood in light of the emphasis placed on fish as a common resource and the fact that non-local boats have a long history of participating in the Finnmark fishery. There has been little recognition of potentially unjust practices resulting from so-called equality policies. The example from the early history of the quota system in Norway when the fjord fishers had not yet started their seasonal cod fishery before the fishery was closed for the part-time fishers, illustrates such an unjust practice. This example also indicates that to live close to the fishing grounds is no longer an advantage. Earlier, fishery people moved to Finnmark to be able to catch as much fish as possible near the fishing ground. The quota system that only allows each boat to harvest a certain amount of fish thus minimizes the advantage of living close to these grounds.

Inequality regarding access to the fishery occurs even though Norway ratified the International Labour Organization (ILO) Convention 169 in 1989, recognizing indigenous peoples' rights to the sea. Carsten Smith, a law professor who, later on, served as Chief Justice of the Supreme Court of Norway, carried out significant judicial research regarding Sami peoples' fishery rights (Smith 1990). This work acknowledged that indigenous fishery rights should be considered in the Norwegian fishery policy. However, Smith's arguments have had little impact as reflected in the minimal changes regarding the inclusion of indigenous perspectives in Norwegian fishery policies (Eythórsson 2003; Nilsen 2003). The Sami Parliament has started to address these questions especially concerning

---

[5] The board members of the Norwegian Sami National Alliance went to Oslo to meet with the fishery minister, Tor Listau (information given by Liv Østmo—one of the board members of Norske Samers Riksforbund /Norwegian Sami National Alliance) at that time.

the quota rights of smaller boats. It should also be mentioned that the discussion about land and water rights have been on the political agenda for many years in Finnmark and the rest of the country. This is evident more recently in the debate over the Finnmark Law. This debate has focused on the nature of the land and water management systems, the number and the appointment process for members of the special commission that manage the law and the area covered by the law. Carsten Smith (Aftenposten, April 13th, 2005) argued that the Finnmark Law should also include sea rights.[6] However, the Finnmark Law was passed by the Norwegian Parliament in May 2005 with limitations regarding the sea.

Gender and Influence on Quota Questions

Decisions about the size of boat quotas and fishery regulations in general are part of a complex process depending on negotiations in the Norwegian–Russian Fishery Commission. This commission decides upon the total allowable catch (TAC) for the following year and the result is approved by the Norwegian Government and presented to the Norwegian Parliament. The Ministry of Fisheries and Coastal Affairs through its Directorate of Fisheries makes the final decisions after the Regulatory Council has presented its recommendations.

The Regulatory Council includes representatives from the Norwegian Fishermen's Union, the Coastal Fishermen's Union, the fish buyers' organization, the Fish Workers' Union and researchers from the Institute of Marine Research. There are also observers from the Norwegian Society for the Conservation of Nature (Friends of the Earth, Norway) and the Sami Parliament. The observer from the Sami Parliament has always tried to focus on small-scale fishers and such efforts have brought about changes in the quota system and the Sami Parliament. The observer from the Sami Parliament has always tried to focus on small-scale fishers and such efforts have brought about changes in the quota system favouring the small-scale fishing fleet. In the 1990s, both of these groups sent women as observers. Nevertheless, the great majority of committee members and observers have always been men. This is also the case for the Norwegian–Russian Fishery Commission. In 2004, both the Norwegian and Russian Commission met with only a few women to negotiate the TAC of the cod stock. These and other fishery-related committees have applied for exemptions from the *Gender Equality Act* in Norway that demands 40% women (or men) in public committees. The argument has been that fishery organizations have few women members and that few women are interested and eligible for such posts (Sloan 2004:87-88). I suggest that the exclusion of women reflects ideas about who are the experts in fishing and who should hold special offices. The former Minister of Fishery, Svein Ludvigsen, argued that discussions about the fishery industry needed to continue with or without the presence of women. He pointed out that only a pool of women environmentalists were available rather than women who worked actively

---

[6] In his feature article, Smith wrote: "Lovutkastets regel om beg rensningen mot sjøen bør sløyfes. Derved vil en videre rettslig og politisk debatt ikke bli hindret. Og domtolene vil på grunnlag av rettsprinsipper kunne fastlegge rettighetene for kyst og fjordfisket." My translation to English is: "The draft law's regulation about the limitation towards the sea should be omitted. In this way a further legal and political debate will not be hindered and the courts could, on the basis of legal principles, determine the rights for the coastal and fjord fishery."

in the fishery. In this way, the Ministry refused to acknowledge women in the fishery as capable experts who can give advice concerning fishery issues.

Many researchers have documented women's contributions to the fishery in Norway (Munk-Madsen 1996; Pettersen 1994; Gerrard 1983, 2003b; Angell 2004; Erlandsen 2004; Sloan 2004). Class is indirectly on the agenda since much of the fishery policy in Norway is related to size and type of boat and equipment, which are preconditions for income level. Sami interest groups as well as the Sami Parliament have tried to increase their representation in different committees in order to influence fishery policy in general and resource management in particular. Here, both ethnic background and geographic belonging are emphasized. This has been intensified since the quota system was launched. The report from the Sami Parliament (2004:8) mentioned above, stated that the quota system has threatened the historical rights of the coastal Sami population because allocations are based on previous years with little available fish in the fjord areas. However, the fishery as a political field aptly demonstrates that there is little space for differences related to gender and ethnicity.

Quota Policy is Followed by Other Changes

Fishery communities—like other communities around the world—have faced other changes that are related to the market-related way of thinking. One observed trend is the centralization of industrial fish production. The filleting and freezing of fish is labour-intensive. Most of the fish factories in Finnmark, big or small, have gone bankrupt because of stock declines and difficult market conditions for filleted fish products. The processing factories that remain are now concentrated in the larger fishery communities. The more modest fish plants in the smaller communities often send the fish away to be processed in the bigger factories. There are also examples of a more globalized system of production. Fish from the Barents Sea are now transported to China for production and sent back to Norway and other European countries to be sold and consumed. The industrial filleting production in the northernmost province of Norway has, in this way, decreased enormously.

A general restructuring of the public economy has also taken place. A very large part of Finnmark was burned to the ground by Hitler's regime in 1944. This led to massive, government-supported construction and resettlement after the war. Over the last few decades of the twentieth century, the Norwegian state transferred money to different sectors of the municipalities. They still continue this practice, but today, the transfer of money is not dependent on sector. At the same time, more responsibilities and tasks have been handed over to lower administrative levels. That means that more work has to be done for the same or less amount of money, since compensation for higher wages is not always provided. The amount of money is dependent, in part, on the number of inhabitants in a municipality.[7] Fewer people in a municipality means a reduction in income as well as public jobs such as teaching, cleaning schools and other public buildings, caring for children or elderly and sick people in their homes or in special institutions. Subsidies to the fishing industry have also decreased. The cut in subsidies is partly due to regulations set by the European Union (EU), which establishes the agreement on European

---

[7] There are still special arrangements for municipalities with fewer than 3,000 inhabitants.

Economic Area (EEA) between EU and the European Free Trade Association (EFTA) countries of which Norway is a member. However, the changes may also be related to a more right-wing or neo-liberal economic-oriented policy. These trade agreements can, in my opinion, also be seen as a reflection of this wider neo-liberal ideology. These and other changes have resulted in fewer employment opportunities and a decrease in the population of the coastal communities in Finnmark. These days, it is not only young women who are moving, but also young men and established families.

Women and Men in Transition – Theoretical Considerations

The changes presented above help shape the context that may contribute to local women's and men's changing practices and identities. Elsewhere, I (Gerrard 2003a), have described such changes in practices and identities as 'being in transition' or 'being in between'. These are perspectives that I also apply here. What do I mean by being 'in transition'? Women and men whom I consider to be 'in transition' or 'in between' are faced with structural changes, which may lead to new ways of thinking and new types of knowledge and practices. They can also lead to role, status and identity changes. When we define situations as changing or 'in transition', we do so because we are faced with situations where we often have to seek alternative actions as far as employment, education, leisure, and settlement are concerned. Under such circumstances, negotiations of values and ideas about what is appropriate behaviour for men and women may be needed and take place. I am inspired by Susan Stanford Friedman's concept that "identity resists fixity" (Friedman 1998:23). This means that identity may change depending on the situation.

Of course, the results of change are rarely neutral and my adoption of a feminist perspective ensures that I consider the differing consequences of the new quota system for women and men, especially for their practices and identities. As a result of these consequences, must we as researchers reconstruct the concepts of a fisher, fishery and fishery culture? How do local women and men interpret the changes? Do they look upon them as improvements indicative of progress or as steps to control their lives? Do these changes strengthen women's positions?

Asking such questions forces us to consider issues of power, which have a long tradition in feminist research. In Norway, Ellingsæter and Solheim (2002:35) have made an important contribution to studies on gender and work. They emphasize the meaning of work and the symbolic dimensions of power. At the same time, they highlight economic and social aspects. That means that they advocate analyses where the symbolic aspects are connected with institutions as well as other social structures. They claim that the concept of power has to be studied within its context, and the context, varies depending both on the time period and the topic under review. Through many studies, feminist researchers interested in fisheries have revealed gender biases and power imbalances in the preoccupation of men's work in policy, cultural images, and research (Gerrard 1983, 1986, 1995; Munk-Madsen 1996; Davis and Gerrard 2000; Angell 2004; Power 2005).

A focus on power requires not only an examination of gender but also ethnicity in the current context of changing fisheries policies. Open discussions in coastal Finnmark about ethnic background and heritage are rare. However, when ethnicity is on the formal political agenda, these kinds of discussions are more likely to be debated among ordinary people. For example, regional and national print media coverage of the Finnmark Law during the spring of 2005 sparked local discussions. My own research in the 1970s on a labour union in the filleting industry in one of the biggest fishing villages in Finnmark revealed that talks about ethnic heritage often took place informally in a joking manner and more so among men than women (Gerrard 1986). Women stayed quiet when ethnic identity was on the agenda, even in informal talks with each other. There are many reasons for such reactions. The Norwegian-speaking people represent the majority in many of the coastal communities. Some of my informants felt that they were stigmatized if they revealed their background. In my eyes, this silence reveals 'hidden' power relations between the Norwegian and the Sami populations.

Concerning gender, traditionally, men's fishing activities influenced when and where women's work was carried out. At the same time, women worked independently and very hard (Gerrard 1975). The result has been an efficient fishery. In addition to fishery-related work, women have had to carry out the many tasks of household work. In my early research, I argued that such patterns represented an important part of the local culture of small-scale fisheries (Gerrard 1975). In the early 1980s, I used the concepts of ground or shore crew when I analyzed the situation of fishers' wives. I emphasized that women's tasks were extremely important in fishery societies, but were not recognized formally in terms of social benefits and rights. In the Norwegian welfare state, paid work provided the basis for allocation of economic rights. Thus, I pointed out that there was a mismatch between women's fishing-related work and the official welfare rights that were developed after World War II including compensation for being sick, for holidays, and pension rights (Gerrard 1983). Given women's and men's unequal positions, we must consider the consequences of the changes in fisheries regarding gendered power relations and women's rights. Are gendered relations and women's rights in the fishery today apt to change for better or worse?

How then do we go about studying these gendered and ethnic power relations? In earlier studies, I focused on the visible actions of women and men in specific contexts and traced what happened with some of the issues women raised on different political levels (Gerrard 1986, 1995). In all societies, visible interpersonal power relations do not cover all aspects of power. Foucault and Bourdieu want us to account for the invisible (Ellingsæter and Solheim 2002). When we intend to study such 'hidden power relations' or what I will call 'power that sits in the walls', then we must search for perspectives that lead us behind what we can immediately observe as researchers. This approach can be extended to discussions of ethnicity.

Focusing on quotas at the same time as studying events locally implies challenges concerning the level of study. While gender studies often deal with the micro level, in many cases, local level studies require considerations of the macro level. Doreen Massey (2001) states that the challenge of global–local studies is to "grasp the powerful nature of social relations at all levels." For feminist and ethnic-related studies that might impact on gender relations and identity, this means we should try to trace the interrelated processes and events on micro and macro levels. However, from experience, we know that this is more difficult than it sounds, especially in empirical research. One reason is that many processes and events being studied are complex and interrelated. Such an approach is also of interest when we are dealing with ethnicity, especially in situations where ethnicity is hidden or receiving little attention. In this way, a focus on gender and ethnicity encourages us to pay attention to such analytical challenges.

One way of solving such analytical challenges, however, is to choose limited and specific events or patterns where we can try to trace the interacting fields reaching the different levels. In the example I will present here, I trace the fishers' practices as well as the political structures and policies with which fishers must act. In my case, I must place particular emphasis on the quota system representing some crucial changes at the national level and decipher how women and men deal with these changes within a specific local context.

Local, but Exceptional Changes:
Together at Sea–Together on Shore

Åshild and Rolf Ove's lives illustrate the link between processes on different levels. They live in Skarsvåg where most of the population has an historical attachment to fishing and work at the fish plant due to the rich fishing areas located near the shore. Today, the public sector and tourism represent important sources for both men's and women's employment. In spite of these changes, the local population, local authorities, and visitors identify the village as a fishing community.

Rolf Ove is in his 50s. He was born and raised in Skarsvåg. Åshild, who is in her 40s, moved into the community as a young woman after finishing high school in the mid-1970s. When she came to the village, she established her home with Rolf Ove who had been a fisher since he finished school at the age of 16. He has fished with his father and been crew on bigger boats. He has had his own boat for many years and today, Åshild and Rolf together own a boat that measures between 10.5 and 11 metres. During the spring of 2004, they bought another slightly smaller boat. Because of the new boat, they have the right to another quota and thus, they have the formal right to catch more fish. Before the quota system was introduced, they would have been able to fish as much as they could manage according to the market situation and their working abilities on one boat.

Åshild and Rolf Ove are two of 20 full-time registered fishers who live in Skarsvåg today. In addition, there are 6 part-time fishers of whom 5 are retired. In Norway, retired fishers are able to fish as part-time fishers or work on shore for other fishers. Four fishers living in the municipality centre, Honningsvåg, 27 kilometres from Skarsvåg, commute every morning during the fishing season. During the spring fishery, the local harbour is full of fishing boats from other places. The fact that many boats come to this and many other harbours in Finnmark is a well-established pattern reflecting the migratory pattern of cod stock that passes by on its way to and from the Barents Sea.

When Rolf Ove started his fishing career in the 1970s, there were between 35 to 40 fishers in the village. While the number of boats (around 20) has not changed significantly, there are now fewer boats measuring 14 metres and above and more boats measuring between 9 and 11 metres. Today, four of the registered boat owners or companies own 12 boats and cod quotas. The number of quotas exceeds the number of boats present in the community. This is possibly due to the new policy that has been adopted which states that one can transfer one quota from one boat to another within the same length group, for example, within the group of boats from 10 to 15 metres. Two of the owners have organized the ownership in private limited companies; two others have individual or sole enterprises where the owners are responsible for all the debts and credits of the boats in contrast to the shareowners who are responsible only for the value of the shares they have put into the firms. From a historical perspective, this latter structure has been the usual ownership pattern for smaller fishing boats in most villages along the North Norwegian coast. Åshild and Rolf Ove belong to this group.

Åshild and Rolf Ove deliver their catch to the local fish plant that is owned by one of the biggest fishery corporations in Norway, Aker RGI Holding AS and its Norwegian and Danish fishery section, Norway Seafoods. The fishers themselves are responsible for gutting and cleaning the fish before delivering it to the local plant. In earlier years, fish plant workers helped them with this kind of work. Now, due to quality regulations and a reduced workforce at the plant, they must do this kind of work on their own boats. When the fish is delivered, one of the fish plant workers weighs the fish before the foreperson registers the catch. The workers ice the fish in boxes and send most of the fish by trucks to the company's filleting factories situated elsewhere in North Norway or Denmark. In this way, the fish plant in Skarsvåg functions as a transit station with a handful of workers, most of them men. This is in contrast to the years before the imposition of the quota system when the filleting production demanded anywhere between 25 and 40 workers including both women and men.

Åshild and Rolf Ove live alone in their house. Two of their children are now adults. The youngest daughter attends high school in the municipality centre. She comes home to visit on the weekends while Åshild and Rolf Ove sometimes see her during the week when they go shopping. Åshild has been responsible for most of the daily care of the children and for household tasks. Late in the 1970s, before the two eldest children entered school, she began her teacher's education, following a decentralized study program for training teachers. After finishing her education, she worked both full- and part-time at the local school. She also worked together with her husband, baiting the long lines and occasionally fishing herself.

The public sector has meant a lot to Åshild and to women and their families in fishing communities. Today, Åshild still acts as a substitute teacher when needed at the local school. In 2004, the local school employed 7 of the 9 women who work in full- or part-time public sector jobs. Since 2002, the number of women who worked in the kindergartens and cared for elderly people has decreased. The reduction in employees is mainly due to changes in policy concerning the public sector but also to demographic changes. Fewer children have been born in the area in the last few years. Some of the seniors who received assistance in their homes, have had to move to the municipality centre when they needed more intensive medical care.

Åshild and Rolf Ove have experienced many changes during their years in the fishing community. There has been a decrease in the population from 220 in 1980 to less than 100 in 2004 (Gerrard 1980, 2005a). The majority of young people, especially young women, have a long tradition of moving away from the fishing villages for different reasons. This trend was strengthened after the moratorium because of a reduction in filleting work and fewer vacant jobs in other sectors. However, the mismatch between the jobs available and the formal education of the young people of today is another reason for moving away. As a consequence, few new households are established and the number of children has decreased. In 2004, there were 4 children below the age of 6 and 17 between the ages of 6 and 16. The older school children, one of whom is the daughter of Åshild and Rolf Ove, now live and attend high school in the municipality centre or in other school centres. In January 2006, there were only about 10 school children.

Åshild has expressed to me several times: "I have always wanted to go fishing". Nine years ago, she took the compulsory safety and security course and subsequently, the follow-up course. In June 2003, she took a leave from her school job and started to fish full- time with her husband. Today, Åshild and two other women in the village are registered fishers. While Åshild works both at sea and on shore, the two other women have restricted their fisheries work to shore-related tasks such as administration, accountancy, baiting and giving advice to their spouse. One of the two women is no longer active and, since her husband's death, she does most of her paid work in other sectors.

Åshild was just as eager as her husband to buy their newest boat and now hopes to be able to earn her living entirely from the coastal fishery. Both agreed that obtaining another boat with a quota was necessary to financially support two people without Åshild's salary from teaching. With another boat with quotas, they are now able to catch more cod.

Rolf Ove is the skipper on their boats and is the most experienced and knowledgeable about the sea. Both talk about the necessity that Åshild must learn to steer the boat and start and stop the engine. When they fish with the long lines, Åshild is responsible on the shore organizing the work and baiting the lines, a type of work she has been doing for many years. She also goes to sea and fishes especially when they are fishing with hook and line. They have a joint household economy. They both have their own salary and file separate tax returns. In this way, they both acquire formal welfare rights. The work of the husband has, in many ways, become easier since the work is now shared. The division of labour, however, follows gender lines that have existed for many years. Perhaps this helps explain why few seem to notice and discuss the fact that there are registered female fishers in Skarsvåg.

As mentioned earlier, few local people speak about their ethnic identity. I have never heard Åshild and Rolf Ove or other inhabitants in the fishing community speak about their ethnic background. This is understandable in a context where there have been few advantages ascribed to a self-identified Sami or a Sami fisher (Nilsen 2003). The local branch of the Fishermen's Union has followed the policy expressed by the union in general in which the Sami fishery policy has seldom been put on the agenda. We must also bear in mind that currently, many voices from the coastal areas in Finnmark are critical of the changes that have been proposed concerning the management of land and water (the Finnmark Law). In a radio program broadcast by the regional branch of the Norwegian Broadcasting in April 2005, young men from the village expressed strong feelings opposing this new law and the Sami people whom they identify as reindeer herders. Some elders[8] and local people who have developed relations with Sami reindeer herders who have stayed in the village over the summer because of the good pastures, expressed more positive attitudes toward Sami people in general.

In 2004, the Sami Parliament accepted Skarsvåg with Gjesvær, Repvåg and Kamøyvær, all fishing villages in the municipality of Nordkapp, as areas covered by the Sami Development Fund.[9] The Fund provides the population with another way of financing new initiatives for employment. In the municipality of Nordkapp, the Fund is a new arrangement, and some people have already mentioned that they will submit an application to support tourism development. It will be interesting to see if these new developments will lead to changes in how the local people understand Sami questions, including the question of identity.

---

[8] Interview with Jørgen Lindkvist, 29. April 2005.
[9] This is a public fund that gives support to business activities of different kinds in certtain areas that, according to political decisions by the Sami Parliament, are considered as Sami areas.

Women Fishers: A Minority in a Male–Dominated Profession

How can we relate Åshild and Rolf Ove's story to broader processes and especially to the quota system? Let me start with the number of registered fishers since registration is one of the requirements to obtain a boat with a quota.

Table 1 shows that Åshild was one of 283 full-time registered women fishers in Norway in 2003,[10] while 130 women were registered as part-time fishers. The table also shows that between 1988 and 1998, the number of female fishers was relatively stable. Following the moratorium and the first years of the quota system, Norway had the largest number of registered female fishers since the gendered registration started. There has been little research on this topic with the exception of Munk-Madsen, 1996. There are many reasons for the stability in the number of female fishers between 1988 and 1998.

Table 1: Women employed as full and part-time fishers in Norway (1983 – 2003)
(Directorate of Fisheries, 2004)

Year	Full-time	Part-time	Total
1983	182	106	288
1984	218	100	318
1985	250	112	362
1986	362	115	477
1987	504	109	613
1988	575	102	677
1989	579	105	684
1990	554	112	666
1991	583	109	691
1992	583	116	699
1993	572	105	677
1994	534	124	658
1995	506	130	636
1996	526	133	659
1997	520	152	672
1998	530	166	696
1999	410	178	588
2000	361	166	527
2001	321	161	462
2002	313	130	443
2003	283	130	413

---

[10] As mentioned earlier, there are no statistics kept for Sami fishers, either women or men.

However, more research is needed to more fully answer this question. Since 1998, there has been a decrease in the number of female fishers. The same trend is evident in Finnmark. While Finnmark had 26 registered female fishers in 1983 and 55 in 1989, there were 24 in 2003.[11] In Finnmark, I know of only one woman who is skipper on her own boat with her own crew. While she has been fishing for many years, Åshild can be considered an exception. Having the option to choose between teaching and fishing, she chose fishing. Now, with her children grown up and living away for much of the time, it is much easier for her to be a full-time fisher. Her life as a fisher is more flexible than being a teacher. Together with her husband, she can now decide where, when, and how to work. With hard work and good weather, she can also earn the money she needs. In this way, Åshild has acted similarly to women in Newfoundland, Canada after the moratorium. Here, women represent about 19% of the fishers in the province in 2000 (Grzetic 2004). Many of them are wives of long-established fishers who work together with their spouses. In Newfoundland, this increase also has links to fisheries restructuring and policy changes. From a researcher's perspective, one can wonder if this trend reflects a real choice, as in Åshild's case, or whether it reflects the limited employment possibilities for women. Women's minority situation becomes even clearer when we compare the numbers of male and female fishers.

Table 2 indicates that the decline in the number of full- and part-time fishers is also evident among men. A decrease of more than 9,000 full-time fishers occurring at the same time as the restructuring of the industry is significant. The greatest decline in the number of fishers, about 7,000 respectively, took place after the quota system was launched in 1990. The number of women with formal ownership of boats is relatively small. In the whole of Norway, 161 women and 7,386 men owned registered fishing boats in 2005. In Finnmark, there are 28 female and 1,065 male boat owners belonging to this category.[12]

On the basis of such low numbers of registered women fishers and few women as registered boat owners, it is evident that access to fishery and quotas has become men's formal property right, meaning that the boat a person owns has a quota and that this person is allowed to fish and sell the fish to a registered fish plant. Feminist researcher Eva Munk-Madsen (1996) argued that a resource that has been common property and open to 'everybody' has, with the new quota system, become closed for most women—or about half of the fishery population. Since few women are registered and few boats are registered in the names of women, few women have a formal right to the quota. Until recently, widows who were not registered as fishers felt that they had to sell their boats with the quota even when they wanted to keep the boats and start to fish because they were not entitled as 'fishers' according to Norwegian law. This was the case even if the woman has performed substantial unpaid work for the boat and the enterprise. Since Åshild is a registered fisher and is active and experienced, she has the right to buy a boat of her own or take over the boats and continue to fish or rent out her boat if her husband passes away.

---

[11] Statistics supplied by the Directorate of Fishery in Norway 16.08.04.
[12] The Directorate of Fisheries provided this information in October and November 2005. It is based on various registers. The information is valid for boats smaller than 28 metres and with an owner share higher than 49.9%.

Table 2: All fishers involved on a full and part-time basis in Norway 1983–2003
(Fishery statistics, Statistics Norway 2004)

Year	Full-time	Part-time	Total
1983	22455	5849	28304
1984	22864	6768	29632
1985	22465	7101	29566
1986	22619	7362	29981
1987	22622	7293	29915
1988	22048	7302	29350
1989	21448	7207	28655
1990	20475	7043	27518
1991	20003	6963	26966
1992	19779	6973	26752
1993	19072	6324	25396
1994	16442	6478	22920
1995	17160	6493	23653
1996	17087	6310	23397
1997	16661	6255	22916
1998	15141	6157	21298
1999	15328	5945	21273
2000	14270	5838	20098
2001	13700	5254	18954
2002	13913	4735	18648
2003	13260	3999	17259

Fishing, Gender and Power

Åshild's and Rolf Ove's case raises a number of questions: Why haven't more women become active fishers? Does a woman's lack of direct involvement as a registered fisher reveal the fisher husband's power over the wife since he does not share fishing rights with her? An investigation into who makes decisions about the fishery and how the fishery policy is defined reveals a much more complicated picture than these questions suggest. Power seems to be embedded in rules and regulations far from local women's and men's control. The quota system also reveals that women have much less access to the fish resources than ever before. Even though few women were fishing before the quota system was launched, they could at least continue to own their boat or rent it out if their husbands passed away. With the quota system, a boat without fishing rights has low value. Since most owners are sole proprietors, the value of the boat is higher when they can sell it with a quota.

Most of the fishery population opposed the quota system when it was introduced. Since its implementation, the fishers' associations, with very few female members, have tried to influence the system. Men, either as bureaucrats or fishery politicians, have tended to keep the questions of resource management beyond women's reach. Indeed, gender has not been a question on the 'malestream' resource agenda. In this way, the quota system can be looked upon as a new and even stronger symbol of men's power in fishing.

The few women who have entered the political arena and placed emphasis on the fisheries, have seldom raised questions about gender. The women who have tried are researchers, members of the environmental movement, the Sami Parliament or left-wing political parties. It is interesting to note that the historical data on the involvement of fishers disaggregated by gender presented in Table 1 was not collected until 2004, following a request by the Sami Parliament. They needed the data for the *Report on Fishing as Industrial and Cultural in Coastal and Fjord Areas* (Sami Parliament 2004), which focuses on women's situations and presents ways to secure women's position in fishing.

Women in fishing communities in general have had little influence on the types of questions asked regarding policy directions at the national level. Whenever women have tried to influence policy, for example, in the committee that gives advice to the Ministry regarding fish stocks, they have made little progress. In August 2004, the mayor of the province of Finnmark, Helga Pedersen[13] who is also a Sami from Tana (a fjord and inland area) attended the annual meeting of the Finnmark Fishermen's Union and suggested some revisions of the system. Using the analyses of male researchers, she suggested that a regional board should have the right to redistribute quotas to fishers from the area in the cases where they were returned to the Norwegian state.[14] In this way, she hoped to reduce the number of quotas sold to owners in other regions. She called upon the fishers to nominate members to a committee to continue to work in this direction. The suggestions were turned down immediately since many fishers considered her suggestions as a step toward transferring more power to regional authorities, thereby breaking with the established pattern where central institutions have the most influence on political decisions. In Norway, there are other examples of formalized regional management in the Lofoten fishery (Jentoft and Kristoffersen 1989). This is a management system that was established long before the quota system was launched. To regionalize the quota system will, as already mentioned, break with the principles of the open commons and probably give more fishing rights to those living in the region—rights that do not seem to count in a quota regime (Nilsen 2003). This example illustrates that there are some women interested in fishery policy but that fishery policy and resource management policy represent arenas in which men still have the power to define the agenda.

Åshild could have chosen to continue her career as a teacher employed by the municipality or in the service sector in general as other fishers' wives and women in fishery communities have done. She enjoys the work at sea. Her enjoyment of working outside in nature was reflected in her teaching as well. For example, she developed courses for the school children built on outdoor life. For her, fishing represents a better alternative than employment in the public or private sector. For Åshild, the advantages in fishing are greater than the stable income and more secure work outside the fishery. Other women in fishery communities have become increasingly attached to the labour market outside fisheries in the last decade and in a way, are supporting their household economy and husbands'

---

[13] In October 2005, she was appointed as Minister of Fisheries and Coastal Affairs.
[14] Examples include cases where the boats are condemned and the owner received compensation from the government.

fishing efforts financially. Even if we cannot always 'count' on the official statistics, they at least display a trend. Table 1 indicates that from 1987 to 1998, there were more than 500 registered female fishers. During the last few years, the number has decreased. During the same period, the market or the men themselves have taken over more of women's fisheries-related tasks that have never been registered or formally recognized. In the 1970s and 1980s, researchers documented that women did the accounting, washing of the boats, and sewing or knitting of the clothing (Gerrard 1983; Jentoft 1989).

Åshild's case demonstrates that where women have a close connection to the fishing boat, they may be able to influence husband's fishing patterns such as when and where to fish. The new ownership model, based on private limited companies that has been put into practice in many other fishing villages, takes a business approach. In Skarsvåg, the two private limited companies have no women on the boards. This way of organizing ownership represents a new way of running a business among Norwegian owners of smaller boats, measuring between 10 and 15 metres. Before the quota system was applied and the policy changed, we seldom found examples of such ownership patterns among the owners of small boats.

To summarize, I see different tendencies. Since the implementation of the quota system, some women have been more active in fish harvesting, often working together with their husbands. Some of them are registered fishers and have formal status. However, in general, women continue to have little formal influence in fishery policy. This is also confirmed by recent studies (Angell 2004; Sloan 2004). It will be exciting to see if the Sami Parliament Report (2004) on fisheries with its emphasis on women will have long-term positive consequences. The other tendency is that the majority of women have a looser connection to fishing work and processing than before the depletion of the fish stocks that ended with the moratorium and the implementation of the quota system. The majority of women choose or are forced to give priority to work outside of the fishing sector, for example, teaching, or other jobs in the public sector since fishery work has been so heavily downsized (Gerrard 2003a).

It is likely that the differences between smaller and bigger enterprises, between limited companies and individual owners may, in the long run, influence men's identity construction, men's relations to each other and the fishery as a cultural community since the preferences and the values in fishing may vary. Today, it is not only the length of the boat that makes an owner 'big' or 'small'; today, 'big' or 'small' are also ways of describing the number of boats and quotas one has.

Towards a New Understanding of Practices and Identities

How are women's and men's differing access to the quota system and unequal positions in decision-making linked to changing practices and identities? Are women and men in a situation of being 'in between' or 'in transition' as I suggested earlier in this chapter? Historically, in Finnmark as well as in other Norwegian fishery areas, big catches, fished on one boat and carried out largely by men with women as the shore or ground crew, represented the norm in the coastal fishery. Such practices with its attendant values also provided the basis for masculine identities in many fishery areas (Gerrard 2004). Today, it is difficult for fishers with a small quota to continue these established values and practices, especially if they are in a situation of having to provide economically for a family and pay debt owed on the family house, the boat, the car and their children's education.

When fishers today buy a second boat in order to get another quota, they break with established local ideas based on the principle: "One or two owners—one boat." Yet, some fishers opt to purchase a second, a third or a fourth boat because, in their view, this is one of the ways they can secure their future as active coastal fishers in cod-dependent areas. In this way, they might also be able to fulfil accepted standards of masculinity.

These new practices also have an impact on fishers' knowledge. Obviously, fishers use, establish and develop knowledge about how to catch fish. Now, fishers must position themselves within the quota regime and turn their attention to fishery policies, rules and regulations, financial options and the market to a much greater degree than in earlier years. The ownership of several boats and quotas is a visible indicator of such a shift. These fishers are entrepreneurs. Entrepreneurship within this context is not unusual. However, the establishment of private limited companies of relatively small boats (10 to 15 metres) built upon the model of the larger shipping companies represents a shift from earlier models. This new model is reminiscent of the neo-liberal, globalized economy described by Connell (1998). This type of boat ownership may also be a means to acquire more rights in order to be able to harvest more fish. Yet many fishers choose to continue with their single boat, one cod quota and only one man on the boat. This decision may reflect fishers' ability to manage with only one cod quota. This is more readily apparent amongst the fishers who do not have much debt or who may have spouses who are bringing an additional salary into the household. In this way, they often manage with less cash than families in towns and cities.

Another reason for rejecting the trend towards multiple boat ownership is the increased workload that accompanies additional boats, quotas and a larger crew. Few fishers share the boats and the quotas with their wives, like Åshild and Rolf Ove. The changes outlined above have lead to a situation with new choices, practices and knowledge. The quota system has 'forced' fishers to turn more and more of their attention and practices "from the sea to the shore". It is more difficult to know if these new practices and knowledge also shape new identities. More quotas in order to harvest more fish can also be understood as a means to fulfil established masculinity standards including being an industrious and successful fisher and having the ability to catch enough fish in order to support a family. Some of the coastal fishers in Finnmark, especially those who are responsible for a crew, are more business oriented compared with earlier periods. The business orientation is not a new orientation among fishers. It has existed for many years in the Norwegian fisheries, especially on the western coast of Norway, but in Finnmark and especially in the small communities in the small-scale fisheries, it seems to be rather new. Thus, fishers seem to be in a situation of having an identity 'in transition or in between'—although this means different things for men and women.

It is important to remember that fishers' identities in most countries are directly linked to men's roles and to masculinity (Power 2005). Women comprise a minority of the registered fishers, and of those women who are registered, not all go to sea. They do however perform important and necessary work on shore. We also know that there are women fishing with husbands who are not registered. Munk-Madsen (1996) documented that some women in the beginning of the 1990s experienced problems registering as full time fishers since they also had to care for children. Instead, they were classified as part-time fishers and only had admission to the maximum quota and thus had to fish less than full-time fishers with a regular boat quota. The examples from Skarsvåg show that this has not been the case there. Most of the women who choose to go fishing and register as a fisher are breaking with established practices. Since women represent a minority group among the active fishers and are spread out on different boats coming from different villages along the long Norwegian coast, there are not many visible signs or markers to identify women as fishers. There are few women to identify with and thus, few role models. It is also important to remember that the ways women carry out their work vary. They are in a situation where they have to create their own meanings about what it is to be a fisher and their ways of acting and identifying as fishers change from one individual to another. Eva Munk-Madsen (2000) has outlined interesting examples of variation and challenges in the construction of identity by female and male fishers.

Åshild, for example, has been, and continues to combine many tasks connected to different sectors of society. She has had varied working roles and practices. She has formal status as a fisher according to the rules and regulations of the adopted quota system. The step from being a teacher to a fisher was not a big one since she had been working with her fisher-husband in addition to being a teacher for many years. Even though today she calls herself as a fisker (in Norwegian) and uses the word fisherman when she presents herself in English, it is obvious that the ways in which she carries out her work are different from other registered women fishers in the village since she goes out fishing and they work on the shore. However, she also spends time together with children, siblings and friends. Her husband does the same, since they are doing many tasks together both on land and at sea. In contrast with Åshild, he has been a fisher all of his adult life and has much more experience and knowledge of some of the fishing tasks.

These examples of women's and men's practices can be related to Susan Friedman's (1998:23) treatment of identity and identity change. She argues that: "Situational approaches assume that identity resists fixity, but they particularly stress how it shifts fluidly from setting to setting." This means that a female fisher might consider herself a fisher when she is fishing, cleaning the boat or baiting or doing other types of fishing-related work. As a full-time fisher she is, however, a newcomer. When she takes care of her children and grandchildren, she identifies as a mother or grandmother. When she gives lessons at school, she might identify as a teacher. Even though the total number of registered female fishers has decreased on a national basis, in Skarsvåg, three women have registered since the quota system was launched. However, on the whole, fewer women are directly engaged in fishery-related activities in these communities. In addition, there is a diversified fishing and ownership pattern. Although women have always performed many fishery tasks on shore, neighbours or people outside fishery rarely recognize women performing such work as fishers. When women begin to work on the boat, this seems to change. The boat is therefore a strong symbolic marker of being a fisher. This is a perception that seems to exist for women and men, as well as for locals and outsiders. In this way, the connection between the man and the boat seems to represent an underlying key to understanding 'hard programmed' gender perceptions of what a fisher is as well as women's weaker power positions in fishery politics. This form of symbolic meaning is in line with Ellingsether and Solheim's way of analyzing the relation between gender, power and work.

The reality today indicates a complex situation for both women and men. Given this complexity and the shifting organization of the Norwegian fishery, we must reconsider or reconstruct the concepts 'the fisher,' 'fishery knowledge,' and 'fishery culture.' We should also reconsider how we understand fishers' identities in a local as well as a wider context. Another question is this: will more coastal fishers—men or women—identify as Sami fishers in public arenas in the future? Perhaps new means to financing fishery activities and other advantages will change the situation. In other places, such as some of the fjord areas in North Troms, the province south of Finnmark, new rights connected to the area combined with the efforts of the younger generation have lead to a revitalization of Sami culture, language and identity. In the coastal community that has been the focus for this chapter, representatives of the younger generation do not yet seem to have adopted the experience and the attitudes as in North Troms. In many coastal communities of Finnmark, it seems likely that women and men with Sami roots will continue to keep their ethnic identities private. However, since identity resists fixity as Susan Stanford Friedman (1998) has stated, this might change when new situations arise.

Conclusion

In response to the implementation of the quota system, women and men in fishery communities have, in some cases, adopted new practices in fishing and other fishery-related tasks. At the local level, changes have led to a reconstruction of the gendered division of work and the redistribution of economic rights and have opened spaces within which women can exercise power at least at the micro level. This is evident in Åshild's case. These changes indicate that even in the context of imposed structural change, women's and men's practices in fishery-specific contexts varies and thus, provides us with an example of how fishery people contribute to cultural changes. On the other hand, the quota system, created and controlled by Norwegian institutions (mostly men), have not made it easy for women or newcomers in general to acquire fishing rights, rights that have traditionally been held in common. In fishery policy, men who are secure within the establishment at all levels have maintained their influence and power. They also have retained static ideas of what they consider as fishery culture and who they define as a fisher.

I have demonstrated that there are challenges in the construction of what it means to be a fisher. Following Friedman's (1998) argument, a fisher belongs to a diversified group with different ways of defining identities and roles. While this has probably always been the case, the particular shape of new economic and political processes, for example, the buying and selling of boats and quotas and their intersection with local restructuring of fisheries, may mean more visible diversifications, or, at the very least, new possibilities for identity formation.

My arguments here show that it is not enough to talk about female and male fishers and a single fishery culture. Einar Eythórsson (2003) has asked if the coastal Sami represent a 'pariah caste' of the Norwegian fisheries. The quota policy has not made it easier for women (or young men) to go into fishing. Perhaps we also can ask whether women can be considered as another 'pariah caste' since so few women have had the opportunity to choose fishing as a way of life?

While male researchers interested in Sami fisheries have only to a minimal degree raised the 'gender question' (*see* Jentoft *et al.* 2003), female researchers with interests in gender have seldom focused on gender and ethnicity in the Finnmark fishery. There are, however, some exceptions including a recent study by Elisabeth Angell (2004) and the film produced by Rossella Ragazzi (2003). Therefore, as male and female researchers, we should, together with the fisher population, try to gain a deeper understanding of how women and men themselves construct identities and/or combinations of identities and how they define fishery cultures. There are likely many practical and political consequences of this approach, since ideas as well as organizational or social and cultural patterns are connected with notions of 'the fisher.' My discussion here shows that a simple notion of what a fisher is does not fit with the reality—for women, men, Sami or Norwegian!

Epilogue

As already mentioned, this article is based on data up to 2005-2006. Many changes have taken place. Since then, the number of fishers continue to decrease both in Skarsvåg and in Norway as a whole. Some committees have closed down and some have come up with new propositions concerning the quota policies. There have also been changes in the quota policy, allowing fewer quotas connected to one fish boat. In February 2008, a Norwegian Official Report titled *The Right to Fish in the Sea Outside Finnmark* (Fiskeri og Kystde partementet 2008) was launched by a committee appointed by a Royal decree, June 30th 2006 under the leadership of Carsten Smith, former Chief Justice of the Supreme Court. This report focused on the coast and fjord fishery in Finnmark. The occasion for this committee was the Sami fishers' situation. The committee recommended that the population in Finnmark living at the coast and the fjords shall be treated as equals. Another of the many propositions was that boats smaller than 15 metric tons belonging to Finnmark, should have the right to fish without the limitations that the quotas set. They also proposed new laws and institutions to carry out the new policy. The report is now discussed in many institutions and has not yet been discussed in the Norwegian Parliament. The Fishermen's Union, to mention one, disagree to many of the committee's propositions. However, in the meanwhile the fishers like Åshild and Rolf Ove, fish as they did when they told me their story some years ago. What changes in fishery policy they can expect in the future is therefore difficult to say.

Acknowledgements: My warmest thanks go to the population in Skarsvåg, the Municipality of North Cape and especially to Åshild and Rolf Ove Pettersen who shared their experiences with me and who gave their consent to use their full names and also use the photographs they have taken. I also thank Nicole Gerarda Power for the enormous and patient assistance she has given me asking critical questions as well as suggesting excellent changes as far as language is concerned. I am grateful to Joanna Kafarowski, Vigdis Stordahl and Robert Paine for inspiration and critical comments. My gratitude also goes to Gunnar Grytås, a journalist in Fiskeriblandet, Odny Aspervik, the Directorate of Fisheries and Inge Eriksen, consultant in the Sami Parliament for valuable information, to Barbara Neis and the research program "Coast under Stress" and Memorial University that gave me excellent working conditions while staying in Newfoundland. I am also grateful for the financial assistance provided by the Department of Planning and Community Studies, University of Tromsø and last, but not least, the Norwegian Research Council that financed the project 'Sustainable fishery culture?'

References

Andersen, S. (2001). 'Bruk og forvaltning av kystressursene i et samisk perspektiv,' pp. 51-56 in *Rapport fra Sametinget fiskeriseminar,* February 2001, Karasjok, Norway.

Angell, E. (2004). *Kjønn og etnisitet i fiskeripolitikken. En analyse av kvinners rolle i samisk fiskerpolitikk og Sametingets posisjon i norske fiskerireguleringer.* Report No. 2004(4). Alta, Norway: Norut NIBR, Finnmark.

Binkley, M. (2000). Getting by in tough times. Coping with the fishery crisis. *Women's Studies International Forum* 23(3): 323-332.

Connell, R.W. (1998). Masculinities and Globalization. *Men and Masculinities* 1(1): 3-23.

Davis, D.L. and S. Gerrard (2000). Introduction: Gender and the resource crisis in the North Atlantic fisheries. *Women's Studies International Forum* 23 (3): 279-286.

Ellingsæter, A.L. and J. Solheim, eds. (2002). *Den usynlige hand. Kjønnsmakt og moderne arbeidsliv.* Oslo: Gyldendal.

Erlandsen, M.M. (2004). 'A Sami, a Woman and a Fisheries Politician,' pp. 112-116 in L. Sloan, ed. *Women's Participation in Decisionmaking Processes in Arctic Fisheries Resource Management. Arctic Council 2002-2004.* Norfold, Norway: Forlaget Nord.

Directorate of Fisheries (2004). *Fiskarar som har fiske som hovedyrke og biyrke fordelt etter hokjønn for kvart fylke i perioden 1983–2003 (16.8.04).* Bergen: Fiskeridirektoratet.

Eythórsson, E. (2003). 'The Coastal Sami: A "Pariah Caste" of the Norwegian fisheries? A Reflection on Ethnicity and Power in Norwegian Resource Management,' pp.149-162 in S. Jentoft, H. Minde and R. Nilsen, eds. *Indigenous peoples: Resource Management and Global Rights.* Delft: Eburon.

Friedman, S.S. (1998). *Mappings: Feminism and the Cultural Geographies of encounter.* Princeton: Princeton University Press.

Gerrard, S. (1975). *Arbeidsliv og lokalsamfunn: Samarbeid og skille mellom yrkesgrupper i et nordnorsk fiskevær. Magistergradsavhandling.* Tromsø: Institutt for Samfunnsvitenskap, Universitet i Tromsø.

Gerrard, S. (1980). Levedyktighet og lokal organisasjon: Om befolkningsutvikling og sosial forandring i et nord-norsk fiskevær. *Tidsskrift for samfunnsforskning* 21(3- 4): 265– 281.

Gerrard, S. (1983). 'Kvinner i fiskeridistrikter: Fiskerinæringas bakkemannskap,' pp. 217-241 in B. Hersoug, ed. *Kan fiskerinæringa styres?* Oslo: Novus Forlag.

Gerrard, S. (1986). *Kvinners makt og avmakt.* Occasional Paper 6/86. Alta, Norway: Finnmark College.

Gerrard, S. (1995). When women take the lead: Changing conditions for women's activities, roles and knowledge in North Norwegian fishing communities. *Social Science Information (SAGE)* 34(4): 593-631.

Gerrard, S. (2003a). 'Må det bo folk I husan?,' pp.143-158 in M.S. Haugen og E.P. Stræte, eds. *Ut i verden og inn i bygda.* Trondheim: Tapir akademiske forlag.

Gerrard, S. (2003b). Utfordringer i formidlingen av kulturorientert fiskerforskning. *Plan.*6: 2-35.

Gerrard, S. (2004). *Arbeid i endring—kjønn i spenning: Noen perspektiver på maskuliniteter, feminiteter og fiskerliv.* Kjønnsmakt i Norden. Konferanserapport. Electronic document: http://kjonn.maktutredningen.no/aktuelt/746. Last accessed: February 28, 2009.

Gerrard, S. (2005a). 'Living with the Quotas,' pp. 47-55 in K. Frangoudes and J.P. Fernandes, eds. *Women in Fisheries and Aquaculture: Lessons from the Past, Current Actions and Ambitions for the Future. Proceedings from International conference AKTEA.* La Laguna, Tenerife: Asociacion de Antropologia La Laguna.

Gerrard, S. (2005b). Living with the Fish quotas: Changing gender practices? *Tidsskrift for Kjønns forskning* 4: 34-49.

Grzetic, B. (2004). *Women Fishes These Days.* Halifax: Fernwood.

Jentoft, S., ed. (1989). *Mor til rors.* Tromsø: Norges Fiskerihøgskole, Universitetet i Tromsø.

Jentoft, S. and T. Kristoffersen (1989). Fisheries Co-management: The case of Lofoten. *Human Organization* 48(4): 355-65.

Jentoft, S., H. Minde, and R. Nilsen, eds. (2003). *Indigenous Peoples. Resource Management and Global Rights.* Delft: Eburon.

Massey, D. (2001). Geography on the agenda in progress. *Human Geography* 25(1): 4-17.

Munk-Madsen, E. (1996). *Fiskerkøn. Afhandling til dr.graden i fiskerividenskap,* Tromsø: Norges fiskerihøgskole, Universitetet i Tromsø.

Munk-Madsen, E. (2000). Wife the deckhand, husband the skipper. Authority and dignity among fishing couples. *Women's Studies International Forum* 23(3): 333-342.

Neis, B., B. Grzetic, and M. Pidgeon (2001). *From Fishplant to Nickel Smelter: Health Determinants and the Health of Newfoundland's Women Fish and Shellfish Processors in an Environment of Restructuring.* St. John's: Memorial University of Newfoundland.

Nilsen, R. (2003). 'From Norwegianization to Coastal Sami Uprising,' pp.163-184 in S. Jentoft, H. Minde and R. Nilsen, eds. *Indigenous peoples. Resource Management and Global Rights.* Delft: Eburon.

Norges offentlige utredninger (2008). Retten til fiske i havet utenfor Finnmark. *Norwegian Public Reports* 2008: 5. Oslo: Fiskeri- og kystdepartementet.

Pettersen 1994

Porter, M. (1993). *Place and Persistence in the Lives of Newfoundland Women.* Aldershot: Avebury.

Power, N.G. (2005). *What Do They Call a Fisherman? Men, Gender and Restructuring in the Newfoundland Fishery.* St. John's: Institution for Social and Economic Research, Memorial University of Newfoundland.

Ragazzi, R. (2003). *Hjemme i verden, Valmimise aasta / At home in the world.* 63 minutes. Tromsø: University of Tromsø.

Skaptadóttir, U.D. (2000).Women coping with change in an Iceland fishery community. A case study. *Women's Studies International Forum* 23(3): 311-321.

Sloan, L., ed. (2004). *Women's Participation in Decision-making Processes in Arctic Fisheries Resource Management. Arctic Council 2002-2004.* Norfold, Norway: Forlaget Nora.

Sami Parliament (2004). *Sametingets melding om fiske som næring og kultur i kyst og fjordom rådene.* Karasjok: Sami Parliament of Norway.

Statistics of Norway (2004). *Fishery Statistics: Fiskarar, etter fiske som leveveg 1983-2003.*

Smith, C. (1990). Om samens rett til naturressurser—særlig ved fiskerireguleringer. *Lov og rett* 1990: 507 –535.

Smith, C. (2005). Rett for samer og nordmenn. *Aftenposten.* April 16. Electronic document: http://www.aftenposten.no/meninger/kronikker/article1016016.ece?service=print. Last accessed: February 28, 2009.

CHAPTER EIGHT

"It's Our Land Too": Inuit Women's Involvement and Representation in Arctic Fisheries in Canada

Joanna Kafarowski

Abstract: Within a socio-economic context, sustainable Arctic fisheries can be attained through the full participation of Indigenous peoples, including women. Even though Indigenous women are involved in fisheries—both directly and indirectly—they are often poorly represented and are seldom included in formal decision-making processes with regard to fisheries management. Focusing on Canada, this chapter investigates gender distribution and Inuit women's access to and participation in decision-making processes in community and territorial organizations related to fisheries.[1]

Introduction

Following the 2002 *Taking Wing: Gender Equality and Women in the Arctic* Conference hosted by the Arctic Council in Finland, the Northern Feminist University in Norway initiated the circumpolar "Women in decision-making processes in Arctic fisheries resource management project." This two-year project was approved by the Arctic Council in 2003 and involved the member nations of Norway, Canada, Greenland, Iceland, Sweden and Denmark (the Faeroe Islands) and three Indigenous organizations: Pauktuutit Inuit Women's Association, the Norwegian Saami Parliament (*Samediggi*) and the Inuit Circumpolar Conference Greenland. Based in Pangnirtung and Iqaluit, Nunavut and Holman, Northwest Territories, the Canadian component of this project investigated Inuit

---

[1] This chapter is based on earlier work published in *Women's Participation in Decision-making Processes in Arctic Fisheries Management*. Forlaget NORA, Norfold, Norway. The author kindly acknowledges funding and support from Pauktuutit Inuit Women's Association; the Department of Foreign Affairs and International Trade (Circumpolar Affair Division); Status of Women Canada (Women's Program); Indian and Northern Affairs Canada (Circumpolar Liaison Directorate); Canadian Circumpolar Institute; Aurora Research Institute; Walter and Duncan Gordon Foundation; the Department of Sustainable Development (now Department of Environment), Nunavut; Department of Health and Social Services (Womens' Initiatives), Nunavut; Nunavut Wildlife Management Board; Kakivak Association and Baffin Fisheries Coalition.

women's involvement, representation and access to decision-making roles in fisheries management at various levels.

Legislative Framework

In Canada, gender equality as it pertains to environmental management is enshrined in both international and national law. This is recognized in the *Brundtland Report* (1987), *Agenda 21* (1992) and the *Rio Declaration* (1992). According to the Rio Declaration: "Women have a vital role in environmental management and development. Their full participation is therefore essential to achieve sustainable development (Principle 20)" and "Indigenous people and their communities and other local communities have a vital role in environmental management and development because of their knowledge and traditional practices. States should recognize and duly support their identity, culture and interests and enable their effective participation in the achievement of sustainable development" (Principle 22).

In 1995, Canada adopted the United Nations *Platform for Action* which calls for the inclusion of a gender-based analysis process. The *Platform for Action* emphasizes the need to take measures to ensure women's equal access to and full participation in power structures and decision-making and the need to increase women's capacity to participate in decision-making and leadership.

Additionally, legislation specific to fisheries management such as the United Nations 1995 *Agreement on Straddling Fish Stocks and Highly Migratory Fish Stocks* which came into force in 2001, calls for transparency in decision-making processes and asserts that all representatives from organizations concerned with straddling and highly migratory fish stocks shall have the opportunity to take part in decision-making processes. According to federal legislation in Canada, Sections 15 and 28 of the *Canadian Charter of Rights and Freedoms* (1985) were enacted to guarantee the equality of women and men and prohibit discrimination on the basis of sex. The *Federal Plan for Gender Equality* (1995) continues to be a pioneering document used by departments of the Federal Government to address gender equality. The objectives of the Federal Plan are as follows:

1. Implement gender-based analysis throughout federal departments and agencies.
2. Improve women's economic autonomy and well-being.
3. Improve women's physical and psychological well-being.
4. Reduce violence in society, particularly violence against women and children.
5. Promote gender equality in all aspects of Canada's cultural life.
6. Incorporate women's perspective in governance.
7. Promote and support global gender equality.
8. Advance gender equality for employees of federal departments and agencies.

The Plan states that federal departments and agencies in Canada are responsible for conducting gender-based analysis and specifies that implementation should be phased in over a five-year period. The Canadian government's gender-based analysis policy requires federal departments to conduct gender-based analyses of policies, legislation and practices. Some government departments have developed a mechanism for conducting gender-based analysis; i.e., the *Policy of the Department of Justice on Gender Equality Analysis* (1997), Indian and Northern Affairs Canada's *Gender Equality Analysis Policy* (1999) and *Health Canada's Gender-Based Analysis Policy* (2000).

While Canada is to be commended for its support of progressive gender equality legislation and its commitment to conducting gender-based analysis in federal departments and agencies, the actual implementation of gender-based analysis remains inconsistent. To date, the Department of Fisheries and Oceans Canada has not developed a gender equality policy nor does it conduct gender-based analysis of its policies, legislation and practices. As outlined later in this chapter, the failure of the Department of Fisheries and Oceans to develop a gender equality policy and promote gender-based analysis is linked to this Department's contradictory attitude toward gender issues, including hiring Indigenous women for senior positions.

Inuit Women's Involvement in Arctic Fisheries in Canada

Subsistence Fisheries

Historically, both Inuit and Inuvialuit women and men located in coastal communities in Nunavut and the Northwest Territories have been equally involved in subsistence fisheries and this has changed little over time. Roles varied somewhat in that men were primarily responsible for going out in the boats, catching the fish and dooking (chiseling during ice fishing). Women's roles were diverse including making, mending, setting and checking nets, cleaning and cutting up the fish and deciding how to distribute the fish. Women helped with dog teams, including hitching the teams up and harnessing them and tying the dogs down when camping. These roles, separate but equal, were interchangeable between sexes to some degree.

> When I was growing up, it was the men who went out to fish and the women were the background players. When my father brought back the fish, my mother would prepare them. She would cut them up, freeze them, dry them, cook them—she would look after them once they came home. My mother would decide where the fish would go. My father would take some to my aunt and some to my uncle. It would be shared out in the family. In my family, there was no clear definition of who did what. My father would help sometimes in the drying and deciding who got what but my mother not so much with the fishing. But she would help my father when she could. Looking after the nets—mostly cleaning them was my mother's role but looking after the gear was my father's role. It was like this in many households in the community.
>
> Iqaluit resident

> My childhood experiences were mostly fun. A lot of girls had specific roles and didn't take the fish out of the net because that was the duty of the boys. The boys also made holes with the chisel.
>
> Holman resident

In Holman, Northwest Territories and in Nunavut, girls were brought up to help their mothers and other female relatives, and boys were expected to hunt and fish and follow the example of male relatives. However, there were still many opportunities for young girls to learn about and experience fishing. Girls who loved the outdoors and learned to hunt and fish often did so in the company of fathers and brothers.

> I used to enjoy going fishing with my dad and it was so much fun. One time I got a big big fish we call *Ihook* (big mature fish). I think it was maybe at Uyagaktuuk (big big lake). I was really young and he made a hole first and left me the fishing hook so I started fishing and I got a bite and it was so big, really heavy. I was trying to pull it up for a long time, just going backwards and then once in a while the fish used to pull me forward and I kept trying and finally I got it up and it was bigger than me.
>
> Holman resident

> I used to follow my father in the kamotiq (sled). He didn't really encourage me at the beginning but he did after a while. I was persistent when he went out hunting and fishing so in the end he would ask if I would like to follow him. I was a tomboy and had five brothers and I was very competitive with them. I didn't do what was customary. My mother would ask me to sew and I would say "no." I would rather go out hunting than sew. My friends and other female relatives would like to sew and they would ask me why I liked to hunt and fish and I said "I just do."
>
> Iqaluit resident

As in many Inuit and Inuvialuit communities, some women assumed a greater share of responsibilities for fishing. This group included single women, widows, or others with few male relatives who were prepared to assist in these activities.

> Women set nets, they had quite a good tide there and they would walk and set them and clear them out at low tide and use the fish. Because of their circumstances, they didn't have someone else hunting and fishing for them. They provided for themselves and their family in that way and they were very diligent fishers.
>
> Pangnirtung resident

Today, Indigenous women in Northwest Territories remain active in fishing as indicated in Table 1. Although Inuit and Inuvialuit women of all ages fish, women over the age of sixty participate most frequently. Women between the ages of 25 and 59 (prime years in the labour force) report fishing most frequently on day-trips or weekends. States one Holman resident: "I see more ladies going into fishing than men cause the men are mainly busy or hunting and we have to do our part to feed our kids." This is particularly so in Holman and in Nunavut communities—coastal regions based on cultural, social, economic and spiritual connections with the sea and the land. Many women state that, today, women's and men's roles in fishing are the same, although responses to this may be different according to the age of the woman.

Today, technological changes have increased women's participation in fisheries. The use of the auger instead of the traditional chisel improves the efficiency of fishing and reduces the time needed to complete the task. Both women and men use the auger in ice fishing.

Table 1. Number of NWT women participating in hunting and fishing.
(Source: NWT Bureau of Statistics July 2000)

Age of women who hunt/fish	Frequency of hunting/fishing		
Aboriginal women	More than day-trips or weekends	On day-trips or weekends	Rarely hunts or fishes
15-29 years	17 %	48 %	28.3 %
25-39 years	13.7 %	48 %	26.8 %
40-59 years	19.4 %	50 %	11.3 %
60 and over	24.2 %	38.6 %	22.3 %
Inuvialuit/Inuit women			
15-29 years	34.2 %	28.8 %	23.4 %
25-39 years	17.8 %	43.2 %	29.5 %
40-59 years	23.3 %	46 %	10.6 %
60 and over	45 %	25 %	30 %
Non-Aboriginal women			
15-29 years	2.6 %	83.7 %	12.8 %
25-39 years	5 %	61.3 %	28.1 %
40-59 years	9.5 %	47.5 %	38.2 %
60 and over	12.4 %	19.1 %	18 %

> Now for me, I still do both. I use the chisel to make holes when I go fishing by myself and but when there's other people with me, my husband or my kids, I use the auger. It's a lot faster but I always tell my kids it's more fun to use the chisel. More work but more fun! When we were kids all you really had to do was actually fish and jig and maybe fetch water once in a while. Now I could drive my own vehicle to get to where I want to go fishing and if I'm not lazy, make a hole. I prefer to chisel rather than auger because sometimes the auger kicks back or spins too fast and I tend to hurt my wrist.
>
> Holman resident

Other technological changes include the use of snowmobiles and all-terrain vehicles. Limited access to these technological advances may prevent some women from taking part in fishing. However, many report that women in this situation are supported by others in the community. States a Holman resident: "I see my auntie across the street take out her sister a lot. Her sister doesn't have any real transportation to go out and I see it with my other relatives too, they take them out fishing or whatever."

Women also support other women through teaching them the best places to fish—where to go and how far, how to use equipment and learn about the weather. Holman women are highly knowledgeable about fishing and are aware of their role in passing down this knowledge to younger women. "Inuit women can do just practically whatever it takes to go fishing and do what you have to do for fishing, like for instance, pitch up the tent, get where you're going, set up camp and make the holes if you have to" asserts a Holman resident. Another Holman woman says that when she is passing on this knowledge: "It's a good job for your mental self. You are socializing and teaching the children how to continue with what we're teaching them."

In traditional Inuit and Inuvialuit societies, the contributions of all members- elders, youth, women and men are considered critical. Gender roles, in which it is expected that women and men fulfill specific tasks and responsibilities that are valued differently (with women's responsibilities assigned lower value), have been imposed due to Western influence. According to an Iqaluit woman: "As Inuit, there's no gender thing. If it was left up to traditional ways, women would be the equals." In both the Northwest Territories and Nunavut, Inuvialuit and Inuit residents must deal with the challenges imposed by often conflicting gender paradigms represented by traditional Indigenous perspectives and Western values. This conflict is particularly apparent in communities such as Iqaluit.

> Few have had to deal with the astounding cultural transition as is being experienced here in Nunavut. Many of the older generation were born into a hunter-gatherers society while the younger generation struggles to find a home in today's cyber age. The traditional Inuit culture is struggling to adapt to more southern norms. As far as women's equity and equality issues, these concepts have not yet made it into the paradigm of many of the average Inuit woman's conceptual reality. If you are not aware there is a problem and therefore do not have the conceptual language to understand the issues or discuss it, how can you be expected to find a solution?
>
> Iqaluit organizational representative

Within subsistence fisheries, the significant and active involvement of Inuit and Inuvialuit women will continue. This is particularly true of women living in remote communities in Nunavut and the Northwest Territories which are more reliant upon a traditional diet.

Commercial Fisheries

Currently, the participation of Inuit women in Nunavut and Inuvialuit women in Holman, Northwest Territories in commercial fisheries is minimal. In Nunavut, few women work in offshore fisheries. This is because of limited available opportunities but also because of the physical hardships of the job, stress imposed by being away from family and community, language barriers and differing cultural attitudes towards work. However, current pressure being exerted on the Department of Fisheries and Oceans and other decision-making bodies to extend Nunavut's quotas, and an increased call for Inuit-owned boats in Nunavut, may result in a more extensive role for Inuit (including Inuit women) in commercial fisheries. In Holman, no commercial fisheries exists although a license to determine commercial viability for Arctic cod, snow crab, Northern shrimp and halibut has been issued and some work into experimental fisheries has been conducted.

Fish Processing

Nunavut has fish processing facilities located in Cambridge Bay, Whale Cove, Rankin Inlet and Pangnirtung. Pangnirtung Fisheries is the largest of these and provides one of the main sources of employment in the community. Approximately 75% of the 55 employees are female with most senior positions occupied by men.[2] Work in the fish processing industry is physically demanding and mentally taxing. Plant employees may be exposed to harsh conditions including extreme cold and heat and perform tasks that are monotonous and repetitive. Job-related injuries are common in the fish-processing industry although this was not specifically researched.

Lack of support services such as childcare affects the performance of female fish processing workers in Pangnirtung and prevents some women from entering the workforce. Women who do not require childcare or who are assisted by family members are able to find work in the plant as long as fish are available. A woman from Pangnirtung contends: "Some women just stay home because there are no sitters for them. But now there is a daycare, there might be more female workers here. More might find work instead of staying home. It can help their minds about finding work. We need to think about our rent, our welfare."

Inuit Women's Representation in Fisheries Organizations

Nunavut

The federal Department of Fisheries and Oceans, the Nunavut Wildlife Management Board, the Hunters' and Trappers' Organizations[3] (HTOs), the territorial Department of the Environment and, to a much lesser extent, the Baffin Fisheries Coalition are responsible for decision-making regarding Nunavut fisheries. According to Table 2, the Nunavut Wildlife Management Board and the Department of Fisheries and Oceans Canada employs the highest percentage of female staff. This percentage includes all women at both the senior and junior levels. At the territorial level, government re-organization in the spring of 2004 resulted in fewer women being employed in the Department of the Environment now responsible for fisheries.

A particularly low percentage of women is represented at the board level of the Nunavut Wildlife Management Board (NWMB) and the board of the Baffin Fisheries Coalition (*see* Tables 2,3). The Baffin region is considered the most traditional of the three regions in Nunavut. Members of the Nunavut Wildlife Management Board and board members of the Baffin Fisheries Coalition are elected by the Hunters and Trappers Organizations from 11 communities. The maximum number of female members of the board of the NWMB from its inception in 1994 has been 2. This disparity in gender distribution is also reflected in the board membership of Hunters and Trappers Organizations (HTO) in Nunavut as outlined in Table 4. As indicated in Table 4, the number of

---

[2] All statistics in this chapter are accurate as of December, 2005.

[3] The local community groups responsible for fisheries and wildlife management in Nunavut are referred to as Hunters' and Trappers' Organizations (HTO) while similar groups in the Northwest Territories are Hunters' and Trappers' Committees (HTC). Responsibilities for these groups in Nunavut and the Northwest Territories are essentially the same.

female board members of the Hunters and Trappers Organizations of the Baffin Region is very low at 6.8% in 2003-2004 and 9.1% in 2004-2005.

Table 2: Women's representation in fisheries-related organizations in Nunavut

Organization	Percentage of Women
Department of Fisheries and Oceans Canada	43%
(Iqaluit: n=12)	
(Rankin Inlet: n=2)	
Department of Sustainable Development	29%
(n=109) (prior to April, 2004)	
Department of Environment	19%
(n=115) (after April, 2004)	
Nunavut Wildlife Management Board	
(Board: n=11)	11%
(Staff: n=8)	62%
Baffin Fisheries Coalition	0%
(Board: n=11)	
(Staff: n=2)	

Table 3: Percentage of female board members of the
Nunavut Wildlife Management Board

Year	Percentage of Female Board Members
1994	0%
1995	0%
1996	11%
1997	11%
1998	22%
1999	22%
2000	22%
2001	22%
2002	12.5%
2003	11%
2004	11%
Number of members: 9 each year, except in 1994, 2002 (n=8)	

Table 4: Percentage of female board members of Nunavut Hunters and Trappers

Region (including total number of board members)	Year	
	2002 – 2003 (% of female board members)	2004 – 2005 (% of female board members)
Baffin (n=102)	6.8%	9.1%
Kivalliq (n=50)	18%	18.8%
Kitikmeot (n=150)	14%	14.2%

Holman, Northwest Territories

The federal Department of Fisheries and Oceans Canada (DFO), the Fisheries Joint Management Committee,[4] the Hunters and Trappers Committees and, to a lesser extent, the territorial Department of Resources, Wildlife and Economic Development are the main decision-making bodies for fisheries in the Northwest Territories. According to Table 5, the Department of Fisheries and Oceans Canada (DFO) in the Northwest Territories employs the lowest number of women (17%) amongst the significant decision-making organizations related to fisheries. This includes all women at both the senior and junior levels. However, it is important to note that the majority of DFO positions in the Northwest Territories are represented in the Coast Guard. In Department of Fisheries and Oceans positions that are directly related to fisheries, women hold 32% of the positions. Two of the senior staff in the Department of Fisheries and Oceans Canada who have the most frequent contact with the community of Holman are female. The role of the territorial Department of Resources, Wildlife and Economic Development is more peripheral regarding fisheries, focusing primarily on sport fishing and enforcement. Women are poorly represented in this department, comprising only 18% of the staff dedicated to working solely on wildlife/fisheries issues. Fifty percent of the staff working for the Fisheries Joint Management Committee (1 position out of 2) is female.

As indicated in Table 6, women are poorly represented at the board level of the Hunters and Trappers Committees located in the Inuvialuit Settlement Region.[5] Over a six year period, less than 20% of all board members of the Hunters and Trappers Committees

---

[4] The Fisheries Joint Management Committee was established in 1986 under the Inuvialuit Final Agreement and is a co-management organization with members appointed by the Inuvialuit Game Council and the Government of Canada. A Chair is appointed by the members. The Inuvialuit Game Council is made up of two representatives from each of the six Hunters and Trappers Committees. There have been zero women members of the Inuvialuit Game Council between 1998-2004.

[5] The Inuvialuit Settlement Region in the Northwest Territories comprises the six communities of Holman, Aklavik, Inuvik, Sachs Harbour, Tuktoyaktuk and Paulatuk and is the area administered under the terms of the *Inuvialuit Final Agreement* (1985).

in Inuvik, Sachs Harbour and Tuktoyaktuk were women. Paulatuk has never had a woman board member during this period and is recognized as the most traditional of these communities. Holman and Aklavik have the highest number of female board members and both communities are recognized for its strong women in leadership positions. There are no female board members on the Fisheries Joint Management Committee.

Table 5: Women in decision-making organizations related to fisheries

Women in fisheries organizations in the Northwest Territories (2004)	
Organization	**% of female staff**
Department of Fisheries and Oceans Canada	
Yellowknife (n=20)	30%
Inuvik (n=8)	37%
Hay River (n=5)	30%
Coast Guard (n=65)	10%
Total fisheries-related staff (n=33)	32%
Total DFO staff (n=98)	17%
Organization	**% of female staff**
NWT Dept. of Resources, Wildlife and Economic Dev. (Wildlife and Fisheries staff)	
Yellowknife (n=30)	27%
Inuvik (n=14)	21%
Fort Simpson (n=6)	0%
Fort Liard (n=1)	0%
Norman Wells (n=5)	20%
Deline (n=1)	0%
Tulita (n=1)	0%
Fort Smith (n=12)	8%
Hay River (n=1)	0%
Fort Providence (n=1)	100%
Fort Resolution (n=1)	0%
Lutsel K'e (n=1)	0%
Total staff (n=76)	18%
Organization	**% of female staff**
Fisheries Joint Mgt. Committee	
(Board: n=6)	0%
(Staff: n=2)	50%

Table 6: Average female representation on HTC boards in the Inuvialuit Settlement Region, 1998-2004

Name of community	Average female representation on HTC boards 1998-2004
Holman	34%
Aklavik	26%
Inuvik	17%
Sachs Harbour	14%
Tuktoyaktuk	12%
Paulatuk	0%

Access to Decision-Making Positions in Fisheries Organizations

Involvement of Young People

Women who hold decision-making roles in fisheries departments or organizations in Nunavut and the Northwest Territories are well-educated and experienced in their respective field. The combination of relevant education and expertise is critical. Fisheries management is a complex area fraught with a myriad of unknown variables. This complexity is under-scored in the Arctic—a region in which so much research is still to be conducted on stock assessment, habitat loss, and effects of climate change. Therefore, Inuit and Inuvialuit women who aspire to professional positions in fisheries would benefit from taking part in this activity in their youth.

High school students from Holman, Northwest Territories learn about fishing from females and males and fish frequently with family and friends of both sexes. In Holman, women as well as men act as role models to young people. Holman students stated that both women and men were equally good at fishing. As one male student stated: "Women and men are both good at fishing because it comes with tradition and culture. If you're Inuit, you're born with it." Another student said "Everyone gets taught the same way."

Although representatives from the Department of Fisheries and Oceans Canada, the Department of Resources, Wildlife and Economic Development or the Fisheries Joint Management Committee (the three main fisheries-related decision-making bodies in the Northwest Territories) makes an annual presentation to the high school about their respective organization, students learn little about fisheries in their formal curriculum. However, completing Grade 12 is paramount for students wishing to pursue a career in fisheries. Due to the technical nature of fisheries research and management jobs, most require post-secondary education. In order to assist Inuvialuit youth to complete high school and become prepared for a career in fisheries, the Fisheries Joint Management Committee (FJMC) began the Student Mentoring Program in 1996. Supported by the Fisheries Joint Management Committee, the Joint Secretariat, Department of Fisheries and Oceans Canada, the territorial government and private industry, this program provides high school students with the opportunity to work co-operatively with these organizations on exciting projects. Over the past three years, the FJMC Student Mentoring Program has employed at least 50% young women. While this program offers unparalleled opportunities for young people to gain experience in fisheries research at an early age, it is important to recognize that even applying and entering such a program poses challenges for many young people from the more remote communities. Separating from family and a strong support network in the community at an early age imposes stresses that cause some promising students to drop out of the program.

Apart from gaining practical experience through fishing with family and friends, young people also benefit from participating in natural resource-related workshops, conferences and other public fora. Hunters and Trappers Committees in the Northwest Territories and Hunters and Trappers Organizations (HTOs) in Nunavut are the single most important decision-making body regarding fisheries at the community level in both Northwest Territories and Nunavut. However, most HTOs in Nunavut do not allow youth members to join the board although they are welcome to attend meetings. Most HTOs impose an age limit of 18 years for board members that relates directly to the age required to obtain a licence to hunt. This is similar in the Northwest Territories, although some communities in the Inuvialuit Settlement Region are considering lowering this age limit to 16. Students from the Inuvialuit Settlement Region who wish to enter fisheries and wildlife as a career usually apply to attend the Natural Resources Technology Program at Aurora College in Inuvik while students in Nunavut apply to the Environmental Technology program at Nunavut Arctic College in Iqaluit.

As indicated in Tables 7 and 8, female students are poorly represented in fisheries management classes at both Nunavut Arctic College and Aurora College, likely due to the fact that this is still perceived as an untraditional career choice for young women. Despite these low numbers, female students who enroll in these classes have a similar completion rate as their male counterparts. Between 1997 and 2004, male students at Nunavut Arctic College successfully completed the class at a rate of 74% compared to 71% of female students who completed the course. At Aurora College, 90% of male students completed the course while 81% of female students did so.

Instructors of these classes report that the presence of female students changes the dynamic of the class in that female students tend to be more mature and committed to completing their program than many male students. All students entering this course are individuals with an interest in the outdoors. Instructors state that the women in this program tend to be somewhat older than many of the male students and that the tenor of the class is positively affected. In these cases, the quality of student questions and submitted work tends to be higher compared to all-male classes. No specific initiatives have been established at either Nunavut Arctic College or Aurora College to encourage the enrollment of more female students nor to support female students currently enrolled in these programs.

Community Involvement

As indicated earlier, women are largely invisible on the boards of most Hunters and Trappers Committees in the Inuvialuit Settlement Region in the Northwest Territories and the Hunters and Trappers Organizations in Nunavut. Both community and organizational representatives assert that sitting on the board is a valuable learning experience for any individual and provides an excellent opportunity for a young person interested in entering fisheries as a career. Additionally, there is a recognition in some communities that many board members have been sitting on the board for many years and there is a need to attract new members.

Table 7: Student enrollment and graduation rates in Freshwater Fisheries class, Aurora College 1997- 2004. (Source: Aurora College, 2004)

Year	Male students enrolled	Female students enrolled	Successful completion by males	Successful completion by females
2004	9	2	9	0
2003	3	5	3	5
2002	10	1	10	1
2001	6	4	5	4
1998	7	3	4	3
1997	7	6	7	4

Table 8: Student enrollment and graduation rates in fisheries management class. (Source: Nunavut Arctic College, 2004)

Year	Male students enrolled	Female students enrolled	Successful completion by males	Successful completion by females
2004	9	2	7	2
2003	0	0	0	0
2002	3	0	1	0
2001	1	2	1	1
2000	5	1	4	1
1999	6	0	4	0
1998	3	2	3	1
1997	0	0	0	0

> We should encourage young people because we can't live forever, our end could come at any time. There should be someone who could carry it on if we pass it on to them.
>
> Holman resident

One challenge in many smaller communities is that there is often a restricted pool of individuals who join committees and boards. However, a perception exists in the community that you are taking a job away from someone else if you sit on too many boards.

One significant barrier for women is that Hunters and Trappers Committees in the Northwest Territories and Hunters and Trapper Organizations in Nunavut are perceived primarily as a place for hunters (mostly male) rather than a place for fishers (female and male equally in most communities). Staff in fisheries-related organizations recognize that these organizations are often traditional, patriarchal organizations. It is possible that more men sit on the boards of these community organizations because their skills as hunters and fishers are more recognized locally.

> Most of the meetings we have, it's been a male-dominated situation. That has been the case in Nunavut. When we go in and meet with Hunters and Trappers Organizations and meet with the fishers, traditionally most of the people we meet with are male. I think it's something we need to work on and try to change. I think there are extremely knowledgeable individuals that I have met and chatted with on fisheries. But I think there's room for further expansion or further development of that, through having more women involved in fish management, fish development and overall organizational management.
>
> Iqaluit organizational representative

Although remuneration is provided for board members, there is no reimbursement for child or elder care costs that may be incurred as a result of meetings. This proves another barrier to women who often are not able to make the significant time commitment required of full-time board members. States an Inuvik-based organizational representative: "Most younger women in the communities have 9-5 jobs in the community and so don't have the time. There are definitely more women than men in full-time government jobs there. It's hard when you work full-time and have a family too." It is likely that even though many women do not hold formal decision-making positions in fisheries in most communities, women are able to influence the decision-making process through other means. Organizational representatives in the Northwest Territories report that although some women are outspoken and have no difficulty participating openly in public meetings, many women participate differently when alone or with other women than when they are in mixed groups. Additionally, women may influence the opinions of their husbands, brothers and sons and thereby affect the decision-making process more informally (Fox 2002). Both community and organizational representatives emphasize that the presence of women on the board only improves the efficiency of the Hunters and Trappers Committees. States a Holman resident:

> It is beneficial for women to be on the HTC. Women have different points of view and different ways of doing things than men that are more efficient.

Many Holman residents comment on the fact that many board meetings are long and drawn-out and that if women were running the meetings, the meetings would be concluded much earlier with all the work done. Certainly, Hunters and Trappers Committees are charged with a great deal of work to oversee and administer in the community. In 2004, many Holman women asserted that:

> Women get right down to the point and are very efficient while men talk forever. Women are more conscious of time and they need to manage time the wisest. They need to find the most efficient way of doing this. They will try newer ways of doing things to be efficient while men used tried and true ways.

Organizational representatives report that increasing the diversity of the board membership has a positive impact on the board's effectiveness. Particularly in the Inuvialuit Settlement Region, the boards of Hunters and Trappers Committees that comprise a combination of elders, experienced hunters/trappers/fishers and business people (including women) are often the most effective. This diversity ensures that a wide spectrum of opinions and experience is represented and that all aspects of an issue are argued. It has been noted that the lack of women on the majority of these boards may be attributed more to the need for board development and training and the need to better advertise new board openings than to the desire to prevent women from entering this male domain of power.

Although women are poorly represented as HTO and HTC board members, they frequently hold the position of Secretary/Managers for these organizations. This is an advisory position but a Secretary/Manager can provide some minimal guidance to board members. Despite the fact that women hold many of these positions, their decision-making powers are limited. Due to low pay, these positions have a high turnover rate with some HTOs and HTCs employing several Secretary/Managers in one year.

Employment in government organizations

The Department of Fisheries and Oceans Canada is the single most significant fisheries-related organization in both the Inuvialuit Settlement Region, Northwest Territories and in Nunavut. As outlined earlier, this department is not noticeably lagging behind other federal government departments regarding maintaining a gender balance amongst its staff. However, its failure to establish a gender-based analysis in programs, legislation and staffing and its lack of support and interest in gender-related fisheries issues is significant. Current and past DFO employees comment differently on the working climate for women. Some women currently in senior positions in both NWT and Nunavut report cases of sexual harassment encountered during their working lives at DFO while other female employees report a positive working experience.

> DFO has policies that encourage women to apply for jobs outside the ordinary kind of job that women usually work at. Lots of the biologists are women, lots of the people who do research are women. I think DFO encourages women to seek employment outside their traditional type of employment and it's a good thing. Women have lots to offer and I think they bring a different perspective to how things are done.
>
> Iqaluit organizational representative

Many DFO staff and community residents comment on the dominant patriarchal culture associated with fisheries and wildlife offices that still exists within DFO. However, this culture in DFO has not prevented some women from being hired into decision-making roles.

According to Article 23 of the *Nunavut Land Settlement Act*, Inuit beneficiaries of the Settlement are hired above other qualified individuals. Because of this regulation, Inuit women beneficiaries will be hired before more qualified non-beneficiaries, male or female. Staff at both the DFO and the Department of Environment in Nunavut assert that these organizations will hire the best-qualified candidate regardless of gender (apart from beneficiary status). Staff in the Department of Environment are very supportive of the role of women (and, in particular, Inuit women) in decision-making positions in Nunavut fisheries. States an Iqaluit organizational representative: "It would be better to have Inuit people running the show, running fisheries development. Because it's closer to them, they're running their own territory and they're not relying on *Qalunaaq*[6] like myself from

---

[6] Inuit term usually applied to non-Inuit 'outsiders' from the south.

the south. So that's where we have to go. I think that's very important and those are the steps we need to take."

Most Northern colleges including Nunavut Arctic College and Aurora College have a bridging or transitional year for students who need to upgrade their academic credentials. Often, this year is not sufficient and many students are not able to successfully make the transition between their own community and school life in Iqaluit or Inuvik. Budget cutbacks for colleges results in program cutbacks, overworked staff and overexpended resources. However, given the high percentage of successful female graduates from this program, a recruitment program aimed at attracting more females students to the Environmental Technology and Natural Resources Technology programs would likely be successful.

Childcare and Support Services

Services are required on various levels to promote women's greater involvement in decision-making processes in Arctic fisheries. Childcare is one of the most significant and necessary of these support services. Traditionally, most Inuit families are supported by the extended family network. However, as more Nunavut residents participate in the waged economy, the demand for regular and consistent childcare increases. As employees of the Pangnirtung fish plant have indicated, lack of childcare services may result in decreased employment opportunities for women (Department of Education and Department of Health and Social Services 2004; Inuit Tapiriit Kanatami 2005). Lack of childcare also affects students who wish to take part in the Environmental Technology program at Nunavut Arctic College or the Natural Resources Technology Program at Aurora College. Young mothers who enter the fisheries management programs often have difficulty successfully completing. Indeed, this is one of the primary reasons that female students drop out of the program. Male students with a young family would also benefit from the availability of childcare. This is critical given Nunavut and the Northwest Territories' high percentage of young residents. Although childcare is a common need for women (and men) in many cultures, Inuit women in Nunavut also benefit from other support services that enable young women to overcome traditional gender roles within small communities, to learn more about non-traditional career opportunities at the high school level, to successfully complete high school, to make the transition from a remote community to college in Iqaluit, to sustain and support students while in college and to assist graduates in finding a job in fisheries management.

Training Opportunities

Given the technical nature of many positions within fisheries, the need to develop and maintain training opportunities has also been identified even after a diploma or degree has been attained. This has been highlighted in communities such as Pangnirtung in which Nunavut's only major fish processing plant is located. Although staff need to be trained to learn the skills necessary to carry out a particular function such as fish gutting, other technical skills such as first aid, HACCP[7] and WHMIS[8] are highly marketable and in great demand.

---

[7] Hazard Analysis and Critical Control Point
[8] Workplace Hazardous Materials Information System

A lot of people ask me do they need training on the job. Well, they don't need training for what they're doing every day. You need the inside training for like saving lives and HACCP. And you know, there's nobody in Occupation and Health, there's nobody in First Aid, there's nobody in HACCP, there's nobody in Quality Control and the list goes on. And especially the chemicals part of it all—WHMIS. I think that's a good opportunity for women to get into the Canadian Food Inspection Agency and start going around and inspecting their own fish plants along with the Worker's Compensation Board.

Pangnirtung organizational representative

Training opportunities to acquire both technical and cross-cultural communication skills are made available at government departments including the Department of Fisheries and Oceans Canada and the Department of Environment. Access to ongoing training opportunities is critical as many new staff in senior positions are recruited from the south with limited experience in working in a cross-cultural context.

Conclusion

Inuit and Inuvialuit women are actively engaged in fishing at the community level in Nunavut and Holman, Northwest Territories. Currently, this is primarily at the subsistence level but this may be due to lack of economic opportunities. In particular, the expansion of commercial fisheries and potential increased fishing quotas in Nunavut may offer future opportunities for Inuit women. Membership on the board of a Hunters and Trappers Organization or Hunters and Trappers Committee offers the greatest scope for assuming decision-making powers within the community. With the exception of a few communities, women have had limited formal involvement on the boards of these organizations in both Nunavut and the Inuvialuit Settlement Region in the Northwest Territories. This limited involvement may result in a lack of diversity of perspectives at the board level and a lack of recognition of women's knowledge and considerable experience where fisheries are concerned. The failure to consistently involve youth members on a formal basis at the board level also results in lost opportunities for young women and men to gain valuable experience that would help to prepare them for an education and future career in fisheries management. However, women may be involved in decision-making processes at the community level on an informal basis through influencing male family members.

Inuit and Inuvialuit women are also poorly represented at the board level of fisheries co-management boards including the Nunavut Wildlife Management Board in Nunavut and the Fisheries Joint Management Committee in the Inuvialuit Settlement Region. All members of the Nunavut Wildlife Management Board and fifty percent of the members of the Fisheries Joint Management Committee (with the exclusion of the Chairperson in both cases) are appointed or elected by the Hunters and Trappers Committee or Hunters and Trappers Organization. Therefore, women's lack of representation on these boards directly impacts on their lack of representation on co-management boards including the Nunavut Wildlife Management Board and the Fisheries Joint Management Committee. Inuit women are represented at the staff level of formal organizations including the Hunters' and Trappers' Committees/Hunters' and Trappers' Organizations and both co-management boards although, in many cases, the staff role is to support the decisions of the board. Women are relatively well-represented in decision-making positions in the Department of Fisheries and Oceans Canada in both territories but are not as visible at the territorial level. However, the implementation of gender-based analysis in a review of federal and territorial programs, policies and practices would be effective in approaching the goal of gender equality as specified in *The Federal Plan for Gender Equality* (1995).

Nunavut Arctic College and Aurora College could provide further support for young Inuit and Inuvialuit women wishing to pursue a career in fisheries. Support services including onsite daycare facilities, counselling services that assist young parents and students who are isolated from their families and communities and additional financial aid would be welcomed. Finally, stronger links between the Hunters' and Trappers' Committees and Hunters' and Trappers' Organizations and the communities, Nunavut Arctic College, Aurora College, Fisheries Joint Management Committee, Nunavut Wildlife Management Board and the federal and territorial government bodies responsible for fisheries need to be strengthened and maintained.

References

CARG (1987). Theme issue, "Arctic Fisheries: New Approaches for Troubled Waters." *Northern Perspectives* 15(4).

Community Growth Solutions (2003). *Holman Community Economic Development Plan.*

Department of Education and Department of Health and Social Services (2004). *Nunavut Early Childhood Development Update Report. 2003/2004.* Electronic document: www.gov.nu.ca/education/eng/pubdoc/ECD_UpRep_0304.pdf. Last accessed: March 26, 2009.

Fox, S. (2002). 'Women's participation in self government negotiations in the Northwest Territories, Canada,' pp. 144-151 in *Taking Wing Conference Report.* Helsinki: Ministry of Social Affairs and Health.

George, J. (2003). Nunavut interests seek bigger share of turbot, shrimp. *Nunatsiaq News.* November 7.

Inuit Tapiriit Kanatami (2005). *State of Inuit Learning in Canada. Prepared for the Canadian Council on Learning.* Ottawa: Inuit Tapiriit Kanatami Socio-Economic Department.

Kellert, S. and S. Ebbin (1993). *Empowerment and equity of Indigenous peoples of North America: Emerging co-operative institutions for fisheries management."* Report prepared for U.S. Man and the Biosphere and The High Latitude Ecosystems Directorate, Yale University.

Reist, J. and M. Treble (1998). 'Challenges facing Northern Canadian Fisheries and their co-managers,' pp. 155-165 in J. Oakes and R. Riewe, eds. *Issues in the North. Volume III.* Occasional Publication No. 44. Edmonton: Canadian Circumpolar Institute, and Winnipeg: Department of Native Studies, University of Manitoba.

Reynolds, J.B. (1997). *Fish ecology in Arctic North America.* Bethesda: American Fisheries Society.

Senate of Canada. (1998). *The Nunavut Report: Report of the Standing Committee on Fisheries and Oceans.* Electronic document: http://www2.parl.gc.ca/HousePublications/Publication.aspx?DocId=1031534&Language=E&Mode=1&Parl=36&Ses=1. Last accessed: March 2, 2009.

Senate of Canada (2004). *Nunavut Fisheries: Quota allocations and benefits.A Report of the Standing Committee on Fisheries and Oceans.* Electronic document: http://www.parl.gc.ca/37/3/parlbus/commbus/senate/com-e/fish-e/rep-e/rep04apr04-e.htm. Last accessed: March 2, 2009.

Standing Committee Report (1999). *Government response to the Seventh Report of the Standing Committee on Fisheries and Oceans (Nunavut Report).*

CHAPTER NINE

Gender Equality and Governance in Arctic Swedish Fisheries and Reindeer Herding

Maria Udén

Abstract: This chapter discusses women's situation in fisheries and reindeer herding in Arctic Sweden in relation to the governance of these activities. From a gender equality perspective, there are discouraging similarities between these two industries; similarities which distinguish them from the general picture of gender equality in Sweden today. The author's interest in the subject stems from insights gained in two research projects located in Sweden's northernmost region, Norrbotten County. The chapter describes local circumstances in some rural communities in an Arctic region, but ultimately addresses challenges at the national level as fisheries, reindeer herding and gender equality are nationally governed.

Introduction

This chapter draws together a description of women's situations in Arctic Swedish fisheries on the one hand, and an account of how gender is regarded in the sector's public governance on the other, with corresponding descriptions from another small scale traditionally-based industry in the Swedish Arctic, namely reindeer herding. The incentive for preparing a dual analysis of this kind stems from insights gained in two different research projects; one on women in reindeer herding and another on women in fisheries. The first (2001-2003) was initiated by women reindeer herders in Sirges, a Sámi village in Norrbotten County, Sweden. The project examined social patterns and reviewed the technical modernization of reindeer herding, in order to support an increase of women's representation as reindeer herders. The second was part of an Arctic Council initiative on women in Arctic fisheries (2003-2004), and in this, women's situation in the coastal fisheries of Norrbotten County was portrayed (Udén 2003, 2004). Both of these projects were thus located in the same geographic region and administrative unit of Norrbotten County in Northern Sweden. However, I do not focus on the regional level. Fisheries, reindeer herding and gender equality are all nationally governed in Sweden and the ethos in matters of gender equality and women's rights is coupled to Sweden as a nation. Thus, my discussion ultimately addresses challenges at the national level.

In Sweden, as in most other countries, women's access to political influence, in other words, political power, is generally considered a cornerstone for gender equality. Even if Sweden was not among the very first nations taking the historical step to introduce equal rights to vote for women and men, Sweden is currently recognized as one of the leading nations in terms of gender equality.[1] For instance, the present Parliament of Sweden (Riksdagen) consists of 45% women and the national government is made up of eleven women and nine men (2004). The second cornerstone of the Swedish gender equality policy is the right of women to gainful employment. Swedish economic and labour market policies of the twentieth century have gradually abandoned the male breadwinner model in favour of a dual breadwinner model. The state has thus engaged in advancing women's integration in the labour market through extended welfare rights and services (Mósesdóttir 2001).

At present, the concept of power and influence, together with conditions for gainful employment, continue to be prioritied gender equality issues, as evidenced by the Swedish Government in their gender equality policy for the current electoral period. The government's list of priorities in this area of politics includes some recently added issues such as the sexualization of the public sphere (i.e., the manner in which women's bodies and sexuality are exposed and exploited in mass media, advertising, pornography etc.), but critical issues remain equal access to positions of power and influence and equal pay for equal work and work of equal value (Swedish Government Office 2002:5-6). It is also evident that the concept of gender equality has broadened, for example, it now includes gender representation and balance on the boards of public and private businesses. As the dual breadwinner model has taken precedence in Sweden, a woman is no longer primarily identified according to her marital status as was the case well into the twentieth century. For instance, the income of married couples used to be taxed as one unit in a manner which made two incomes redundant, but, in 1971, the income tax for married couples became individually determined. The widow's pension, which compensated for a loss of a deceased husband's/breadwinner's support, was formerly included in the Swedish public pension system, but the range of this pension was limited from 1989 onward (Clayhills 1991; Kyle 1987).

Norrbotten County Profile

Norrbotten County covers the northernmost part of Sweden. It faces Västerbotten County in the south, Norway along its mountainous western border, Finland in the east and the Gulf of Bothnia in the southeast. The county covers one fourth of Sweden's land area or 98,000 km^2 and had 252,874 inhabitants as of late 2003 (County Administrative Board 2005). The economy is based on natural resources extracted in the region but utilized elsewhere, and include: electricity (from hydropower), timber, metals and minerals. Industrialization arrived late to this Arctic part of Sweden, and developed in areas of interest for large-scale natural resource exploitation. One quarter of the county is made up of parks and protected areas. After a period of positive migration nets in Norrbotten during the early twentieth century, net negative migration developed during the latter decades. This was partially counteracted by relatively high birth rates. Nevertheless, the downward trend has been particularly noticeable amongst young women. The population statistics for 2004 indicated 86 women per 100 men aged 20-34 years in Norrbotten County, and in some municipalities as few as 72 women per 100 men in this age group (County Administrative Board 2005:17).

---

[1] In Sweden, women were given the right to vote from 1921 onwards. In Finland, women were able to vote in 1906, in Norway 1913 and in Denmark and Iceland 1915 (Rönnbäck 2004).

Norrbotten County is inhabited primarily by Sámi, Finnish, and Swedish populations with other ethnic groups, including the Rom,[2] represented. The exact ethnic composition of the population is not known, as ethnicity is not publicly registered in Sweden. However, being a recognized Indigenous population, the Sámi people and Sámi issues have been accorded attention by the government, and the need to establish accurate demographic data for the Sámi has attracted attention. Officially, the Sámi population in Sweden is estimated to be 20,000, compared to an estimated total of 80,000 Sámi in Scandinavia and North West Russia (www.sametinget.se). However, these statistics are problematic. Amft (2000) has commented upon the implementation of sometimes arbitrary criteria which have marginalized individuals and groups of Sámi people and, in particular, women.

Differing power dynamics between the various groups in Sweden are revealed in formal and informal regulations regarding language use. Dominant societal norms and former government policy emphasizing the benefits of being (or becoming) 'Swedish' led to originally Finnish and Sámi-speaking families increasingly turning to Swedish as their first language during the latter part of the twentieth century (Elenius 2001). Today, recent legislation accepts and promotes Indigenous rights to use national minority languages. For example, the Sámi language has special status in Arjeplog, Gällivare, Jokkmokk and Kiruna as does Finnish in the municipalities of Gällivare, Haparanda, Kiruna, Pajala and Övertorneå. This arrangement indicates the ethnic composition of pre-industrial settlement patterns in Norrbotten.[3] Among the coastal municipalities of Norrbotten—Piteå, Luleå, Kalix and Haparanda—only Haparanda is mentioned in the minority language law (Finnish).

Fisheries and Reindeer Herding in Norrbotten County

Just under two percent of those employed in Norrbotten work in the primary sector, which includes forestry, farming, fishing and reindeer herding (County Administrative Board 2004). From this, we understand that neither fisheries nor reindeer herding affects the total employment rates in the county other than marginally. However, in a number of local communities in the most sparsely populated areas, these industries are crucial. Additionally, there are secondary industries fully or partially dependent on primary industries including food production, transport services, and tourism. It should be noted that fisheries, and reindeer herding, tend to provide various forms of income. For example, reindeer herders may also fish in mountain lakes and fishers as well as reindeer herders may hunt as a part of a traditional lifestyle. In fishery households, gardening and picking berries are also part of a local subsistence-based economy. Tourism as related to both fisheries and reindeer-hunting is also being developed.

According to the register at the National Board of Fisheries in April 2004, 65 individuals had a license for commercial fisheries in Norrbotten. Of these, six fish in the mountain lakes in the state-owned western parts of the county, while the others fish in the Gulf of Bothnia and the Baltic Sea. The coastal population of the Gulf of Bothnia traditionally fish along the river outlets, along the coast and in the archipelagoes of the

---

[2] The Romani people, known colloquially in English as 'gypsies'.
[3] See the Swedish Government's proposition 1998/99:143 and further, the Swedish Parliament's decisions bet.1999/200:KU6, rskr.1999/ 2000:69.

Gulf of Bothnia. Fisheries developed into the main or sole source of income for a specialized fisher corps only after World War II. This group of specialized fishers expanded in the 1970's with the introduction of trawlers. This expansion took place at the same time as it decreased in Sweden as a whole (Karlsson 1981; Rova 1999). The register kept at the National Board of Fisheries shows that, in 2004, almost half of all fishers in Norrbotten were fifty years of age or older, and only one professional fisher in Norrbotten was younger than thirty. This may be perceived as a problem; in any case, it signals the state of the fisheries in the county. Concerns about the renewal of the reindeer herders as a group are also currently being expressed among reindeer herders (Udén 2004). Nonetheless, far more youth are in evidence amongst reindeer herders than in fisheries today.

The Norrbotten fisheries has remained relatively small-scale. Usually the fisheries is made up of one-person enterprises, providing a source of income primarily for their owners. Following shifts in the rural economy, settlement patterns have changed with the focus moving from islands in the Norrbotten archipelagoes to the mainland. To a large extent, Norrbotten fisheries remains a coastal fisheries adapted to the climate. Thus, incomes derived from fisheries decline during the winter when the Gulf of Bothnia is covered with ice. To counteract this, the larger vessels of the Norrbotten fleet are at times used for trawling and fishing in the Baltic Sea during winter (Karlsson 1981; Rova 1999). Combining fishing with another source of income is again customary. In my own investigation (Udén 2004), a random sampling of Norrbotten's licensed fishers (or approximately 33% of the sea and coastal fishers in this county), were interviewed regarding their businesses and, in particular, women's representation in their businesses. The results suggested that 50% of Norrbotten fishers have other gainful employment or income besides that provided by their fisheries companies. Gustavsson and Johnsson (2000) examined a large number of Swedish coastal fishery household income declarations from 1997, and concluded that the average income from the fishery businesses represents only 33% of the total income in households of this type.

In Sweden, reindeer herding is a Sámi privilege. As with the commercial fisheries, the industry is regulated in terms of individual rights, but the issue is more complicated as access to grazing land is determined according to membership in a Sámi village.[4] Sámi village members have the right to let their reindeer graze within their village's territory, but they do not individually have rights to the territories. There are fifty-one Sámi villages in Sweden, 32 of which are located in Norrbotten County. In 2001, the total number of reindeer in Sweden was 220,000, out of which 121,000 grazed in Norrbotten. On average, the single owner in Norrbotten owns fewer reindeer than in the more southern counties as there are also many reindeer owners. In 2001, out of 4,500 reindeer owners in Sweden, 3,888 lived in Norrbotten (Jordbruksverket 2003).

The majority of Sámi villages in Sweden stretch out in a southeast/northwest direction following the river valleys from mountain springs to the coast. This pattern corresponds to traditional annual migration cycles. Migration from one area to another is fundamental for the maintenance of larger herds. Keeping one's herd fenced during winter is, at times, an option, but is generally avoided. Today, reindeer herding communities are semi-nomadic and partly residential in villages and small towns near the winter grazing areas. Herders still move with their reindeer all year-round, and may stay in cottages in the grazing areas for weeks at a time. Snowmobiles and other modern vehicles make it

---

[4] Another expression for a Sámi village is grazing community.

possible to travel longer distances on a daily basis than was possible a generation ago (Beach 1993; Amft 2000; Udén 2003; Labba and Jernsletten 2004).

In comparison to the Swedish median, the average income level in coastal fisheries and reindeer herding is low. For reindeer herder households, the average income in 1997 was 121.000 SEK (or approximately $20,000 CAD), out of which 14 % was derived from the business and 86 % from other employment. The minimum subsistence income for self- employment in 1997 was calculated to be 36.300 SEK plus housing expenses. The average declared income from coastal fishery activities and reindeer herding were thus below the calculated subsistence level in Sweden (Gustavsson and Johnsson 2000).

Women in Norrbotten's Fisheries and Reindeer Herding Industries

In Sweden, as in many other countries, fisheries and reindeer herding are male-dominated sectors. The situation in Norrbotten is no exception. Of the sixty-five commercial fisheries license owners currently operating in Norrbotten, only one is a woman. Västerbotten County, Norrbotten's nearest neighbour to the south, had 22 license owners in early 2004 and none of these was a woman. It is estimated that 10% of the full-time active reindeer herders in Sweden are women (Kråik 2002). All reindeer owners are not herders, but, as a rule, the largest owners are also herders. In 2001, 34% of Norrbotten's 3,888 reindeer owners were women, but only 18% of the reindeer in the county were owned by women (Jordbruksverket 2003).

From ownership we can now turn to employment. Gustavsson and Johnsson's (2000) investigation of coastal fishery household income declarations shows that the incomes from fisheries in these households were declared by the male members of the household. The incomes from fisheries declared by the women of these were insignificant. Thus, the wives of fishers are generally not employed in their husband's fisheries, or earn any income from it. At the local level in Norrbotten, Lundgren *et al.* (2004) asserted that active fishery companies in the municipality of Kalix in Norrbotten County (eleven companies in total) provided 14.5 full-time positions. Only one position was held by a woman. This was in the largest fishery company, which, at the time of the investigation, had four employees including the owner. Women's employment in reindeer herding has not been recorded, with the exception of self-employment. From a local perspective, it is evident that no established labour market for women exists in this industry.

Women cannot expect gainful employment in fisheries or reindeer herding companies unless they are the owners of the business themselves, and few women own a business in either of these sectors. Nonetheless, I maintain that women are engaged in both of these industries through informal, usually family-based arrangements, and through short-term labour arrangements. In my own fisheries study, first-hand information from fishers suggested that women perform informally organized work in 70 to 90 per cent of the Norrbotten fisheries, most usually within household relations. Fishers' spouses and other women work for the fisheries as assistants, performing clerical tasks, maintenance, catch handling and processing. Except for family arrangements, women are regularly employed when labour is temporarily in-demand, for example, in the handling and processing of the fall vendace catch.[5] After the vendace is caught, it must be handled and the roe processed

---

[5] Vendace (*Coregonus albula*), is a crucial species to the Norrbotten fishers. The vendace is estimated to comprise 50%-70% of the Norrbotten fisheries' total income (Karlson 1981;

within hours. This stimulates an intensive but short-lived need for labour during a few weeks in the fall.

Women's informal participation in reindeer herding is a well known phenomenon and has been presented by a number of scholars including Amft (2000), Eikjok (1989), Joks (2001) and Ulvevadet and Klokov (2004). In reindeer herding, various responsibilities are performed by female family members and not least by spouses. Traditional tasks associated with housekeeping (which is more complicated and economically vital in a rural society built on primary production than it is in urban areas), and bookkeeping recur in these discussions. Women are primarily responsible for the time-consuming task of transporting people and goods to and from the grazing areas.

Women are deeply engaged informally in reindeer herding through financing their spouses' business. The history and extent of the practice whereby women guarantee household incomes and business finances has not been investigated in depth and people do not openly discuss first-hand experience of these arrangements. Thus, the scale, strategies and structures of this economic support are not known. Nevertheless, there are strong indications from both industries that such solutions exist. The practice has been reported by Amft (2000) in Swedish reindeer herding and there is an open discourse on this in herding communities. Regarding fisheries, the picture is less clear. Lack of cash, for instance, is not necessarily a problem to coastal fisher families. On the contrary, I have encountered an attitude that cash is never a problem in these families. Nevertheless, according to informal data, women in Norrbotten have regularly guaranteed credit for investments in their spouses' fisheries businesses. A comparison throughout the Arctic region shows that women's economic support of spousal businesses appears to be recurring in local fisheries communities (Sloan 2004). If women guarantee the basic standards of living and/or investments in Norrbotten fisheries and reindeer herding, what are the potential economic consequences of this support? One result is that today, reindeer herders' and fishers' spouses have gainful employment outside the fisheries. Earlier, spouses were mainly occupied within the realms of the household and their husband's business (Amft 2000; Gustavsson and Johansson 2000; Udén 2004). Keeping in mind that, according to Gustavsson and Johansson (2000), the fishery generated 33% of the fishery households' declared incomes, it is interesting to note that women's earnings from gainful employment were significantly greater than that of men. Gustavsson and Johnsson (2000:47) concluded: "Income from employment comes in all essentially from women."

Public Governance

Fishery and reindeer herding are governed by the Ministry of Agriculture, Food and Consumer Affairs in Sweden which oversees all primary sectors and is also responsible for Sámi issues. The National Board of Fishery is the central authority in all fishery issues, while the Swedish Board of Agriculture is the highest authority for the governance of reindeer herding. At the regional level, the County Administrative Board is responsible for fishery and reindeer herding issues. Each County Administration Board manages information and control, handles subsidiaries and compensation to the fisheries and, in

---

Rova 1999; Udén 2004). The value of vendace to the fisheries is generated from its roe, a luxury product which is processed to consumer-ready quality within the fishers' own companies.

the northernmost counties, to Sámi villages and individual herders within their respective territory. In the early 1990s, a number of governance tasks related to fisheries were re-allocated to the national level from the County Administration Boards. One of these tasks was the approval of fishery licenses. However, the Norrbotten County Administration Board still manages the large state-owned areas in the northwestern section of the country.

Meetings and discussions with public officials in reindeer herding and fisheries management provides evidence that many of these officials believe that women cannot be fishers or reindeer herders. There is no need for complicated strategies of investigation to come to this conclusion: officials who hold this view simply state that women are not fit for the job. When I encountered such statements, officials usually meant to inform me, as an 'outsider,' about the nature of fisheries or reindeer herding. The conditions under which the majority of women participate in Norrbotten's fisheries and reindeer herding are such that they cannot aspire to membership in fishers' federations or fisheries labour unions, or become full time reindeer herders. Somewhere between the attitude that 'women cannot be' fishers or reindeer herders, and the invisibility of women doing informally organized work, women's voices regarding the governance of these industries and the natural resources they rely on, have been muted. This is expressed in informally communicated attitudes, but also in more formal instances and in documents. One example of the latter is a booklet on the subject "Who owns the fish?" (Johansson 2004), that was recently published by the Fisheries Unit at Norrbotten County Administration Board. This booklet features ten people from Norrbotten who offer their perspectives on fisheries—only one of whom is a woman!

The Household Model

Over time, the patriarchal, household-based model has been abandoned by the Swedish state. In relation to the households that people live in, women and men are increasingly regarded as individuals rather than family members. Husbands, fathers, or other adult men are no longer the guardians of adult women, as was the case early in the twentieth century (Kyle 1987; Clayhills 1991; Rönnbäck 2004). As it has become less acceptable to advocate gender segregation in Swedish working life, and gender equality is today one of the goals of national Swedish as well as European Union policy, it has astonished me how, without toning down or rephrasing, officials publicly hold the view that women 'cannot' be fishers or reindeer herders. Furthermore, I have noted that practices around women's informally organized but unpaid labour are not only known, but accepted as integral to the industries by public officials. This is especially the case for fisheries. Among fishers as well as public servants, a common way to explain the household-based organization of women's work and distribution (or lack thereof) of income from this work is that "it is arranged within the family." In fact, the household model is still referred to in the laws and regulations for both fisheries and reindeer herding. Labba and Jernsletten (2004:131) explain this household-based organization of reindeer herding units as follows: "The reindeer industry is made up of approx. 950 units. According to the government, a husbandry unit may be defined as an economic association managed by a reindeer owner and his/her household/family. A husbandry unit is regarded in legal terms as an enterprise, with either one or several owners."

The term 'husbandry unit,' as used by Labba and Jernsletten refers to the household model (Swedish: *hushåll*) in the regulating law (SFS 1971:437). The fundamental position of the household concept affects women's and men's positions in reindeer herding quite differently. This happens through the grading of Sámi village membership into classes and the fact that, in a family, only one person can attain the highest class of membership (Class 1). Today, it is not obvious that this classification system is gender-biased, even if it can be noted that fewer women than men are Class 1 members (full-time reindeer herders). However, as Amft (2000) has indicated, a historical review reveals the gendered structure of this system quite clearly. In an earlier law that preceded the existing one,[6] the gender bias inherent within this system was explicitly imprinted. The class system was structurally the same as it exists today with the significant exception that members of Class 3 were explicitly described in the text as 'wife and children' or 'widow and orphan' after deceased higher class members. This stratification leaves nothing unsaid as the members of the lowest class were clearly defined as women and children and the law described a patriarchal family.

This situation is mirrored in the fisheries particularly regarding the laws and regulations for acquiring a fisheries license. According to FIFS 1995:23, in certain cases, the household of the applicant shall be considered in the decision to approve or deny a license application. Normally, the regulation states that licenses shall only be given to those whose main income is fisheries. In particular cases, commercial fisheries licenses can be approved in order to provide a secondary income to full-time employment, namely, in local communities where income opportunities are rare. If, in such a case, the applicant's secondary income from fisheries is expected to contribute at least 20% of the household's income, then a license can be granted.[7] The other household members' incomes must be low if the license owner has full-time employment in addition to fishing, but the income from fisheries still contributes 20% of the total household income. The image of the fisher as breadwinner stands forth in this regulation of fisheries license approval. How pertinent is this law today?

How is Women's Situation Perceived in Local Communities?

We have seen in fisheries and reindeer herding that women's participation is organized informally, often 'within the family,' or as short-term employment. We have noted that this practice has certain support in legislation, by the state's definition of these two industries as family matters, and the family as a patriarchal unit. Among public officials at various levels, this is regarded as natural and women's ability to attain other than a secondary position in the respective industries is not questioned. Today, neither Arctic Swedish fisheries nor reindeer herding are, overall, highly profitable industries. In the short-term, therefore, the economic consequences of women's lack of participation in decision-making roles in reindeer herding and fisheries are not obviously dire. Modest enterprises may not generate enough income to actually pay a regular salary to anyone but the owner, or at least, the salary level would be very low. One reason is that high taxes and insurance (about 40-50%) are deducted and this money then disappears from the household, at least in the short term. It may appear to be the better choice, to let fisheries earnings stay in the company, allowing choices for the family lifestyle at another level than

---

[6] SFS 1928:309.
[7] FIFS 1995:23, 3§.

the individual. Nevertheless, when women's labour stays within the private, informal sphere, the benefits of the Swedish welfare state that relate to employment are denied to them (i.e., retirement pensions, insurance against illness and unemployment, and parental leave). Within the household, these consequences may appear less important, especially in the short-term. However, in situations of divorce or other dramatic life changes, this is no longer the case. Women's rights to her husband's business, and her ability to maintain these rights in case of the husband's illness or death, are limited. Among Sámi women and men, this is acknowledged as a problem.

The effects of the household concept being the accepted paradigm are most obvious within reindeer herding; both because the law itself (and its history) is so explicit on this point, and because women from organizations such as Sáráhkká and Sámi Nisson Forum challenge this norm. Maria Kråik, the former chairperson of Sáráhkká, stated: "The issue of equal rights for the entire Sámi population, irrespective of gender, cannot be understood without a historical perspective. To understand present frustration among Sámi women, it appears necessary to consider the development of the rights to practice reindeer-herd management" (Kråik 2002:138). The work of these organizations has been complemented by actions taken by women as individuals. For example, in 2001, the status of two sisters who were both reindeer herders and leaders of their own companies was downgraded from Class 1 membership upon their respective marriages. The two sisters challenged this decision requesting that it be overturned. This case was documented in the national media.[8]

In the coastal fisheries, the gendered division of labour is not usually openly challenged. Thus, on the surface, relations between women and men in the fisheries sector may appear more harmonious and less problematic than in reindeer herding. Thus far, within fisheries, women have infrequently challenged the status quo at least not to the point where it becomes publicly observable. Women have preferred other strategies, for example, daughters in fisheries households have simply left the communities. "There is nothing for them here," as I was often told during the course of the fisheries study. Regardless of the positive factors discussed, interviews conducted with fishers' wives in Norrbotten County revealed certain aspects of the role of fisher's wife as problematic (Udén 2004). As outlined below, varying generations have somewhat different experiences, but this is rather a matter of nuance.

One older woman remembered that in her youth, working in the local fishery processing plant during the summer was something that 'everybody' did. She and her husband spoke about the intense commerce that was always found on the landing stages in earlier times, compared to how slow and empty it is today. Fresh fish was sold directly to households in the villages near the coast by salesmen who bought herring from the fishers and drove around the villages to sell the fresh catch the same day. When they were young, the wife and children accompanied the husband on the boat during the summer and they would stay in a hut on an island in the Archipelago. This arrangement became more difficult when the children grew older partly because social activities became more important. As the children went to school, the wife took up gainful employment outside the household primarily due to her need for social contact. "One must be psychologically very strong to be alone at home all the time," she says. Nevertheless, during the years of paid work, she always made sure to be available for the vendace processing.

---

[8] The case was documented in the Norrländska Socialdemokraten, 6 November 2001.

Another woman stated that she and other women believed that life as a fisher's spouse was particularly difficult for the younger generation of women. She stated that it was not as easy for younger women to give up their own professions in order "to be available in the roe hut" or to take part in other fisheries activities when needed, as it had been for the fishers' mothers. In her situation, she shared a daughter with a partner from whom she had separated. She was sure that her former spouse would be happy if their daughter would want to go into fisheries and, in time, take over his business. But, at the time of the interview, this did not appear likely to happen. Even sons are not interested in becoming fishers today, the interviewee said. "Nevertheless," I remarked, "some do choose this life—what is their reason?" "It is a free and adventurous life," the interviewee answered. "Sometimes you cannot plan the future to the full, as you cannot really know what the catch will be. There is excitement in this." Also, she emphasized that contact with nature added to the quality of life in a fishery household.

In both reindeer herding and Norrbotten fisheries, family ties provide the basis for entering and remaining in the business. As described by Amft (2000), a son is raised to carry on reindeer herding following the death of his father and while sons are preferred, daughters may also follow this tradition. According to my research (Udén 2003), the attitude to women as active reindeer herders varies. Some male herders are more comfortable with the idea than others. Regarding fisheries, there are fishers who would not recommend that their daughters become fishers. Likewise, there are those who would gladly encourage their daughters to follow in their footsteps. In the interviews reported in Udén (2004), one Norrbotten fisher stated that he does not appreciate working with women and saw to it that he did not have to. Still another spoke warmly about how there is now one woman in Norrbotten who has a commercial fishery license and that he supported her taking steps to apply for this license. It may be that attitudes toward the participation of women in fisheries relate to the long history of fisheries in the families. Certain fishing locations in the coastal areas have been in the same family for centuries, handed down from generation to generation. During interviews conducted for this study, it was more usual that a male fisher related to his father as a fisher, than to the maternal grandfather (or the mother). Nevertheless, several fishers referred solely to their mother's families as those who had introduced them to the fishery. For example, one fisher stated that his father and paternal grandfather were self-employed in sectors other than fishing, while the mother's family—maternal grandfather and uncles—were fishers. Signs of positive change regarding the acceptance of women in fisheries do exist and, to a certain degree, this is supported in the local community.

Similarities in the difficulties of the fishery and the reindeer herding life for women are apparent. For families in both areas, schooling can be a significant challenge. For example, it has been difficult to reconcile semi-nomadism with the public school system schedule. Presently, the recent waves of closure of rural schools puts one more obstacle in the way of combining family life with a traditional industry. Marit Myrvoll (2004) cites a Finnish Sámi woman: "When local schools are nonexistent or closed, the children may have 2 to 2½ hours of transportation to school every day. Of course, this can be strenuous for the children. This situation also influences the parents' choice of home location and jobs. Who knows, over time, perhaps all of the families will live in the centre of the community" (Myrvoll 2004:104). In Norrbotten today, there are cases where women and children of a fisheries household have an apartment in town, while the father/husband lives on an island. This separation of the family is a further consequence of the closure of local schools in the archipelago.⁹ Living close to nature is necessary and important in fisheries as it is in reindeer herding. The interviews with Bay of Bothnia fisheries women made the link to and relationship with nature evident, and this sentiment has similarly been expressed by Sámi women. States Myrvoll: "The close relationship with nature is still alive, and must be kept vital as long as reindeer husbandry is practised. Living in relationship with animals and nature, people have learned to master life and adapt to insecurity. As one informant stated: "Who knows what kind of winter will come?" (Myrvoll 2004:102). However, this connection with nature is also challenging as household members must adapt to an increasingly urbanized and non-traditional society.

Conclusion

According to current Swedish policies, adults are expected to earn their own income and have equal opportunities to participate in political processes. Gendered structures in Arctic fisheries and reindeer herding contradict national gender equality goals of "equal access to positions of power and influence" and "equal pay for equal work and work of equal value." The application (or lack thereof) of these concepts has been examined in relation to the household model. But what does the significance given to the household concept and the gendered organization of labour within the fisheries and reindeer herding households really imply? We see how Sámi women organize to speak about their rights and how male fishers encourage women to enter this profession. Additionally, in comparing the two industries, we must remember that reindeer herding is an exclusive Sámi right in Sweden while fisheries is not distinguished on ethnic grounds. Fisheries and reindeer herding are maintained by two different groups with a different cultural history and they also inhabit varying geographical areas. The results put forth in this chapter shows that official doubts regarding women's abilities are not necessarily shared by the fishers and reindeer herders themselves, and that preconceptions regarding the family economy that is expressed in legislation is often not reflected at the household level.

References

Aasjord, B. (2003). 'Where have all the fishes gone?,' pp. 36-44 in *Taking Wing: Conference on Gender Equality and Women in the Arctic,* Helsinki: Ministry of Social Affairs and Health.

Amft, A. (2000). *Sápmi i förändringens tid.* Ph.D. Dissertation. Sami Studies, Umeå University.

Beach, H. (1993). *A year in Lapland: Guest of the reindeer herders.* Seattle and London: University of Washington Press.

Clayhills, H. (1991). *Kvinnohistorisk uppslagsbok.* Stockholm: Rabén and Sjögren.

County Administrative Board (2004). Fakta om Norrbottens län 2004. Länsstyrelsen i Norrbottens Län, rapport 2/2004.

County Administrative Board (2005). *Talking of women and men in Norrbotten County.* Electronic document: http://www.bd.lst.se/publishedObjects/10001015/eng05.pdf. Last accessed: March 2, 2009.

Eikjok, J. (1989). *Kvinner og menn mellom to verdener: samisk kvinne- og mannsidentitet i en dring.* Tromsö: Hovedoppgave i samfunnsvitenskap, Universitetet i Tromsö.

---

⁹ According to statements made by the younger woman interviewed for my study.

Elenius, L. (2001). *Både finsk och svensk*. Ph.D. Dissertation. Umeå University.

FIFS—Fiskeriverkets Författningssamling (1995). *Om kontroll på fiskets område*. 23.

Gustavsson, T. and B. Johnsson (2000). *Kustfiskebefolkningens ekonomi*. Report 2000:1. Göteborg: The National Board of Fishery.

Johansson, Å. (2004). *Vem äger fisken? 10 åsikter om fisket i Norrbottens Län*. Länsstyrelsen i Norrbottens Län, rapportserie 1/2004.

Joks, S. (2001). Boazosámi nissonolbmot: guovddálzis báike-ja siidadoalus muhto vajálduvvon almmolalclcat. Dieðut 2001:5. Guovdageaidnu: Sámi Instituhtta

Jordbruksverket (2003). *Renägare och renskötselföretag: Rennäringens struktur 1994-2001*. Rapport 2003:14. Electronic document: http://www.sametinget.se/2362. Last accessed: March 2, 2009.

Karlsson, Y. (1981). Fiskförädling i Norrbotten. Luleå: Länsstyrelsen i Norrbottens län. *Planering savdelningens rapportserie* 1981:5.

Kråik, M. (2002). 'Sámi women's equal rights—Yesterday and tomorrow,' pp. 156-161 in *Taking Wing: Conference on Gender Equality and Women in the Arctic*. Helsinki: Ministry of Social Affairs and Health.

Kyle, G. (1987). *Handbok i svensk kvinnohistoria*. Stockholm: Carlssons.

Labba, N. and J-L. Jernsletten (2004). 'Sweden,' pp. 131-142 in B. Ulvevadet and K. Klokov, eds. *Family-based reindeer-herding and hunting economies, and the status and management of wild reindeer/caribou populations*. Arctic Council 2002-2004. Tromsö: Center for Sámi Studies.

Lundgren, A., Å. Redin and D. Müller (2004.) *Laxfisketurism och kustfiske vid Kalix älv*. Umeå and Kiruna: Spatial Modelling Centre.

Mósesdóttir, L. (2000). 'Gender Mainstreaming: The Swedish Case,' pp. 32-50 in U. Behning and A.S. Pascual, eds. *Gender Mainstreaming within the European Union*. Brussels: European Trade Union Institute.

Myrvoll, M. (2004). 'Finland,' pp.__ in B. Ulvevadet & K. Klokov, eds. *Family-based reindeer herding and hunting economies, and the status and management of wild reindeer/caribou populations. Arctic Council 2002-2004*. Tromsö: Center for Sámi Studies.

Rova, C. (1999). *Natural Resources and Institutional Performance*. Luleå University of Technology: Licentiate thesis 1999:55.

Rönnbäck, J. (2004). *Politikens genusgränser*. Ph.D. Dissertation. Stockholm University and Atlas Akademi.

Sloan, L., ed. (2004). *Women's Participation in Decisionmaking Processes in Arctic Fisheries Resource Management. Arctic Council 2002-2004*. Norfold: Forlaget Nora.

Swedish Government Office (2002). Jämt och ständigt—Regeringens jämställdhetspolitik med handlingsplan för mandatperioden. *Regeringens skrivelse* 2002/2003: 140. Electronic document: http://www.regeringen.se/content/1/c4/20/69/2c27bade.pdf. Last accessed: March 2, 2009.

Udén, M. (2003). *Nya möjligheter för kvinnor i renskötselföretag*. Final report on project funded by EU Objective 1 Sápmi Norra. Delivered to The Swedish Sámi Parliament 2003-12-22. Electronic document: http://epubl.ltu.se/1402-1536/2004/23/LTU-TR-0423-SE.pdf. Last accessed: March 2, 2009

Udén, M. (2004). 'Swedish case study,' pp. 89-100 in L. Sloan, ed. *Women's Participation in Decisionmaking Processes in Arctic Fisheries Resource Management. Arctic Council 2002-2004*. Norfold: Forlaget Nora.

Ulvevadet, B. and K. Klokov, eds. (2004). *Family-based reindeer herding and hunting economies, and the status and management of wild reindeer/caribou populations. Arctic Council 2002-2004*. Tromsö: Center for Sámi Studies.

CHAPTER TEN

Beyond the Pale: Locating Sea Sami Women Outside the Official Fisheries Discourse in Northern Norway

Elina Helander-Renvall

Abstract: The purpose of this paper is to analyze the power of Sea Sami[1] women in fisheries. My goal is to look through women's eyes and determine why they are located outside the official discourse through which they could empower themselves and claim their rights. This paper shows that Sea Sami have, for generations, been subjected to pressure from the Norwegian government and obliged to deny their ethnic background and culture. The Sea Sami political and cultural awakening began in the 1980s and is ongoing. Power in the Norwegian fisheries belongs to men and Sami people have not been able to gain access to the corridors of power regarding the fishery. Although Sami men are involved in the fishing industry and discourse, Sami women are marginalized as modernization has resulted in an emphasis on large-scale fishing in which only men take part. The Sea Sami have not managed to reach Norwegian authorities and power regimes with their demands regarding fishing and the management of natural resources. In this sense, both men and women seem to be powerless. However, women have various options to influence their situation. For example, Sea Sami women empower themselves through making ethnic clothes, handicrafts and other cultural activities that were once censured by the Norwegian government. Participating in these cultural activities becomes a political act which also affects other members of their community.

Introduction

In this paper, my purpose is to analyze the status of Sea Sami women regarding the fishery through women's eyes. To what extent do women have power? How do they voice their opinions? Why and how have women been excluded from the official discourse? I have studied traditions and customary rules among the inland Sami in the Tana municipality and among the fjord Sami in Norway (Helander 2001, 2002). I have also collected material about Sami traditional knowledge including women's knowledge. This informa-

---

[1] Today, most Sea Sami live along the northern and north-west part of the Norwegian coast but historically, they were found further south to present-day Trøndelag.

tion was primarily gathered in Sirma, Båteng, Horma, Hillagurra and Vestertana in the municipality of Tana in Northern Norway although some of the informants live outside of these villages. Most of the material for this study was collected in 1998-1999 during which I interviewed 80 people. During 1999-2001, I also collected data in Utsjoki, Northern Finland. In 2002 and 2005, the collected material was complemented by additional interviews and discussions mainly with residents of Porsanger and Tana in Northern-Norway and Utsjoki, Northern Finland. In total, I interviewed 127 people between the ages of 20 and 93. The great majority of the informants wished to remain anonymous. In addition to interviews, I also have reviewed local newspapers and other written documents addressing the subsistence activities, rights, identity politics and power issues of the Sami people.

The Sea Sami women's situation in contemporary times has been minimally researched. Often, when research is conducted on Sami women or gender issues, emphasis is placed on reindeer herders rather than fishers. The Sami political field including women's organizations marginalizes the roles and contributions of Sea Sami women and it is difficult to find information in print about current issues facing these women. This dearth of information about women means that they remain outside of a discourse through which they could empower themselves and claim their rights. Why are Sea Sami women situated beyond the 'official' discourse of fisheries in Norway? Many of them are not formally engaged in politics since there is limited space in society for their involvement. As Trinh Minh-ha writes: "Something must be said... The story depends upon every one of us to come into being. It needs us all, needs our remembering, understanding, and creating what we have heard together to keep on coming into being" (1989:119).

Sea Sami people and other Sami groups have a long-established colonial relationship with Norwegian society and their present condition is the outcome of colonial views expressed and implemented by the dominant society (Helander and Kailo 1998). Subject to assimilation by the Norwegian people and state authorities (Paine 1957; Eidheim 1971; Nielsen 1986; Pedersen 1987), the identity of the Sea Sami has been stigmatized over time (Eidheim 1971) and many have not dared to openly express their 'Saminess' (Nielsen 1986). Indeed, many have been ashamed to be Sami (Nilljut 1994). As Kateri Damm (1993a:14) states: "the erasure of another's identity can be a very damaging and oppressive action based on ignorance, racism and racial power relations." Oppression of the Sea Sami has been continuing for such a long time that many have hidden or forgotten their language and cultural ways and values (Eidheim 1971; Nielsen 1986). To have a Sea Sami identity has had negative connotations and many have become passive and uncritical (Ballari 1983). It takes a long time to recover from the impacts of oppression and racism. The wounds are deep and healing takes time (Helander and Kailo 1998). There is a risk that the oppressed adopts the views of the oppressor and the results of recovery are unsure (Ballari 1983; Helander and Kailo 1998; Kuokkanen and Riihijärvi 2005). Since the 1980s, however, an awakening has taken place among many Sea Sami and they are now culturally and politically more active than before (Nielsen 1986; Eythórsson and Mathisen 1998; Helander 2002).

The imposition of colonialism, the pressure to assimilate into Norwegian society and the modernization of the fishing sector adversely affected the position of Sea Sami women and their overall position in society became weakened. Sea Sami women became invisible within Sea Sami families and within the wider political sphere. A Sea Sami woman from Northern Norway, Maddja Nilljut claims: "I don't know what are the women's traditions. What I now see are Norwegian traditions. I have to go back to my childhood[2]" (1994:4). In traditional Sami societies, women's work was highly valued. They played a powerful role as mediators of traditions (Helander 1985). Sea Sami women had their own sphere of activity and knowledge. The former resource-based economy—fishing, farming, hunting and gathering—strengthened women's position. Many of the tasks were shared and men and women had diverse opportunities to use their skills and knowledge in their daily lives. By reviewing her childhood and upbringing, Nilljut recognizes what she once was and what she still carries within her to be 'reconstructed' to fit into her contemporary life (*cf.* Anderson 2003:9). This kind of conscious reflection is one of the struggles that women have to go through if they want to uncover the factors that keep them oppressed (McCarl Nilsen 1990).

Research Made for Women

Foucault (1980) focuses on and defends 'locally situated' research activities. This shift away from 'universal' knowledge holds doors open to the relativity of truth claims. The meta-narrative is displaced by the local and contextual. I myself belong to the 'inland' reindeer and fishing society, as I own reindeer in a local reindeer village (Kaldoaivi) in Northern Finland as well as through my family background. My own living conditions and lived experiences are principally, and in many ways, of the same type as those people interviewed and researched along the Norwegian coast.

My Sami background and engagement inspires me to think deeply about the Sea Sami situation in reference to work conducted by Sandra Harding (1987) on feminist methods. According to Harding, "the beliefs and behaviours of the researcher are part of the empirical evidence" (Harding 1987:7). Thus, in feminist science, the researcher's location is critical. Other aspects discussed by Harding include attention being paid to women's experiences and that research on women is made for women meaning that the purpose of feminist research is not "separable from the origins of research problems" (Harding 1987:8).

I also conduct research from the 'inside out,' trying to look outward from the Sami cultural core (Cohen 1985). The Sea Sami and Mountain Sami cultures are different when compared; however, these dissimilarities are limited. For instance, the language that Sea Sami speak is influenced by Northern Sami (my mother tongue). Both groups, the Sea and Mountain Sami, are in a minority position vis-à-vis the majority society. Also, the Sea Sami co-operate traditionally through the so-called *verdde*-institution with the Mountain Sami. '*Verdde*' means a guest-friend or a partner to exchange services and goods with, to live periodically with and to have other kinds of close connections. I have lived for some months in a Sea Sami community (Vestertana) in the Tana fjord to take part in local life and conduct my research there. Women from other areas are also included in my research activities. As I write this article, I am reminded by Damm (1993b:113) that writing is "a means of recognizing and acknowledging the strength, the beauty, the value and the contributions of Native peoples… In words, the healing continues." I acknowledge the strength of Sea Sami women and try to learn how they express their strength from the vantage point of their position as a colonized, assimilated, oppressed, non-visible and almost forgotten group. Empathy is one means of gaining access to other people's

---

[2] Helander-Renvall, translation from Norwegian.

knowledge (Belenky *et. al.* 1997). This mechanism means that you see, feel and connect with the other (Belenky *et. al.* 1997:122). On the other hand, however, I am aware that through analysis and descriptions of the researched material, I am making an interpretation concerning the Sea Sami reality.

The Sea Sami in Vestertana

Vestertana, a Sea Sami community, is located in the north-western part of the municipality of Tana in the province of Finnmark, North Norway. Vestertana is situated on Tana Fjord which, running in a south to north direction, flows into the Barents Sea. The majority of the inhabitants are Sea Sami whose mother tongue is Sami. Sami is also the language of communication between the people of the area. The population of Vestertana declined from 40 in 1985 to 23 in 2000 (Helander 2001:460). The majority of the population was over 50 years of age, when the investigation was made in 1998-1999.

It is common to regard as Sea Sami those who live along the north Norwegian coastal area. Fjord Sami are those who live in the fjords along the coast. As identified by Eidheim (1971) and Nilsen (1999), the Sea Sami identity has been stigmatized and ethnic boundaries and ethnic self-identification, as described by Barth (1969), have been blurred. However, since the 1980s, many Sea Sami have been transforming their 'Norwegian' identity into a positively affirmed 'Sea Sami' identity (Eythórsson 1991; Helander 2001; Nilsen 2003).

At the end of the seventeenth century, the Sea Sami kept reindeer, cattle and sheep. During the eighteenth century, the Sea Sami were semi-nomadic with a number of dwelling areas. They made seasonal use of coastal resources and the inner parts of the fjord area as well as outlying areas. Permanent stationary settlements at the old summer dwelling places began to develop in the Sea Sami part of Tana Fjord during the eighteenth century. Sami culture has always been characterized by hunting and fishing and the Sea Sami, in particular, are involved in fishing, hunting, trapping and other subsistence activities.

The subsistence lifestyle of the Sea Sami is based on having priority access to natural resources in their settlement area. It is also important that a certain number and variety of species are available. Sami people have always combined different resources as the basis for their economic activities. Research in this area shows that historically, the Sami people have had exclusive use of the resources of this area. It is obvious that the Sami people in Tana municipality, including Vestertana, believe that the natural resources of their area belong to them (Helander 2001).

In his description of the coastal and fjordal areas, Ragnar Nilsen (1998) emphasizes the concept of fjord rationality which manifests itself as an adjustment to changing circumstances and diminishing resources. According to Nilsen, it is rational to accept a social situation which maintains traditional subsistence activities and expands on them but which only provides a low income. Fjord rationality rejects the pursuit of big catches and competition with other fishermen as a means of raising one's own living standard. This type of rationality would be complemented and extended by women's point of view, but unfortunately, Norwegian society at large does not embrace this perspective.

In his work, Nilsen (1998) writes about catch-limiting strategies in the fjord and on the coast. Fjord fishers use selective gear such as nets, the '*juksa*' and the fishing line, picking out only mature fish which guarantees that fish live long enough to maintain reproduction. Selective fishing, conducted to a restricted degree in the inner parts of the fjord, represents an activity which is far-sighted, economical, and conducive to sustainability. Despite these practices, Sami fishers have encountered challenges in obtaining official support from Norwegian society for their way of life.

Many of the informants in this study are aware of the necessity to maintain modest living standards especially considering the diminishing resources in the outlying areas and in the fjord. However, it is important to note that these circumstances were imposed by the regulations of Norwegian society as asserted by Nilsen (1999). The fjordal fishermen in the north—at least until recently—"have been forced into an economic situation with lower incomes and more difficult conditions in their jobs and lives than the Norwegian population in the same peripheral areas" (Nilsen 1999:93). The informants also emphasize the importance of other values in life besides mass consumption and business sense. Rationality, therefore, is a way to 'stay in the game' (Freeman 1988:5), and manage in a system based on Sami subsistence activities but influenced by outside circumstances. Vestertana, as a marginalized fjordal Sami community, has a clear cultural boundary that both separates and distinguishes it from Norwegian society. The practical implications of this are continually manifested to the fjord population. Fairness is one of the issues that arises frequently. There are conflicts between the two cultures, Norwegian and Sami, with their different perspectives on many issues, including fisheries management.

Sea Sami Sources of Livelihood and Resource Use

The Barents Sea is one of the world's richest fisheries and Tana Fjord has been recognized for the abundance and variety of its fish stocks. This has enabled the Sami living in the area to maintain their traditional sources of livelihood and culture. Fjord fishing has been combined with other sources of income (Helander 2002). In the twentieth century, Sea Sami culture developed around fishing, farming and supplementary occupations in outlying areas. Additionally, a number of Sea Sami have been employed seasonally or permanently outside the region in the fishing industry, road building, construction or administration. Salaried employment and the introduction of new sources of livelihood have, for many Sami, represented a means through which they can maintain dual occupations as well as remain in the region.

This contrasts with Norwegian society which is dominated by sectorization and the ideal of full-time employment. Research carried out by Gjerde and Mosli (1985) demonstrates that in municipalities investigated in Troms and Finnmark, there were differences between Sami and Norwegians regarding their occupations. More Norwegians are engaged in capital-intensive primary occupations compared to people with a Sami background. Local Sami communities also have fewer jobs available in the public and private service sectors.

Women's and Men's Tasks

Traditionally, Sea Sami women in Vestertana have been responsible for most tasks in agriculture, including the care of animals, milking, cleaning, barn work and gathering feed for the winter. Men took part in hay gathering until they had to leave to fish in the fjords. Men assumed more significant roles as the mechanization of agriculture increased in the north. Small-scale farming has been based on local resources. For instance, lichens, heather, fish heads and seaweed were gathered by women and used as food for cows.

When the snow came, women and men transported the gathered feed from the forests and hills to farms. The farming, childcare, production of the skins and 'soft' handicrafts were part of women's activities. In recent decades, women still produced wool at home while men built boats, vessels, and sledges. They also made fishing nets and other tools for fishing. Men cut down trees and helped women to fetch seaweed and algae. Both women and men gathered berries (Thomassen 1999; Helander 2001).

Traditionally, both women and men took part in fishing activities. Today, commercial fishing is conducted by men while small-scale fishing involves, in some cases, both men and women. Men take care of the nets and other fishing tools. Both women and men cook food although women have the responsibility for food and the preparation of meals. Many tasks are interchangeable and can be performed by both women and men. Some men's work is physically challenging and may be difficult for women to carry out. Generally speaking, in the household and in the community, men and women hold equal status although men are more visible. Women can sell what they have produced and keep the income for themselves. If they have a salary from work outside the home, they can use part of the money for their own personal use. Through gossiping and influencing their men in the household, they may even have a certain impact on the status men hold in the local society and within wider circles.

Jorunn Eikjok (1990) has pointed out that shortly after World War II, the traditional complementary roles of the Sami women and men changed (*cf.* Kailo 2001) and the responsibility for farming and fishing was transferred to men. Capitalist development within Norwegian society stimulated the creation of separate fields for private and public life. The rapid modernization of the Norwegian society impacted negatively on women's opportunities to influence various issues. Helander claims that there has been equality between Sami women and men, but "modernization, overprotective national policies and non-Sami legislation have meant that Sami women are now oppressed within society" (Helander and Kailo 1998:183). As Finnish feminist Kaarina Kailo states, "the world knows of no 'post-colonial' state that would willingly have granted women and men equal access to the rights and resources of the nation-state" (2001:184).

Men are in the Highest Positions

Asta Balto's research (1986, 1990) indicates that the upbringing of Sami children is gender-specific: boys and girls are brought up differently. Boys are educated to be tough and girls are nurturing caretakers. Nancy Vibeke Olsen (1997) claims that young Sami women are socialized in the household while young Sami men are socialized in nature. From their older male relatives, the boys receive a practical understanding regarding the natural environment. Knowledge of nature becomes a part of a man's life (Stordahl 1994). Sea Sami men are more involved in practical fishing activities and acquire a deeper understanding of and training in skills and problems within the fishing industry. Sami women's values, activities and knowledge are less visible (Hirvonen 1996; Eikjok and Birkeland 2004).

Generally speaking, men's tasks in fishing are still valued more highly than the work of women (Eikjok 1985, 2000). I have noticed that much time is spent discussing men's activities and the situation of children, but women mainly discuss their own issues only with other women. Women's status is in many cases gained through working part or full-time outside the home. This maintains the compatible roles that existed in more traditional societies. Modern life means that young women leave their home communities in order to educate themselves and find a job elsewhere (Olsen 1997). Sea Sami women may enter political life but men still dominate this public sphere (Stordahl 2002).

As indicated, men have greater formal authority in the Sami political sphere and hold positions of power in the Norwegian fishing industry (Aasjord 2002). Few women discuss politics at home when their husbands are there. Fishing remains the single most important activity determining Sea Sami identity (Helander 2002) and particularly, the male identity. All people cannot identify with the fishing industry. As a result, some find it difficult to maintain a distinct Sami identity based on fishing, especially considering the low status accorded Sea Sami culture by mainstream Norwegian society (Eidheim 1971; Nilsen 2003). However, fishing provides an avenue for men to status, money and power. According to Joan Acker (1991:167), "men are almost always in the highest positions of organizational power" (*see also* Stordahl 2002). It seems that the sector where Sea Sami women have the most power is the private sector (*see* Nilljut 1994; Erlandsen 2004). "A fisherman's wife is expected to be independent, function as the man's 'ground personnel' and look after the children and the home" (Nilsen 2004:113).

Routine, daily activities of individuals produce and reproduce the structural features of social systems (Giddens 1984). Men sustain local society by working within the highly valued fishing sector and by taking part in the public, political work affiliated with this field. The fishing industry in Norway is a male-dominated sector meaning that formal political power and resource management are in men's hands (Aasjord 2002). Therefore, their economic and political work is visible even if income from both men and women is necessary for the maintenance of Sami households.

Norwegian Fishing Sector Controlled by Norwegian Men

Fish stocks have diminished over the years, but there are still enough to catch in the Tana Fjord. The choice and size of the catch have always depended on local and national demand. The development of the Norwegian fishing industry with regard to economic growth, viability and national regulatory measures threaten the rights of the Sea Sami as an indigenous group and an ethnic minority. Power in the fjord area is exercised by power regimes outside the local communities as well as by Norwegian rules and regulations (Helander 2001, 2002; Aasjord 2002). Due to their post-colonial position *vis-à-vis* Norwegian society (Helander and Kailo 1998), Sami knowledge is not considered valid when making political decisions or when managing fisheries (Eythórsson 2003; Nilsen 2003). The Norwegian fishing sector is controlled by Norwegian men (Sagdahl 1998; Björklund 1999; Aasjord 2002).

During my research (Helander 2001), it became clear that the Sami have a specific environment-based, ecological way of looking at things. As well, the Sami want to participate in the management of natural resources in their own areas (Helander 2001). They fear that Norwegian society, with its system of regulations and its different fishing culture (capital-intensive, large-scale) will deplete natural resources for future generations (Helander 2002). The official Norwegian fisheries policy "aims at safeguarding the profitable development of the fishing industry. Sustainable management of the resources is a precondition for the achievement of that goal" (St.meld. nr. 51, 1997-98). The structural changes undergone by the Norwegian fishing industry since the 1980s have been aimed at rationalizing the industry in view of the existing over-capacity—there were far too many boats, firms, and processing plants. During 1990-1997, the number of small boats under 10 metres decreased by 40 per cent while, during the same period, there was a considerable increase in the number of larger boats. In Sami communities, boats under 10 metres are still commonly used. In the municipality of Tana, the number of boats between 5 and 9.9 metres has decreased since the early 1980s, whereas the number of boats under 5 metres has remained the same. As a rule, fishing with small boats takes place close to home.

More About Fishing and Fishing Policy

Traditionally, fishing in the Barents Sea has been carried out in coastal waters. Fishing in the Vestertana area and other fjords of Northern Norway have been characterized by the use of small boats and modest investments in fishing gear. Various species of fish are harvested all year-round depending on their availability. Until the 1950s, there were many species of fish in the fjord including halibut, salmon, cod, sei, herring, haddock and sea trout (Helander 2001). Over time, a fisher acquired expert knowledge about fjord fish stocks with regard to the location of the fish and the yearly cycles of their movements, as well as the currents and wind conditions in the fjord.

Changes in the fishing industry and the re-building of Finnmark after World War II caused the Sea Sami to become more dependent on national market trends. The fjordal population was not ready for changes that were taking place such as the introduction of large boats, modern gear, loans and investments opportunities as well as contacts with influential traders, bankers and other people representing banking and administration. The material development of Northern Norway became more established and Sea Sami culture was considered to be outdated and reactionary (Halonen and Turi 1983; Nielsen 1986; Nilsen 2003.) As a result, the Sea Sami had very low status with no Sami organizations of their own. Additionally, the authorities did not necessarily regard the Sea Sami as a specific ethnic group and their means of livelihood were excluded from official support systems. The Sea Sami as a group began to feel that they were left out of issues and affairs addressing other Sami peoples and Norwegian society (Nielsen 1986; Helander 2001; Nilsen 2003). In many regions, it seemed that traditional Sea Sami culture was on the verge of collapse.

Steinar Pedersen's (1994, 1997) description of local regulations concerning natural resources in Finnmark's coastal and fjordal regions indicates that residents have always been careful about the protection and regulation of resources. As the fishing industry became more mechanized, a conflict between modern versus traditional practices began to ensue. Over time, the antagonism between the local population and non-resident fishers became accentuated owing to the nature of the laws as well as the general development of society which promoted the interests of Norwegians rather than Sami (Nilsen 1998, 1999, 2003).

Einar Eythórsson (1993) outlines the management of natural resources in Finnmark and the effect of Norwegian regulations on the situation of fishers. He describes many cases in which local people use traditional knowledge and generations of experience to understand and protect their resources in the fjord. For example, local people support conservation measures and oppose the use of more advanced technological gear. In addition, they are concerned about boats from other areas undermining the subsistence of the fishers of the fjord area.

The quota regulations have had a negative impact on the Sami population in the fjord. Owing to seal invasions of the recent past, the Sami have had to live through a number of difficult years as far as fishing is concerned. Many fishers who had been registered as part-time fishers ran the risk of being excluded from the system, and many of the fjord fishers in the northernmost counties lost their fishing licences (Helander 2001). People in the fjord area including Vestertana, have always had a critical attitude toward Norwegian laws as many regulations have a negative impact on their chances of remaining in their home areas and earning their living there. From the Sami point of view, these laws and regulations represent a distant power infrastructure which causes harm. Sami residents in this region have learned to be suspicious of the centralized management of fisheries and administration of environmental issues, the complicated and ever-growing Norwegian bureaucracy, the Norwegian fisheries laws, the large fishing fleets from distant ports, the dominant concepts about 'acceptable' knowledge systems, the fluctuation of prices and markets and the negotiation of fishing quotas at the international level.

As indicated, the introduction of certain regulations in 1990 ended fishing in some traditional Sea Sami areas in the fjord. The Sami Parliament finds this paradoxical at a time when the country has undertaken a commitment concerning the rights of indigenous peoples both nationally and internationally. At the same time, however, Norway introduced a system in which the undermining of the rights of an indigenous people is in fact legal (Samisk fiskeriutvalg 1997).

The establishment of the Sami Parliament and the work it has done have heightened the public profile of the Sea Sami. The Sami Parliament raised the issue of challenges to Sea Sami fishing with the Norwegian Fishery Department. According to Smith (1990), the special stipulations concerning fishing should be linked to Sami areas of settlement and the evaluation of fishing regulations should be done from a comprehensive perspective, taking into consideration the effects of the different dimensions of Sami policy. This means an overall evaluation of the specific measures taken to help the Sami to maintain and develop their culture including traditional subsistence activities. Such an evaluation will also provide a way to assess the extent of the pressure to which Sami culture is exposed.

The Sami Parliament has emphasized that fishing as a source of livelihood should contribute to "the safeguarding and development of Sami culture and Sami social life overall" (Samisk fiskeriutvalg 1997) and emphasized the importance of fishing for Sea Sami culture: "It is necessary that the Sea Sami should be able to practise fishing in a manner that takes into consideration their subsistence base as well as their need for adjustment, especially with regard to the opportunity to combine different sources of livelihood" (Samisk fiskeriutvalg 1997:134). The Sami Parliament has proposed the introduction of a management zone for the coastal and fjordal systems, based on considerations of fishery policies and supporting multi-level management of resources (Sametinget 1998). It also requested changes in legal regulations concerning the ecological harvesting of maritime resources, priority of access to resources granted to traditional fishing on the coast and in the fjord area, the regional management of resources and that those with multiple sources of income should be eligible to receive fishing licences.

The work done by the Sami Parliament and other Sami political actors and the research conducted by Carsten Smith and others has had positive effects on the Sea Sami. Nilsen reports that, "small-boat fishing in the North-Troms and Finnmark regions became somewhat more free than in the rest of the country" (2003:180). But it must be emphasized that the Norwegian fishing policy still proceeds as before and the coastal Sami have not managed to get access to the Norwegian power structure and change their own situation regarding the fishing business (Nilsen 2003). The Sea Sami question has also not been resolved through the newly created *Finnmark Act* (*Finnmarksloven*) which deals with land issues in Finnmark.

Rethinking Politics

Women do not have formalized access to the public discourse and domain where the Sea Sami identity is defined and decisions are made regarding fishery and other issues although women have some opportunities to influence their men at home and in different meetings. Women's status is still measured by their skill in doing housework, handicrafts and childcare and their abilities in the modern world. Those who have skills in handicrafts receive particular acknowledgement. Avtar Brah (1993:25) claims that as women, it is important "to be attentive to how we are positioned in and through these relations of power amongst ourselves and *vis-à-vis* men." Are women deliberately kept outside the political sphere? Are they detached from politics? Or do we have to think about politics in a new way to understand women's position in modern society? Below, I will try to answer these questions.

As shown by Gerrard and Balsvik (1999), women play a significant role in maintaining coastal societies but their work remains unrecognized. This may lead to a situation in which women start doubting their own abilities. A Sea Sami woman, Maddja Nilljut (1994:5), explains: "Earlier, women had faith in their own skills and knowledge. They believed that they really had skills regarding their work. Even if they were not allowed to praise themselves, they had positive thoughts about their own skills. In old times, you did not need to think all the time, as you do these days, that you should have done something in a different way" (1994:5).[3] A relatively young Sea Sami woman from Billedfjord whom I call Kerstin, says: "We have not dared to believe that we are something. We have not had good self-confidence.[4]"

Research conducted in North America by Belenky *et al.* (1997) contends that some women have difficulties in expressing the power of their opinions. French feminist Marguerite Duras claims that women have been 'kept in the dark' for centuries and that women are only now emerging (*see* Belenky *et. al* 1997). Women have had to find their own words to explain their experiences in public fora. A Sea Sami fisheries politician, Berit Ranveig Nilsen (*see* Erlandsen 2004:114), states that:

> If I make a controversial political decision, it seems it is easier to attack me as a female politician than if I had been a male politician. I rarely experience support for the work I do or for the decisions I make in official forums or through the media... However, worst of all is the silence. After all, if people disagree with you, it should be possible to discuss things, but when they simply keep quiet, then the door is literally closed.

It is no wonder that Berit Ranveig Nilsen wears and uses Sami style clothing to gain more attention for her words (Erlandsen 2004:114). According to Foucault (1980:98), as

---

[3] Helander-Renvall, translation from Norwegian.
[4] Interview 2005. 'Kerstin' is a pseudonym.

subjects created by power, women are also subjected to power. Frye (1983:105) states: "The powerful normally determine what is said and sayable." Nilsen (2004) claims that Sami women make different demands than men, for instance regarding the workplace, schools, and stores.

When discussing the situation of women, we have to consider what is meant by 'power,' and how it is expressed. Are we talking about domination over others or empowerment issues? Many feminists have been inspired by the texts of Foucault. He claims that power comes from everywhere and exists in all social interactions. Furthermore, Foucault (1980, 1983) believes that power is local, dynamic and connected to action and resistance. He also stresses that cultural practices and knowledge systems are linked to power. In *Women and Power in Native America*, Laura F. Klein and Lillian A. Ackerman assert that the concept of 'power' is defined as "an active reality that is being created and redefined through individual life stages and through societal history" (1995:12). Thus, power also exists outside the official political realm.

An Italian feminist, Teresa de Lauretis (1987) writes about groups or subjects inside marginal social fields and defines power as micro-political practice and resistance that can impact local affairs. When dealing with politics, she emphasizes the personal identity and experience of women. According to her (1986; 1987), for women, identity is political. It is strategy of emancipation. Other feminists including Hartsock (1983), Kailo (2001) and Anderson (2003) discuss engagement, interconnection and resistance as they understand power to be oppressive and related to patriarchy. Research carried out in Indigenous communities indicates that power that is exercised over an individual or group of people exists in many forms and is often reinforced through cultural, educational and social practices. For Hartsock (1983), it is imperative to redefine power and connect it to women's experiences. Finnish scholar Kari Palonen identifies politics as an aspect of human action (*see* Harle 1996). According to Palonen, "there is no politics without an adversary and resistance" (Harle 1996). The relationship between oppressed and oppressor as an action-against relation is political. The political can be included in any social relation, situation or behaviour. "It does not matter if the actors perceive the political or agree upon its existence" (Harle 1996:412).

Women are excluded from power in fisheries. Exclusion is a part of the political action-against 'other' (*ibid*.:413). A Sea Sami woman from Porsanger, whom I call 'Ingunn[5]' told me, that one reason for the passivity of Sea Sami women in the official political field is 'Norwegianization,' meaning that the former assimilatory policy of Norwegian society still has its effects. In the 1980s and 1990s in the area in which she lives, the Sea Sami started to organize. The question is, are Sea Sami women political? If we accept that power can be dynamic, limited in time and space but existing everywhere, functioning in local and cultural practices, and connected to resistance and various re-definitions of reality (Castells 1997:8), can we analyze the Sea Sami women's political life in a new way?

Collective action frames help us to understand why people take political action. Collective action frames "empower people by defining them as potential agents of their own history" (Gamson 1995:90). Gamson (1995) defines three aspects of collective action. One component is agency. Agency refers to the consciousness that something in one's

---

[5] Interview 2005. 'Ingunn' is a pseudonym.

situation can be changed by oneself. Society has many ways of discouraging people from taking action and making changes. Basic survival and mere daily existence can be so overwhelming that there is no time or opportunity to take part in social or political activities. A Sea Sami woman, 'Kerstin,' told me that she has family, work and so many leisure activities that she does not have time to be involved in politics. In this context, we also have to keep in mind that routine activities reinforce the patterned life of people in a society: "only rarely do people have an opportunity to engage in activity that challenges or tries to change some aspects of this pattern" (Gamson 1995:95). In hierarchical societies, people may feel that they themselves are without power. Feelings of injustice and a 'we/them' dichotomy are also included among the components of collective action frames (Gamson 1995).

Women's Avenues of Power

If they stay in the villages, Sea Sami women can exercise power through their men or through men in their local society. Women also incorporate themselves into the Norwegian power system through gaining higher education, participating in Norwegian politics, networking, etc. However, there are few Sea Sami women in politics and involvement in Sami political organizations does not offer much access to power (Stordahl 2002). Finally, women have difficulties in reaching top positions inside the Norwegian fishing sector (Aasjord 2002). This situation is improving according to Elisabeth Angell (2004).

However, women have other avenues and means regarding power. One of these avenues is the empowerment of personal womanhood and identity. The second is local engagement in cultural and socio- economic issues including the empowerment of men. These two can be combined. The traditions and local knowledge of women are of great importance in this context. Sea Sami women use their traditions, knowledge, creativity and strength to transform their thoughts and experiences into being: old ways into modern identity; old skills into new skills; old social relations into new ones and old cultural styles into new styles. Capitalism, imperialistic domination and colonialism result in cultural oppression, but "with a strong indigenous cultural life, foreign domination cannot be sure of its perpetuation" (Cabral 1994:53). For example, Sea Sami women in Porsanger, Northern Norway use ethnicity and difference in a proactive manner against power and oppression (*cf.* Brah 1993; Erlandsen 2004). Cultural identity, in this sense, is emancipatory process, and it is a matter of becoming and of transformation. In the words of Starhawk (1982:9), we can call this process a 'weaving dance' which entwines together those aspects of life that are of relevance for women.

Traditions play a key role in women reclaiming and redefining their identity and power. As Emma LaRocque (1996:14) asserts, indigenous women "are challenged to change, create, and embrace 'traditions' consistent with contemporary and international human rights standards." Here, I will provide an example from Porsanger as told to me by a Sea Sami woman Ingunn. The women in her area became culturally and politically conscious in the 1980s and 1990s. For several years, women researched their traditional clothing, called *'kofte'* in Norwegian and *'gákti'* in Sami. Women referred back to the eighteenth and nineteenth centuries to learn about local Sea Sami clothing and interviewed their parents and grandparents, did reading, visited museums and conducted other research to understand the elements and colours of the original Sea Sami *kofte*. In addition, they studied the Sami sacred symbols from rock paintings in the Alta area to include some symbols in their Sami *kofte* and belt.

Although it took time for women to learn these things, when they had made their ethnic clothing for themselves, they started using them, producing them and selling them. In 2005, fifteen women were actively involved in the production of kofte and are teaching others to make Sami handicrafts. This undertaking took tremendous courage as Sea Sami had to hide their Sami identity for so many years. Producing, owning and using Sami clothing is a political act for these women. The skills and activities of the local Sea Sami women created new opportunities and contexts for political and economic action. Now, many young people who have migrated from Porsanger make contact with them to obtain their own Sami kofte. The handicraft activities of this Sea Sami group are also mass-marketed. The Sea Sami women translated their ideas from the darkness of private experience into shared community-based activity. In this sense, the concept of power can be understood as capacity, re/empowerment, sharing, transformation and change.

Conclusion

Due to the colonial relationship between Norway and the Sami people and the fact that the Sea Sami have been subjected to powerful assimilatory measures, the Sea Sami do not wield power inside the Norwegian fisheries.The situation of these people has improved somewhat over the last century in part due to the establishment of the Sami Parliament and Norway has, to some degree, altered its policies toward the Sea Sami. It is important to note that the Sea Sami have not yet attained their goals concerning fishing. But their issues will be investigated and discussed through years to come. There is also an increasing recognition and awareness that Sea Sami identity and culture have been stigmatized and that Sami people are now awakening from the historical burdens placed upon them.

While Sea Sami men are involved and visible in fishing activities and politics, women are not yet part of the official fishing discourse and activities. However, women use other strategies in order to reach their goals and to empower themselves and local communities. Overall, Sea Sami women's contributions to the fisheries are not validated by Sami society. If we rethink fisheries and the political sphere in which it occurs as an aspect of human action and as cultural practice, it is clear that Sea Sami women are engaged in fishery politics in their own way. It is critical that women's activities are made visible and part of the formal structures regarding fisheries. Women need a space in politics where they can think and act as women.

References

Aasjord, B. (2003). 'Where have all the fishes gone?,' pp. 36-44 in *Taking Wing: Conference on Gender Equality and Women in the Arctic,* Helsinki: Ministry of Social Affairs and Health.

Acker, J. (1991). 'Hierarchies, Jobs, Bodies: A Theory of Gendered Organizations,' pp. 162-179 in J. Lorber and S. A. Farrell, eds. *The Social Construction of Gender.* London and New Delhi: Sage Publications.

Anderson, K. (2003). *A Recognition of Being. Reconstructing Native Womanhood.* Toronto: Sumach Press.

Angell, E. (2004). 'Gender and ethnicity in the Sámi Parliament's fisheries policy,' pp 102-111 in L. Sloan, ed. *Women's Participation in Decisionmaking Processes in Arctic Fisheries Resource Management. Arctic Council 2002-2004.* Norfold: Forlaget Nora.

Ballari, K. (1983). 'Kulturelle tiltak for vedlikeholdelse av lokalsamfunn. Kultursammenhenger i na eringsproblematikken,' pp. 33-36 in L. Halonen and E. Turi., eds. *De samiske kyst- og fjordområder. Rapport fra seminar 23-25 April 1982 i Lakselv.* Dieðut 1:33-36. Guovdageaidnu: Sámi insttuhtta.

Balto, A. (1986). *Samisk barneoppdragelse og kjönnssosialisering. En studie i foreldrenes og andre voksnes forståelsesformer.* Hovedoppgave i sosialpedagogikk. PFI. Universitetet i Oslo.

Balto, A. (1990). Samiske barnelek i 1980-årene. Eksemplar fra en barnehage. *Tradisjon. Tidsskrift for folkeminnevitenskap* 20: 89-96.

Barth, F., ed. (1969). 'Introduction,' pp. 9-38 in *Ethnic Groups and Boundaries. The Social Organization of Culture Difference.* Bergen; Oslo: Universitetsforlaget.

Belenky, M.F. et al. (1997). *Women's Ways of Knowing: The Development of Self, Voice, and Mind.* New York: Basic Books.

Björklund, I., ed. (1999). *Norsk ressursforvaltning og samiske rettighetsforhold. Om statlig styring, allmenningens tragedie og lokale sedvaner i Sápmi.* Oslo: Ad Notam Gyldendal.

Brah, A. (1993). Re-Framing Europe: En-gendered Racisms. Ethnicities and Nationalisms in Contemporary Western Europe. *Feminist Review* 45: 9-28.

Cabral, A. (1994). 'National Liberation and Culture,' pp. 53-65 in P. Williams and L. Chrisman, eds. *Colonial Discourse and Post-Colonial Theory. A Reader.* New York: Columbia University Press.

Castells, M. (1997). *The Power of Identity.* Oxford: Blackwell Publishers.

Cohen, A.P. (1985). *The Symbolic Construction of Community.* Chichester and New York: Ellis Horwood Limited & Tavistock Publications.

Damm, K. (1993a). 'Says Who? Colonialism, Identity and Defining Indigenous Literature,' pp. 10-26 in J. Armstrong, ed. *Looking at the Words of our People: First Nations Analysis of Literature.* Penticton: Theytus Books.

Damm, K. (1993b). 'Dispelling and Telling. Speaking Native Realities in Maria Campbell's *Halfbreed* and Beatrice Culleton's *In Search of April Raintree,'* pp. 93-114 in J. Armstrong, ed. *Looking at the Words of our People: First Nations Analysis of Literature.* Penticton: Theytus Books.

deLauretis, T., ed. (1986). *Feminist Studies/Critical Studies.* Bloomington: Indiana University Press.

deLauretis, T. (1987). *Technologies of Gender. Essays on Theory, Film and Fiction.* Bloomington and Indianapolis: Indianapolis University Press.

Eidheim, H. (1971). *Aspects of the Lappish Minority Situation.* Oslo; Bergen: Universitetsforlaget.

Eikjok, J. (1985). 'Mytat sámi nissonolbmo birra,' pp. 25-36 in *Sámi nissonolbmuid dili birra. Den samiske kvinnens situasjon. Seminar May 31-June 1,1985.* Ohcejohka:Sámiráddi.

Eikjok, J. (1990). *Indigenous Women's Situation: Similarities and Differences.* Paper presented at Common Struggles or the Future. International Conference and VI General Assembly of the World Council of Indigenous Peoples. Tromsö, Norway, August 8-12, 1990.

Eikjok, J. (2000). Indigenous Women in the North: The Struggle for Rights and Feminism. *International Working Group for Indigenous Affairs* 3: 38–41.

Eikjok, J. and I. Birkeland (2004). 'Natur, kjönn og kultur: om behovet for dokumentasjon av samisk naturforståelse I et kjönnsperspektiv,' pp. 58-71 in L. M. Andreassen, ed. *Samiske landsskapsstudier.* Rapport fra et arbeidsseminar. Diedut 5. Guovdageaidnu: Sámi insttuhtta.

Erlandsen, M.M. (2004). 'A Sámi, a woman and a fisheries politician,' *p*p. 112-116 in L. Sloan, ed. *Women's Participation in Decision-making Processes in Arctic Fisheries Resource Management. Arctic Council 2002-2004.* Norfold: Forlaget Nora.

Eythórsson, E. (1991). *Ressurser, livsform og local kunnskap. Studie av en fjordbygd i Finnmark.* Hovedfagsoppgave i samfunnsvitenskap. Tromsö: Institutt for samfunnsvitenskap, Universitetet i Tromsö.

Eythórsson, E. (1993). *Fjordfolket, fisken og forvaltningen.* FDH-rapport 1993:12. Alta: Finnmark Distriktshøyskole.

Eythórsson, E. (2003). 'The Coastal Sami: a 'Pariah Caste' of the Norwegian Fisheries? A Reflection on Ethnicity and Power in Norwegian Resource Management,' pp. 149-162 in S. Jentoft, ed. *Indigenous Peoples. Resource Management and Global Rights.* Delft: Eburon.

Eythórsson, E. and S.R. Mathiesen. (1998). 'Ressursforvaltning og lokal kunskap i kyst samiske områder,' pp. 40-55 in B. K. Sagdahl, ed. *Fjordressurser og Reguleringspolitikk. En utfordring for kystkommuner?* Oslo: Kommuneforlaget A/S.

Foucault, M. (1977). *Discipline and Punish: The Birth of the Prison.* New York: Vintage.

Foucault, M. (1980). *Power/Knowledge: Selected Interviews and Other Writings 1972-1977.* New York: Pantheon.

Foucault, M. (1983). 'Afterword: The Subject and Power,' in H. Dreyfus and P. Rabinov, eds. *Michel Foucault: Beyond Structuralism and Hermeneutics.* Chicago: University of Chicago Press.

Freeman, M.M.R. (1988). *The Significance of Animals in the Life of Northern Foraging Peoples and its Relevance Today.* Edmonton: Department of Anthropology and Boreal Institute for Northern Studies, University of Alberta.

Frye, M. (1983). *The Politics of Reality: Essays in Feminist Theory.* Freedom, Trumansberg, NY: The Crossing Press.

Gamson, W.A. (1995). 'Constructing Social Protest,' pp. 85-106 in H. Johnston and B. Klandermans, eds. *Social Movements and Culture.* Minneapolis: University of Minnesota Press.

Gerrard, S. and R. Balsvik (1999). *Globale kyster. Liv i endring—kjönn i spenning.* Kvinnforsks skriftserier 1. Tromsø: Kvinnforsk.

Giddens, A. (1984). *The Constitution of Society. Outline of the Theory of Structuration.* Cambridge: Polity Press.

Gjerde, A. and J.H. Mosli. (1985). *Samiske Naeringers Plass i Samfunnsplanlegginga.* Dieðut 5/85. Guovdageaidnu: Sámi Instituhtta.

Halonen, L. and E. Turi, eds. (1983). *De samiske kyst- og fjordområder. Rapport fra seminar 23-25. April 1982 i Lakselv.* Dieðut 1. Guovdageaidnu: Sámi insttuhtta.

Sametinget (1998). *Handlingsplanen for samiske kyst- og fjordområder 1997-2001.* Karasjok: Sametinget.

Harding, S., ed. (1987). 'Introduction. Is There a Feminist Theory?,' pp. 1-14 in S. Harding, ed. *Feminism & Methodology. Social Science Issues.* Bloomington and Indianapolis: Indiana University Press.

Harle, V. (1996). Otherness, Identity, and Politics: Towards a Framework of Analysis. In *The European Legacy.* Vol. 1(2): 409-414.

Hartsock, N. (1983). *Money, Sex and Power: Toward a Feminist Historical Materialism.* Boston: Northeastern University Press.

Helander, E. (1985). 'Sámi nissonat giela ja kultuvrra seailluheaddjin,' pp. 37-42 in *Sámi nissonolbmuid dili birra. Den samiske kvinnens situasjon. Seminar May 31 - June 1, 1985.* Ohcejohka: Sámiráddi.

Helander, E. (2001). *Rätt i Deanodat. Norges Offentlige utredninger. Samiske sedvaner og rettsop pfatninger.* NOU 2001:34. Pp. 459-496. Oslo: Justis- og politidepartementet.

Helander, E. (2002). 'A marginalised minority remains marginalized? On the management of fjord resources,' pp. 203-221 in K. Karppi and J. Eriksson, eds. *Conflict and Cooperation in the North.* Umeå: Kulturens Frontlinjer.

Helander-Renvall, E. (2005). *Biological Diversity in the Arctic. Draft Report to the Convention on Biological Diversity.* Electronic Document: http://arcticcentre.ulapland.fi/docs/BiologicalDiversityintheArctic.pdf. Last accessed: March 2, 2009.

Helander, E. and K. Kailo, eds. (1998). *No Beginning, No End. The Sami Speak Up.* Circumpolar Research Series 5. Edmonton: Canadian Circumpolar Institute.

Hirvonen, V. (1996). 'Research ethics and Sami people—from a woman's point of view,' pp. 7-12 in E. Helander, ed. *Awakened Voice. The Return of Sami Knowledge.* Dieðut 4. Guovdageaidnu: Sámi insttuhtta.

Kailo, K. (2001). 'Gender and Ethnic Overlap/p in the Finnish Kalevala,' pp. 182-222 in H. Bannerji et. al., eds. *Of Property and Propriety: The Role of Gender and Class in Imperialism and Nationalism.* Anthropological Horizons Series. Toronto: University of Toronto Press.

Klein, L. F. and L.A. Ackerman, eds. (1995). *Women and Power in Native North America.* University of Oklahoma Press.

Kuokkanen, R. and M.K. Bulmer (2005). 'Suttesája—From a Sacred Sami Site and Natural Spring to a Water Bottling Plant? The Effects of Colonization in Northern Europe,' pp. 209-224 in S.H. Washington, H. Goodall, and P. Rosier, eds. *Echoes from the Poisoned Well: Global Memories of Environmental Injustice.* Landham, MD: Rowman and Littlefield/Lexington Books.

Thomassen, O.A. (1999). *Lappenes forhold. Nedskrevet 1896-1898 av skolelaerer Ole Thomassen om Lyngen og Porsanger herreder.* Kåfjord Kommune: Samisk Språksenter and Nordkalott-Forlaget.

LaRocque, E. (1996). 'The Colonialization of a Native Woman Scholar,' in C. Miller and P. Chucryk, eds. *Women of the First Nations: Power, Wisdom and Strength.* Winnipeg: University of Manitoba Press.

McCarl Nilsen, J., ed. (1990). 'Introduction,' pp. 1-37 in *Feminist Research Methods. Exemplary Readings in the Social Sciences.* Boulder and London: Westview Press.

Nielsen, R (1986). *Folk uten fortid.* Oslo: Gyldendal Norsk Forlag.

Nilljut, M. (1994). 'Hvilke kvinnetradisjoner og erfaringer har kvinner fra samiske fjord og kystom råder?,' pp. 3-5 in *Samiske kvinners livsvilkår-Sámenisoniid eallindilli. Rapport fra Konferansen. March 8-9, 1994.* Vadsø: Finnmark Fylkeskommune.

Nilsen, R. (1998). *Fjordfiskere og ressursbruk i nord.* Oslo: Ad Notam Gyldendal.

Nilsen, R. (1999). 'Naeringstilpasninger og ressursbruk i samiske fjordströk,' pp. 89-105 in Björklund, ed. *Norsk ressursforvaltning og samiske rettighetsforhold. Om statlig styring, allmenningens tragdie og locale sedvaner i Sápmi.* I. Oslo: Ad Notam Gyldendal.

Nilsen, R. (2003). 'From Norwegianization to Coastal Sami Uprising,' pp. 163-184 in S. Jentoft, H. Minde, and R. Nilsen, eds. *Indigenous Peoples. Resource Management and Global Rights.* Delft:Eburon.

Nilsen, R. (2004).

Olsen, N.V. (1997). *Håper det e' plass til mae også. Samiske ungdoms utvikling og forvaltning av en samisk identitet i et flerkulturelt samfunn.* Hovedoppgave i sosiologi. Tromsö: Institutt for Samfunnsvitenskap, Universitetet i Tromsö.

Paine, R. (1957). *Coast Lapp Society I. A study of the neighbourhood in Revsbotn fjord.* Tromsö: Tromsö Museum.

Pedersen, S. (1987). Fornorskning på nye måter. *Syn og Segn* 3: 202-208.

Pedersen, S. (1994). 'Bruken av land og vann i Finnmark inntil förste verdenskrig,' pp. 13-133 in *Bruk av land og vann i Finnmark. Bakgrunnsmateriale for Samerettsutvalget.* NOU 1994: 21. Oslo: Justis- og politidepartementet.

Pedersen, S. (1997). 'Samisk bruk av sjöressursene i Finnmark,' pp. 20-44 in *Innstilling fra Samisk fiskeriutvalg. Oppnevt av Fiskeridepartementet 21. juli 1993. Innstilling avgitt 10.* Oslo: Fiskeridepartementet, Oslo.

Sagdahl, B.K., ed. (1998). *Fjordressurser og Reguleringspolitikk. En utfordring for kystkommuner?* Oslo: Kommuneforlaget A/S.

Samisk fiskeriutvalg (1997). *Innstilling fra Samisk fiskeriutvalg, oppnevnt av Fiskeridepartementet 21. juli 1993. Innstilling avgitt 10. april 1997.* Oslo: Fiskeridepartementet.

Smith, C. (1990). Om sameness rett til naturresurser—saerlig ved fiskerireguleringer. *Lov og rett* 1990: 507-534.

Starhawk (1982). *Dreaming the Dark. Magic, Sex and Politics.* Boston: Beacon Press.

St.meld. nr. 51. (1997-98). *Perspektiver på utvikling av norsk fiskerinaering.* Oslo: Fiskeri- og Kystdepartmentet. Electronic document: http://www.regjeringen.no/nb/dep/fkd/dok/regpublstmeld/19971998/Stmeld-nr-51-1998-.html?id=191859 Last accessed: March 2, 2009.

Stordahl, V. (1994). *Same i den moderne verden. Endring og kontinuitet i et samisk lokalsam funn.* D.Phil. Dissertation. Institutt for samfunnsvitenskap. Universitetet i Tromsö.

Stordahl, V. (2002). 'The Sámi Parliament in Norway: Limited Access for Women,' pp. 129-134 in *Taking Wing:Conference on Gender Equality and Women in the Arctic.* Helsinki: Ministry of Social Affairs and Health.

Trinh Minh-ha, T. (1989). *Woman, Native, Other. Writing Postcoloniality and Feminism.* Bloomington and Indianapolis: Indiana University Press.

CHAPTER ELEVEN

Women in Sámi Fisheries in Norway —Positions and Policies

Elisabeth Angell[1]

Abstract: Sámi women's participation in fisheries has, until now, received little attention. In Norway, women's tasks in fishing communities are mainly on land—in fish processing plants or preparing and organizing their husbands' fishing activities. The first part of this chapter explores how Sámi women relate to fisheries work while the second part focuses on the Sámi Parliament and its political and financial support to Sámi women in fisheries. I will examine the extent to which this policy is gendered. This article is based on research conducted with women associated with a 2003-2004 project supported by the Sámi Parliament, interviews with Sámi women, a review of Sámi Parliament documents on fisheries and gender policies, interviews with staff and politicians in the Sámi Parliament and relevant literature.

Background

The Sámi are an indigenous people who live in the northern regions of Norway, Finland, Sweden and Russia. The majority of the Sámi live in Norway.[2] The coastal Sámi population in Norway is scattered in many villages and districts but the largest concentration of coastal Sámi is located in the inner fjord areas in Finnmark and North-Troms. In many rural communities, the Norwegian and Sámi live side by side. In the coastal areas, the Sámi language has lost much ground during the twentieth century, mostly because of an active policy of 'Norwegianization'[3] (Nilsen 2003; Minde 2003). Because of the stigma attached to Sámi identity, many people have chosen to assume a Norwegian identity,

---

[1] With thanks to Eva Josefsen, Sveinung Eikeland, Einar Eythórsson, Per Selle and Joanna Kafarowski for helpful and inspirational comments. This is an enlarged and adapted version based on Angell, Elisabeth 2004 a and b.

[2] The numbers vary dependent on the criterion, and in Norway, the numbers vary between 30,000 and 100,000 (Bjerkeli and Selle 2003). Another calculation about the number of Sámi is 50,000 in Norway, among 20,000 in Sweden, approximately 6,000 in Finland and about 2,000 in Russia (http://www.sametinget.se/sametinget/view.cfm?oid=1826). The shaded areas in Figure 1 illustrate the main areas of the Sámi people.

[3] This policy of 'Norwegianization' was active up to the 1950s (Minde 2003) and included discrimination and prohibition of, for example, the Sámi language, music and culture in school

but since the 1980s,[4] there has been a certain revitalization of Sámi culture and identity among the coastal Sámi.

The Sámi Parliament,[5] established in 1989, is an elected national body representing Sámi in Norway. The Sámi Parliament controls the Sámi Development Fund (SDF) which supports economic development including fisheries in Sámi areas. A considerable part of the coast in the north of Norway is included in the geographical area covered by the SDF (see the shaded areas in Fig.1).

This study of Sámi women's participation in Norwegian fisheries is based on data derived from interviews with four Sámi women, statistics collected from the databases of the Norwegian Fisheries Directorate and Statistics Norway and relevant literature. The first half of this chapter draws from application for financial support to the Sámi Development Fund during 1999-2003 while the second half of the chapter utilizes contemporary Sámi Parliament policy documents and interviews with governmental representatives.

Economic Challenges of Life in the Coastal Areas

To provide a picture of the fisheries in the Sámi areas in Norway, I use data for the three northernmost counties (Finnmark, Troms, and Nordland) indicating the total number of individuals employed in fisheries-related industries. These counties include the main Sámi regions although the majority of the population is Norwegian. Figure 2 demonstrates how heavily reliant this region is on the fisheries, including the catching of fish, fish processing and aquaculture. In Finnmark, almost 16 percent of those employed was in the fisheries sector in 1986, and only 8 percent in 2003. However, the decrease in fisheries employment is particularly dramatic in Finnmark, both for fish processing and the catching of fish, and aquaculture is poorly developed. In Troms, the fisheries industries are also diminishing and in Nordland, fisheries employment is more stable, even if the total number of employees is decreasing.

During the last twenty years, there has been considerable out-migration from the coastal areas especially among younger women, but during the last ten years, also among younger men. They move to seek better possibilities for jobs, leisure and higher education and rarely move back to the communities. There is also a low birth rate and the overall population is decreasing (Angell and Lie 2002). The average educational level in the

---

and public arenas. The coastal areas where the Sámi were a minority group were especially vulnerable to this policy.

[4] The discussion and conflict around building a hydroelectric power plant on Sámi traditional territory in Norway, in 1980, in Alta, put the question of the status of the Sámi people in Norway on the agenda. After that, there was an active revitalization of the Sámi culture and identity which changed the official Norwegian attitude to the Sámi (Bjerkeli and Selle 2004). Two government committees were appointed; the Sámi Rights Committee (*see* note 15) and the Sámi Culture Board.

[5] The Sámi Parliament is a consultative and advisory body for the Norwegian government, and can raise any issue that impacts on the Sámi people. There are several examples of delegation of management, i.e., economic development, cultural and language support. The Plenary Session is the Sámi Parliament's highest authority and an executive body, the Sámediggi Council, based on the parliamentary principle, is elected by the majority. There are Sámi Parliaments in Sweden and Finland as well but their status is somewhat different (Josefsen 2003).

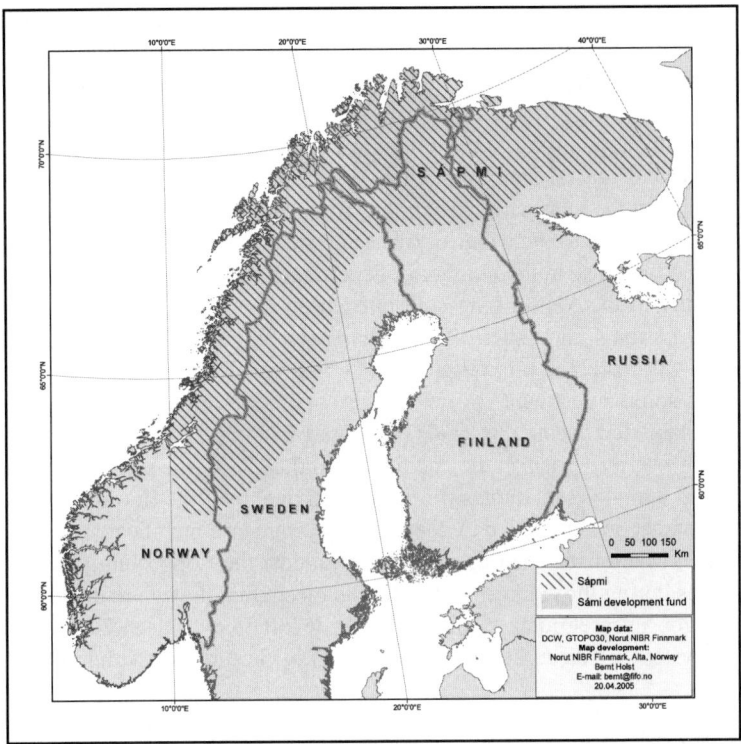

Figure 1[6]. *The shaded areas illustrate the main Sámi areas in Norway, Finland, Sweden and Russia, collectively referred to as 'Sápmi.'*

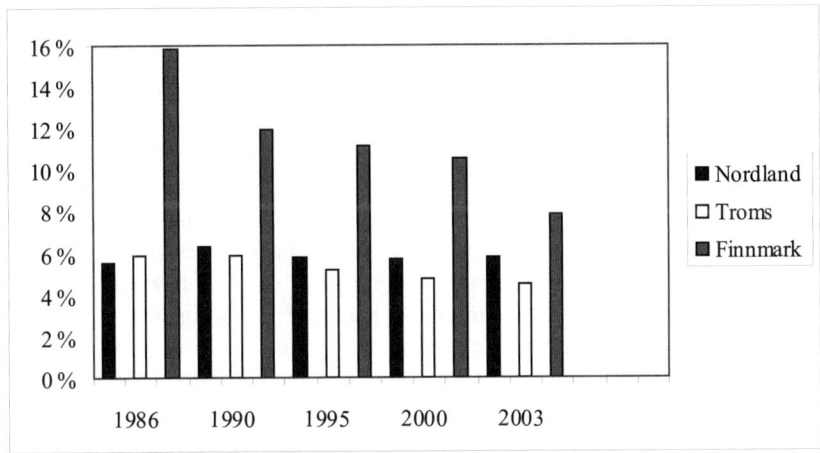

Figure 2. *Individuals employed in fisheries industries in Nordland, Troms and Finnmark. Source: PANDA, SINTEF Industrial Management Economics and Logistics based on data from Statistics, Norway.*

---

[6] The shaded area in the map illustrates the zone for the Sámi Development Fund (SDF). It illustrates the municipalities included in the zone, but in some of these municipalities there are only some villages included in the SDF's zone, not the whole municipality.

coastal areas is lower than in more central or urban regions. The labour market in the coastal communities is strongly gender-segregated and the most important industries for women are fish processing, tourism and the public sector including education, health and social services. The jobs primarily assumed by men are mostly in fisheries, aquaculture, fish processing, manufacturing, transport and public services.

In three typical coastal Sámi municipalities (Tana, Nesseby, and Kåfjord), there has been an increase in the number of fishermen. By the end of the 1980s, fjord fishing was almost non-existent due to the fisheries crisis and seal invasion.[7] During the 1990s, there was a significant increase in the number of fishermen as the cod gradually returned to the fjords. The entry of the Alaska king crab into the fjords in East-Finnmark caused problems for the traditional cod fisheries, but it has been an important source of income for the fishermen who received a crab-quota. New fishermen are entering the fisheries and there has been some investment in new vessels. In other Sámi communities, the number of fishermen stabilized during the 1990s (Eythórsson 2003).

Following a serious crisis in the cod fishery in the 1980s, the government introduced a system of individual vessel quotas for the coastal fleet in 1990. In order to qualify for a vessel quota, a boat owner needed to document a certain volume of catch during the three previous years. Many fishermen in the Sámi areas lost their opportunity to fish because of poor catches during those three years and this meant that they were unable to fulfil the requirements for a vessel quota (Smith 1990; Lillevoll 1998; Nilsen 2003). (The effects of the quota system are discussed in greater length in Gerrard, this volume.)

Fisherwomen in Sámi areas

Women in the fisheries

In Norway, women participate in the fisheries mainly on land as fish processing workers and as organizers and supporters of their husbands' fishing, taking care of children, the elderly and running the household. In the past, women's work was essential for subsistence living in the local communities and to sustain fisheries activities (Gerrard 1975, 1983, 1994; Munk-Madsen 1996). Theories about fishing and fisheries management are mostly focused on the catch. Gerrard (1994) argues that this is a too limited view and that an understanding of resource use and management is widened when fish processing and related household responsibilities are included. A broader view includes women as actors in the fishing industry and fishing communities. I agree this is an insightful approach and that the household adaptations and alternative job possibilities for women are relevant for a cogent analysis of fisheries communities. It is difficult to quantify women's work in the fisheries by available statistics; in most cases, the statistics are not gender-specific. Another challenge is how to measure women's unpaid and unregistered work.

There are available statistics on women in the fishery registry as full and part-time fishers. Table 1 demonstrates that the number of registered fisherwomen is low compared to men. The women who reported fishing as a main occupation represented only 2-3 % of

---

[7] The cod fisheries collapsed in the late 1980s. Due to the lack of cod the authority stopped the cod fishing and introduced a quota system. During the 1980s, there were several years when a lot of seals went into the fjords, ate the fish, spoiled the nets and created great problems for the fishermen.

the total number of fishermen, in the period 1990 to 2003. The highest percentage of full-time fisherwomen is in Nordland. The total number of fisherwomen in Norway decreased by almost half during the 1990s, and in the north, there was a decrease of 37-49%. It is the same trend as for men, even if the decrease for women (-49%) was higher in percentage points than for men (-35%). The share of fisherwomen registered with fisheries as a subsidiary occupation increased during the 1990s in both Finnmark (93%) and Troms (44%), but the numbers are small. For men, both the share of full-time and part-time fisherman decreased. Even if the number of fisherwomen is marginal, it is interesting to note how those women create their place in the most masculine part of the fisheries—the harvest of the fish. There are no available statistics about Sámi fishermen or fisherwomen in particular, but these counties include the areas where most Sámi fishers live, even if the majority of these do not define themselves as Sámi.

During the fisheries crisis in the late 1980s and the early 1990s, Pettersen (1994) examined women's contributions to the reorganization process. In some cases, women took a more active part in the fisheries, including working on board the fishing boat. Another strategy was to find a job outside the household, and, in these cases, women's income acted as a buffer in the family economy. In some instances, women assumed the role of breadwinners. Fishermen households have become more like other households and are economically dependent on a double income. Even if women's efforts have always been important in the fisheries, their work has largely been ignored.

Table 1. The percentage of women in the fisherman-registry from 1990-2003.
Source: The Norwegian Fisheries Directorate 2003, http://www.fiskeridir.no

| County | The percentage of women in the fisherman registry | | | | | | | | Changes 1990-2003 Women | | Changes 1990-2003 Men | |
| | 1990 | | 1995 | | 2000 | | 2003 | | | | | |
	Full-time	Part-time	Full-time	Part-time	Full-time	Part-time	Full-time	Part-time	Full-time	Part-time	Full-time	Part-time
Finnmark	2.2%	2.3%	2.4%	2.4%	2.5%	4.7%	2.0%	6.2%	-44%	93%	-40%	-33%
Troms	1.8%	1.2%	2.3%	2.3%	1.7%	3.4%	1.7%	4.1%	-49%	44%	-46%	-59%
Nordland	2.8%	2.2%	3.3%	1.5%	2.8%	1.9%	2.5%	1.3%	-37%	-72%	-29%	-51%
Norway	2.7%	1.6%	2.7%	2.7%	2.5%	2.9%	2.1%	3.3%	-49%	16%	-35%	-44%

Pettersen (1994) also indicates that when women have paid jobs, it makes their work more visible. The main tendency is that the 'fisherwomen' have become removed from the fisheries industry during the resource crisis. Her findings also documented that a small group of fisherwomen did the opposite, and became more active in the fisheries. However, when work opportunities in the fisheries industry are limited, women are the first to be marginalized. The fisheries crisis made women in fisheries more visible, both economically and in the public political arena (Gerrard 1995, 2000).

Munk-Madsen (1996) emphasizes how the closing of the fisheries commons was an effective way of excluding women, and that these changes further strengthened men's position. Quotas are allocated to the men, who are registered as boat owners. The gender equality debate is on the agenda both in agriculture and reindeer husbandry, but still, it is almost ignored in the fisheries. When recruitment is discussed, in Sámi as well as Norwegian arenas, the focus is on how to recruit young people (read young 'men') to the fisheries, rather than women.

Financial Support to Sámi Fisherwomen

The Sámi Parliament has developed some mechanisms to stimulate industries in Sámi areas; it supports both traditional (i.e., fishing, handicrafts, reindeer hunting) Sámi and modern industries (i.e., tourism). The most important mechanism is the Sámi Development Fund (SDF) which allocates financial support to small businesses within its geographical area without discrimination on an ethnic basis. The support from the SDF is directed toward small-scale businesses and local employment in rural areas. In the five-year period from 1999 to 2003, annual allocations were approximately NOK 30 million ($6 million CAD). In all, women have annually received 10-25% of the allocations from the SDF, while between 16- 25% of the total support has been allocated to fisheries-related projects. Of this amount, 4-23%[8] has been allocated to fisheries projects initiated by women. In most years, approximately about 10%[9] of fisheries support has gone to women.

Although women receive less support than men, most of the applications for fisheries projects from women have been accepted. During the initial five-year period, there were only 18 applications from women, of which only three were turned down. Nine of the 15 approved projects involved partial financing of fishing vessels, one applied for two projects and eight applied for the financing of vessels. In four of these cases, the decision in the SDF board was based upon the desire to involve more women in fisheries, even though the projects did not necessarily satisfy standard requirements for support. After a revision of the guidelines from 2005,[10] gender equality is further emphasized. The objectives state that: "The financial support shall contribute to a reinforcement of women's participation and gender equality in industry." Gender equality is thus incorporated in the criteria for allocation: "Diversity, gender equality and a comprehensive perspective shall be important fundamental principles for the Sámi Parliament."[11] In addition, exceptions from the general requirements for allocation are made possible in cases of women's enterprises. The new rules formalize the use of discretion to facilitate the allocation of business support to women. These rules have not been in place long enough to draw conclusions with regard to how allocation has been influenced. One of the aims of the revised guidelines is to reinforce gender equality and gender perspectives and it is critical to see if the Sámi Parliament is successful in achieving this. A new practice demands a willingness to change the priorities within the industry. It remains to be seen whether the changes will also result in an increase in the support to women in fisheries, with more women wishing to enter the industry. The fact that few women venture into the fisheries sector is not

---

[8] This amounts to between NOK 132,000 (=CAD 26 400) and NOK 785,000 (=CAD 157,000).
[9] There is also great variation, the exceptional year is 2000, when 23% of the fisheries support went to women, and the opposite is 2003 with only 4%.
[10] Guidelines for Sámi Development Fund, 2005, http://samediggi.no/artikkel.asp?AId=140&MId1=5&MId2=132&MId3=158&Back=1
[11] English translation by Elisabeth Angell, 2005.

unique to Sámi areas, but parallels the general picture of fishing as a male-dominated industry.

Women as Skipper

I found it interesting that as many as eight women in the geographical area covered by the SDF received support for investment in a vessel during a period of decreasing fisheries. I wondered if this was a new trend for women to be given priority by the SDF. It was my intention to find out how projects that included support to women for the financing of fishing vessels developed. I interviewed four out of the eight female applicants to the SDF who had been successful. All four women lived in rural areas with few alternative sources of employment and all had bought small fishing boats between 9 and 11 metres in length. The support from the Sámi Development Fund had been essential for all of these women.

For one of the women, it was an attempt to establish herself as a fisher, but she found that she was not fit for that kind of work. She fished for a year and a half with her husband, and then sold the vessel. Another woman got pregnant and now has two small children. She has quit fishing since it is complicated to combine fishing with caring for small children. Her partner now operates the fishing boat, but she is planning to assume a more active role in fishing when the children grow older. These women are optimistic about the future prospects of fishing as an occupation. Neither of these two women have fishermen in their family, as fathers, grandfathers or brothers.

Two of the four women are active 'fishermen.' I call them Berit and Anne. Both grew up in fishermen's families as fathers, grandfathers, uncles or brothers were fishermen. Berit is married and has lived most of her life in her village in Finnmark County, except for a few years in another village while she was young. She and her husband have five adult children, three sons and two daughters. All the sons have their own families and live in the village. They are all fishermen with their own vessels and work part or full-time. The daughters have moved out of the village. Berit has been registered as a full time fisherman for the last four years, but even before, she sometimes went fishing with her husband who has been fishing for many years. She was a farmer and a housewife before she started fishing and their household was a traditional fisherman/farmer type. When one of their sons took over the farm, it gave her an opportunity to change occupations. Berit and her husband are now fishing together. They bought a new vessel with support from the SDF, and transferred the quota from the old vessel to the new one. The capacity in the new boat is greater, but they have not received an increased quota. Berit took over the old boat, and she is fishing in Group II, without a vessel quota. Her husband is crew on her boat, and she is crew on his. On board, he usually is at the helm, and she usually disentangles the nets. The other work they usually do side by side. They fish with nets and hand lines, mostly in the local fjords, but in low season, during autumn and spring, they fish in other areas, several kilometres away from home. A few years ago, a fish processing plant re-started in their village and this made it much easier to deliver the high quality fish. Lower quality fish they dry themselves.

Berit said: "It is a madness that you have to buy the quota. Those quotas were distributed for free by the authorities, and now the fishermen sell them for millions. We are strongly against this system and don't want to participate in this madness, that's why we have not bought more quotas, but we might eventually be forced to do it." She is accepted as a 'fisherwoman.' She has encountered older fishermen who think gender equality has gone too far when the women take over the boat, but she has also experienced positive discrimination, in receiving a better price for her catch. Berit is registered in the Sámi electoral register and her parents were Sámi. She said: "It is difficult, but I consider me as Sámi, even if I don't speak the Sámi language."

Anne grew up in a village on the west coast. Her father and one older brother are fishermen. Anne was widowed at an early age, and, after her four children grew up, she moved to the remote village of her new husband in the Troms County, more than ten years ago. None of her children have chosen fishing as an occupation. Anne has been registered as a part-time fisherman for almost ten years, and has been fishing since 1991. Her husband has a full-time job with every second week free, which makes it possible for him to fish with her in his free time. Mostly they fish in the local fjord, but, during early spring when it is high season in Finnmark, they fish there. Normally, they work side by side, but there is some distribution of tasks. He usually takes care of the engine, does most of the heavy lifting and cuts open the fish. She takes the fish out of the net, and washes the fish. They have installed a new clean-and-wash-line on board which makes the job easier and improves the quality of the fish. On land, they share all the jobs. They fish in Group II, for part time fishing and have no vessel quota, but they are considering buying a quota for full time fishing. According to Anne: "Now, it is a big problem to deliver the fish. We have to bring the fish 120 kilometres one way to the nearest fish buyer. It is a long way and makes it impossible to deliver more than two times a week. We also dry some fish ourselves, but it is difficult to get a good price for high quality dry-fish. The number of fishermen in our fjord has decreased over the last few years. I think the problem of delivering fish is the main reason. Some years ago, a truck came two times a week and collected fish from the local fishermen. It was a good system. But now, new regulations demand special sanitary facilities for the inspector. This was too expensive to fulfil, and therefore, they stopped it. I really regret this. It gives us a lot of extra work, when we have to bring the fish to the fish buyer."

She likes to be a 'fisherwoman' on the boat."Everybody is surprised to meet a woman on the boat, and sometimes they don't immediately understand that I'm a woman. I feel fully accepted on board. One kind of discrimination I have met is from the fish-buyer, he has tried to give me a lower price, and argued that the catch was low quality because I'm a woman." Anne originally came from another part of the country, and does not consider herself as Sámi. Neither of her parents are Sámi. However, she has moved to a part of the country which is included in the Sámi Development Fund zone.

Fishing as a Temporary Career

The four women interviewed live in peripheral rural areas with few alternative job opportunities and they have to create their own jobs. They had some experience in the labour market but their economic options had to be adjusted due to their care obligations for their children. Before the women started fishing, they had flexible job conditions, partially waged-work, partially self-employed, usually based on local resources combined with partial or full-time unpaid work at home with care for children. The work of their partners also varies. In fishing, the women still have flexible job possibilities. All of the women interviewed made it clear that it is necessary to fish with somebody else because fishing is a physically demanding occupation. All these women fish together with their partner so that fishing is a household project. But for the two active fisherwomen, this was an important self-employment project as well.

In areas characterized by a mixed economy combining fishing and agriculture and other available sources of income, this adaptation is often organized as household production. The mixed economy represents a traditional coastal Sámi way of living and is more common in coastal Sámi villages than in the corresponding Norwegian coastal villages (NOU 1988). It requires complementary job-sharing between the sexes that transcends the barrier between work and leisure, paid and unpaid work (Munk-Madsen 1996). This can be characterized as gendered job-sharing. During the last twenty years, it has become common to have paid work outside the household. Women, in particular, take full- or part-time jobs in the service sector. For men, apart from coastal fishing, it is common to gain income through construction work, seasonal work on fish farms, teaching, animal husbandry or service industries. This is an important way to provide regular income to the household and it allows for more independent and flexible work for the spouse.

Two women participating in my study identified themselves as 'fisherwomen,' In one case, the husband has a paid job outside the household and this makes it possible for the woman to be a fisherwoman. In the other case, the spouses are fishing together full-time. In both cases, the couples have adjusted their working lives to take part in a mixed economy. Traditional small-scale coastal fishing today is sometimes an adaptation of elder or retired persons who move back to their childhood village after years spent in a career in the city (Eythórsson 2003). Fjord fishing is rarely now a lifetime occupation. Fishing usually requires a more stable income which means bigger and more mobile fishing vessels. In some cases, it is possible to provide a decent household income from coastal and fjord fisheries in combination with other sources of income. The resource base fluctuates and small-scale fishing is rarely a stable, full time occupation. In any case, small-scale fishing requires access to fish (quota), opportunities to deliver the catch, infrastructure, access to capital for investment in vessel and gear and active local communities for recruitment to the fisheries and to provide services to the fishermen. For the two fisherwomen in this study, small-scale fishing is not a lifetime career, as their fishing must be juggled with the demands of the household, alternative job possibilities and their husband's work. The two women who continue fishing have a family background within the fishery; they know the industry and are familiar with being self-employed. The women feel fully accepted as 'fisherwomen,' but have experienced both positive and negative discrimination.

The support from the Sámi Development Fund was essential for the investment in fishing vessels. Limited access to capital may be an obstacle to participation in the fishery. Therefore, the support from the SDF is an important source of funding. This support may also encourage the involvement of other financial partners. Some of the women who received support consider themselves as Sámi while others do not, which indicates that the relationship to Sámi identity is variable. The SDF supports the population in its area without discrimination on ethnic grounds. Only two of the participants in this study self-identified as Sámi.

These women share the same challenges as many other small-scale fishermen, both in and outside the Sámi area. Berit emphasizes that the main challenge is obtaining a sufficient quota. It is evident that fishermen who maintain flexible and less intensive fishing regimes have become marginalized by the vessel quota-system. The quota system favours fishermen who have made significant investments in vessels and equipment which require intensive fishing to be able to pay debts. Fishermen, regardless of sex, who base their income on mixed resources, flexibility and limited investments, have become second-class fishermen who risk further marginalization (Munk-Madsen 1998). Munk-Madsen also argues that the allocation of individual rights to fish resources to women and men, as women's and men's quotas, is preferable to vessel-quotas as it would serve to establish the link between the control over the fishery capital and control over the fishery resources. This would be challenged as fishing capital is considered male property. The main constraint in fishing to Anne is the problem of delivering fish, since the local fish processing plant has been closed. This underscores the necessity of establishing relevant infrastructure for the industry in peripheral areas, particularly for small-scale businesses. However, both women continue to be pioneers through their involvement in this strongly masculine arena.

Fisheries Policy and Gender in the Sámi Parliament

Sámi interests in Norwegian fisheries policy

After the introduction of the vessel quota regime in 1990, the Sámi Parliament raised questions of principle and argued that the new regime was discriminatory against traditional Sámi fisheries and in conflict with international conventions on minority and indigenous peoples. This introduced the indigenous peoples' perspective into the Norwegian debate. In response, the Norwegian Ministry of Fisheries engaged a lawyer to write a report (Smith 1990) on the Government's legal obligations related to Sámi fisheries. This report established that the Norwegian government is obliged to consult the Sámi Parliament on proposals for changes in the regulation system and it recommended that the Sámi Parliament should be consulted on issues of allocation of quotas in Sámi areas.

The Sámi Parliament has been represented in the Regulatory Council[12] since 1992. The Ministry of Fisheries appointed a 'Sámi Fisheries Committee' in 1993 to consider Sámi fishery interests. The Committee acknowledged that Sámi interests had lost ground as a result of the introduction of quotas (Davis and Jentoft 2001). The Sámi Rights Committee[13] also brought Sámi fjord fishing into a global discourse on the historical rights of indigenous peoples to natural resources in their areas. To date, the Ministry of Fisheries has been of the opinion that Sámi interests can best be safeguarded by means of general regulatory measures. Since the 1990s, the Ministry of Fisheries has, to a certain extent, accepted that an ethnic dimension may exist in fisheries policy (Eythórsson and Mathisen 1998) even though the fisheries resources in coastal regions were not included in the new *Finnmark Act*.[14] The Sámi Parliament is now working with Finnmark County

---

[12] The Regulatory Council gives advice to the Ministry of Fisheries about the distribution of quota, after the total allowable catch (TAC) is decided in the Norwegian–Russian fisheries negotiations.

[13] The government appointed a committee, The Sámi Rights Committee, to report on Sámi rights, after the confrontation between the Sámi and Norwegian government over the issue of the Alta power plant (*see* note 3). This committee worked for 17 years and made several reports. The main report is the Norwegian Government Report 1997a, others are i.e., Norwegian Government Report NOU 1997b, Norwegian Government Report NOU 1993.

[14] This *Act*'s aim is to regulate the management of the land and water in Finnmark County, which is the main Sámi area and follow up on the Sámi Rights Committee report. The *Act* was passed by Norwegian Parliament in the spring 2005, and established a new body, the 'Finnmark-property,' which is the formal owner of the land and water in Finnmark. A board comprising three people elected from the Sámi Parliament and three elected from the Finnmark County Council govern the 'Finnmark-property.'

on a proposal for the regional management of the fisheries resources. Both the Ministry of Fisheries and the Norwegian Fishermen's Association have so far rejected the idea of regionalization, arguing that fish is a national resource, which should be managed at the national level. It is too early to say if the new government will change this policy. There are great expectations of the new Minister of Fisheries Affairs as she is the first Sámi Minister in a Norwegian government.

The Development of Sámi Fisheries Policy

Norwegian fisheries policy is regularly debated in the Sámi Parliament and fisheries policy was on the agenda for Sámi organizations even before the establishment of the Sámi Parliament. A review of Sámi Parliament resolutions between 1989 and 2003 indicates that social issues and community livelihood have provided the framework for the development of the Sámi Parliament's fisheries policy. To a large extent, the fisheries policy issues raised by the Sámi Parliament have been initiated through proposals by Norwegian authorities, in the form of amendments of regulations, legislation, parliamentary white papers, etc. Nevertheless, the Sámi Parliament has made use of opportunities to raise new initiatives, such as a 'Sámi Fisheries Zone' even if the Ministry of Fisheries has not been supportive of the proposals. When the Norwegian government has implemented measures related to Sámi fisheries, it has often come about as a result of proposals from the Sámi Parliament. After the establishment of the Sámi Parliament, many people in Sámi areas were optimistic about increased opportunities to influence the government on fisheries and other economic issues. There is now more or less a national acceptance of the importance of fisheries for the Sámi settlements, but this has not so far resulted in specific initiatives to protect small-scale fisheries in the Sámi areas. These fisheries have become marginalized during the last several decades, despite the existence of the Sámi Parliament (Andersen 2003).

A clear trend is the increasingly legal nature of the Sámi fisheries policy. In the presentation of issues from the Sámi Parliament throughout the 1990s, there is greater emphasis placed on indigenous peoples' rights to resources. During the last 20 years, indigenous peoples' rights to natural resources (including fish), have been placed on the international agenda, and a worldwide recognition of these rights seems within reach.

In September 2004, the Sámi Parliament held a broad internal fisheries debate, based on a 'white paper' prepared by the Sámi Parliament Council. This document gave an overall picture of fisheries policy and priorities for the industry and reviewed recent developments regarding the Sámi Parliament's fishery policy, including the fishery zone and regional management of fisheries resources. A chapter about women in fisheries was included and there were some proposals to strengthen the position of Sámi women in fisheries. The Sámi Parliament endorsed the document from the Council.

The Rural Industry Model

Fisheries policy is often formulated on the basis of two main approaches: sectoral interests and societal interests, which may appear to be in conflict. Sectoral interests place greatest emphasis on what is best for the fishing fleet, the companies, export and profitability, while societal interests primarily address the effects on communities and employment. This may be further refined as two principal models: A 'rural industry' model, which is based on a decentralized structure of small units with regard to vessels and production, and an 'industrialized model' in which the emphasis is on large trawlers, large processing plants, and financially efficient operations. Norwegian fisheries policy has, in general, been developed as a compromise between these two models (Hersoug 2000). However, the current Norwegian fisheries policy and the ongoing structural program have contributed to a further marginalization of the small-scale fisheries. The Sámi Parliament has based its philosophy for fisheries policy on a 'rural industry' model, and gives priority to small-scale fishing and the mixed economy. In the Sámi Parliament's work in the Regulatory Council, the desire to maintain viable coastal communities and small-scale fishing has been an important objective.

Women's Formal Position

The representation of women in the Sámi Parliament had been, until the election in 2005, low and declining even further. In the first term of the Sámi Parliament (1989-1993), 33% of the elected members were women. Since then, the number of women has fallen steadily[15] and in the period 2001-2005, only 7 out of the 39 representatives (18%) are female. This trend changed in 2005, after a major mobilization for new members. More than half of the members in the current Parliament (2005-2009) are female (22 of the 43 representatives) and the new President in the Sámi Parliament is the first female President. The Sámi Parliament Council has had a more equal distribution of the sexes. At times, women have held central positions, which have involved responsibility for the development of the Sámi Parliament's fisheries policy. From 1997 to 2001, a female member of the Sámi Parliament Council was responsible for the development of the fisheries policy. The first two representatives of the Sámi Parliament in the Regulatory Council were women. It has often been women who have put issues of equality and gender roles on the agenda. It is therefore relevant to 'count heads.' At the same time, female representation does not provide a guarantee that issues regarding equality and gender roles will be addressed. It is also relevant to ask whether women's responsibility for implementing gender equality in fisheries policy should be greater than that of the male representatives.

Absence of Gender Perspectives

When the Sámi Parliament was founded, gender equality and women's rights had been on the Norwegian political agenda for two decades. One would have expected this to provide a good starting point for the integration of such matters of the Sámi Parliament from the start, but there is little or no evidence of this in the fisheries policy issues handled by the Sámi Parliament. Equality in Sámi society pertains to ethnic and cultural rather than gender equality. Ethnic equality has to do with the dominant culture and its relations to the Sámi. From the Sámi point of view, this debate has taken the form of a critical discourse *vis-à-vis* the Norwegian authorities. A gender equality discourse would focus on internal Sámi relations, and direct the attention to gender and power in the Sámi society. Those two perspectives do not need to be concurrent (Josefsen 2004).

The renewal of the *Program of Action for Sámi Coastal and Fjord Areas 1997-2001* was the first Sámi Parliament document related to fisheries in which women were actually mentioned. This document states that the principal focus should be on women, but despite this, women are only sporadically mentioned in the sections where fisheries are

---

[15] From 1993 to 1997, there were 12 female members and from 1997 to 2001, there were 10 (Hætta 2002).

discussed—in connection with fish processing and a specified initiative for recruiting women and young people to the fisheries. Gender equality and women's issues are also omitted from policy documents dealing with commercial fishing. Women are addressed in the sections on strategies for agriculture and mixed farming and fishing. In documents from recent years, a change is detectable. There are a few examples of women being mentioned in connection to demographic issues, where the usual approach is how to recruit women to Sámi communities. In the Sámi Parliament's Fisheries Policy, gender equality and women's perspectives are minimized, even if it is a topic in the recent 'White Paper' on Fisheries Policy. However, in decisions regarding fisheries issues, there are, in practice, few traces of women's issues or gender equality in general.

When fisheries issues have been handled in the Sámi Parliament (both in the Plenary Session and in the Council), it has generally been men who have made the proposals, even if women have held important positions. Women have been empowered to take action in connection with the Sámi Parliament's fisheries policy, but it appears that they have only had limited influence. In Sámi politics, women have often promoted gender equality, but this has not been the case regarding the Sámi Parliament's fisheries policy. As demonstrated earlier in this chapter, support from the Sámi Development Fund was fundamental for women in the fisheries. This is interesting since women and gendered perspectives have received little attention in the Sámi Parliament's fisheries policy documents.

A review of the significant documents dealing with the gender equality policy,[16] indicate that primary industries and the mixed economy are only addressed generally. The gender-segregated labour market is described in terms of men's domination in the primary industries while women are more involved in the public sector. Attention is drawn to the fact that jobs in the primary industries in which men are clearly in the majority have higher status than secure and well-paid jobs in the public sector in which most employees are women. References to fisheries are absent in these gender equality policy documents, as references to gender or women are absent from the fisheries policy documents.

It is important to consider the institutional power of the Sámi Parliament when examining how (and if) it incorporates gender equality and women's interests into its fisheries policy. It is evident that the potential of the Sámi Parliament to address change in this area has not been realized. According to Josefsen (2004), the Sámi Parliament does not challenge the gendered structures and processes within its own organization. The gender equality policy has only received sporadic attention by the Sámi Parliament and this policy is not well linked to the political issues which are understood as important in Sámi policy. The fisheries policy is an example of this.

Admittedly, it should be stressed that the Norwegian fisheries policy is not well-oriented toward gender equality either, so the Sámi Parliament is not exceptional here. It would appear that the Norwegian authorities have integrated more gender equality issues into reindeer husbandry and agricultural policies than into fisheries policy. In the process of creating a platform for itself in Norwegian fisheries policy the Sámi Parliament could have been further marginalized if it had taken upon itself the task of promoting the indigenous people's dimension as well as the gender equality interests. However, even internally, the Sámi Parliament has not placed particular emphasis on the question of women

---

[16] The *Plan of Action for Gender Equality* and the *Annual Gender Equality Policy Report* to the Sámi Parliament in 2001, 2002 and 2004.

in fisheries. It appears to be difficult to incorporate the two equality-perspectives, gender and ethnic, simultaneously. In the current fisheries discourse in the Sámi Parliament, gender equality is clearly subordinate to cultural or ethnic equality.

Influence through the International Ethno-political Debate

Fisheries policy, with an emphasis on fjord fishing, has become an integral part of the ethno-political debate (Eythórsson and Mathisen 1998) which has undergone significant changes during the relative short lifetime of the Sámi Parliament. As indicated earlier, gender equality has been largely absent from the ethno-political fisheries debate. At the national level, official documents in one way reflect the international debate on indigenous peoples in which people's right to resources are a critical element (i.e., ILO 169, Agenda 21 and *Convention on Biological Diversity*). This ethno-political fisheries debate has perhaps been most vigorously conducted in academic circles rather than at the national policy level. The Sámi Parliament has introduced the indigenous perspective in the national fisheries policy, but without a clear breakthrough on the level of policy. This is evident from the fact that fishing resources were not included in the new Finnmark Act.[17] These international and academic debates can be an important prerequisite for eventually achieving political support for domestic policy. However, there is also the potential risk that the debate will be confined to the academic sphere and therefore, may have little effect on policy development. In the field of fisheries, there is a question: has the Sámi Parliament been more successful in influencing international arenas in the United Nations system and other international discourses on indigenous people's rights to natural resources than in influencing Norwegian fisheries authorities?

It is interesting to note, despite the fact that gender equality and ethnic equality have marginalization in common; these two principles are not connected in the ethno-political debate in the Sámi Parliament, in the community or in academia. Academics who argue for expanding the fisheries policy to include women[18] assert the rights of other marginalized groups such as the Sámi to follow a traditional way of life. To date, gender equality and fisheries are not debated together in the Sámi Parliament and there has been little connection made at the official or local levels between the ethno-political fisheries debate and the gender equality fisheries debate.

Conclusion

There are some women in the Sámi areas of northern Norway who enter the male domain of the fishing boat. Although these women do men's traditional work, it is often done side-by-side with their husband. The main challenges these fisherwomen face are common to all small-scale fishermen—the access to fish (quota) and the necessary infrastructure for delivering the fish and are thus not specific to gender or ethnicity. Due in part to the modest number of submitted funding applications from women, only a small proportion (approximately 10%) of the support from the Sámi Development Fund (SDF) to fisheries-related projects has been allocated to women. These women all live in remote

---

[17] See note 16.
[18] For example, Siri Gerrard (1983, 1994) and Eva Munk-Madsen (1996, 1998).

Sámi areas with few employment options. The financial support from the Sámi Parliament makes it easier to create their own work and settle in Sámi communities.

The fishing industry and the fisheries policy have traditionally been, and still are male-dominated. The concept of gender equality has been lacking in the Sámi Parliament's work on developing its fisheries policy. The Sámi Parliament has advocated a 'rural industry' model, based on small-scale-business and a mixed economy. Yet, the model has been developed according to a traditional, male-dominated perspective. Only in recent years, following the significant migration of women out of the region, has the Sámi Parliament pointed out the importance of also making local communities attractive for women. As yet, little is being done to achieve this. If the Sámi Parliament is to make progress in the integration of gender equality in its fisheries policy, clear political guidelines, administrative monitoring and a willingness to take action to bring about change are necessary. Maybe the new Sámi Parliament, with a majority of women, and a female President will bring this progress.

There are many parallel arguments between those who have attempted to expand the fisheries policy to include women and those used by the Sámi Parliament to stress equality for indigenous peoples in the national fisheries policy. An expansion of the fisheries policy, directed towards women will also include those who conduct mixed economy enterprises and those who fish part-time. Further co-operation from the Sámi Parliament will support these endeavours. It is significant that no clear connections have existed between the ethno-political fisheries debate and the more female-oriented fisheries debate. The work of the Sámi Parliament has concentrated more on ensuring that the indigenous peoples' dimension is integrated into Norwegian fisheries policy than on getting gender equality issues integrated into its fisheries policy.

References

Andersen, S. (2003). 'Samisk tilhørighet i kyst- og fjordområder,' pp. 246-264 in B. Bjerkli and P. Selle, eds. *Samer, makt og demokrati: Sametinget og den nye samiske offentligheten*. Oslo: Gyldendal Akademisk.

Angell, E. and I. Lie (2002). *Arbeidsmarkedet i tre Finnmarksregioner*. Report No. 2000(2). Alta, Norway: Norut NIBR Finnmark

Angell, E. (2004a). *Kjønn og etnisitet i fiskeripolitikken. En analyse av kvinners rolle i samisk fiskerpolitikk og Sametingets posisjon i norske fiskerireguleringer*. Report No. 2004(4). Alta, Norway: Norut NIBR Finnmark

Angell, E. (2004b). 'Gender and ethnicity in the Sámi Parliament's fisheries policy,' pp. 102-111 in Sloan, L., ed. *Women's Participation in Decision-making Processes in Arctic Fisheries Resource Management. Arcic Council 2002-2004*. Norfold: Forlaget Nora.

Bjerkli, B. and P. Selle (2004). Den nye samiske offentligheten. *Norsk statsvitenskaplig tidsskrift* 4:365-390.

Bjerkli, B. and P. Selle, eds. (2003). 'Sametinget—Kjerneinstitusjon innenfor den nye samiske offentligheten,' pp. 48-86 in B. Bjerkli and P. Selle, eds. *Samer, makt og demokrati: Sametinget og den nye samiske offentligheten*. Oslo: Gyldendal Akademisk.

Davis, A. and S. Jentoft (2001). The challenge and the promise of indigenous peoples' fishing rights —from dependency to agency. *Marine Policy* 25: 223-237.

Eythórsson, E. (2003). *Petroleumsvirksomhet i Lofoten —Barentshavet og samiske forhold. Utredning*. Alta, Norway: Norut NIBR Finnmark

Eythórsson, E. and S. R. Mathiesen (1998). 'Ressursforvaltning og lokal kunskap i kyst samiske områder,' pp. 40-55 in B.K. Sagdahl, ed. *Fjordressurser og Reguleringspolitikk. En utfordring for kystkommuner?* Oslo: Kommuneforlaget A/S.

Gerrard, S. (1975). *Arbeidsliv og lokalsamfunn: Samarbeid og skille mellom yrkesgrupper i et nordnorsk fiskevær.* Magistergradsavhandling. Tromsø: Institutt for Samfunnsvitenskap, Universitet i Tromsø.

Gerrard, S. (1983). 'Kvinner i fiskeridistrikter: Fiskerinæringas bakkemannskap?,' pp. 217-241 in B. Hersoug, ed. *Kan fiskerinæringa styres?* Oslo: Novus Forlag.

Gerrard, S. (1995). When women take the lead: Changing conditions for women's activities, roles and knowledge in North Norwegian fishing communities. *Social Science Information (SAGE)* 34(4): 593-631.

Gerrard, S. (1994). 'Kvinners forvaltning—havets husholdning,' pp. 123-124 in E.O. Otterstad and S. Jentoft, eds. *Leve kysten?* Oslo: Ad Notam Gyldendal.

Gerrard, S. (2000). The gender dimension of local festivals. The fishery crisis and women's and men's political actions in North Norwegian communities. *Women's Studies International Forum* 23(3): 299-309.

Hersoug, B. (2000). 'Er fiskeripolitikk distriktspolitikk?,' pp. 67-103 in E.H. Teigen, ed. *Bygdeutvikling: Historiske spor og framtidige vegval.* Trondheim: Tapir Akademiske Forlag.

Hætta, O.M. (2002). *Sametinget i navn og tall, høsten 2001—høsten 2005.* Karasjok:Sametinget.

Josefsen, E. (2004). *Sametinget som likestillingspolitisk area.* Report No. 2004(9). Alta, Norway: Norut NIBR Finnmark.

Josefsen, E. (2003). *The Sámi and the national parliaments. Channels of political influence.* Guovdageaidnu: Gáldu Resource Centre for the Rights of Indigenous Peoples. Electronic document: http://www.galdu.org/govat/doc/politicalinfluenceevajosefsen.pdf. Last accessed: March 2, 2009.

Lillevoll, T.A. (1998). 'Open commons for fjord fishery in coast Saami areas?,' pp. 121-144 in S. Jentoft, ed. *Commons in a Cold Climate, Coastal Fisheries and Reindeer Pastoralism in North Norway: The Co-management Approach.* Man and the Biosphere Series 22. Paris and New York: UNESCO and The Parthenon Publishing Group.

Nilsen, R. (2003). 'From Norwegianization to Coastal Sami Uprising,' pp. 163-184 in S. Jentoft, H. Minde, and R. Nilsen, eds. *Indigenous Peoples: Sametinget og den nye samiske offentligheten. Resource Management and Global Rights.* Delft:Eburon,

Minde, H. (2003). 'Urfolksoffensiv, folkerettsfokus og styringskrise: Kampen for en ny samepolitikk 1990-1960,' pp. 87-123 in Bjørn Bjerkli and Per Selle, eds. *Samer, makt og demokrati: Sametinget og den nye samiske offentligheten.* Oslo: Gyldendal Akademisk.

Munk-Madsen, E. (1998). 'From common property to all-male property—quota distribution in Norwegian fishing,' pp. 145-165 in S. Jentoft, ed. *Commons in a Cold Climate, Coastal Fisheries and Reindeer Pastoralism in North Norway: The Co-management Approach.* Man and the Biosphere Series 22. Paris and New York: UNESCO and The Parthenon Publishing Group.

Munk-Madsen, E. (1996). Fra alminding til mandsejendom—kvotefordeling i norsk fiskeri. *Sosiologi i dag* 1996(3): 83-104.

Norwegian Government Report (1988). *Næringskombinasjoner i samiske bosettings områder.* NOU 1988:42. Oslo: Justis- og politidepartementet.

Norwegian Government Report (1993). *Rett til og forvaltning av land og vann i Finnmark. Bakgrunnsmateriale for Samerettsutvalget.* NOU 1993:34. Oslo: Justis- og politidepartementet.

Norwegian Government Report (1997a). *Naturgrunnlaget for samisk kultur.* NOU 1997:4. Oslo: Justis- og politidepartementet.

Norwegian Government Report (1997b). *Urfolks landrettigheter etter folkerett og utenlandsk rettbakgrunnsmateriale for Samerettsutvalget.* NOU 1997:5. Oslo: Justis- og politidepartementet.

Pettersen, L.T. (1994). 'Hovedsaken er at kjerringa er i arbeid. Husholdsstrategier i fiskerikrisen,' pp. 65-76 in E.O. Otterstad and S. Jentoft, eds. *Leve kysten?* Oslo: Ad Notam Gyldendal.

Smith, C. (1990). Om sameness rett til naturresurser—saerlig ved fiskerireguleringer. *Lov og rett* 1990: 507-534.

CHAPTER TWELVE

Gender, Human Security and Northern Fisheries[1]

Gunhild Hoogensen

"In no society are women secure or treated equally to men. Personal insecurity shadows them from cradle to grave…And from childhood through adulthood they are abused because of their gender" (UNDP 1994:1).

Abstract: This chapter provides a preliminary exploration of the relevance of the human security concept to gender and fisheries. The chapter is divided into two parts, the first addressing the concept of security itself and the ways in which the notion of security has and can be broadened to accommodate non-military approaches to security. It is argued that a relational approach to security is important, and this relational approach is derived primarily from gender literature. A relational approach to security makes visible relations between dominant and non-dominant peoples, groups, communities within larger entities such as states, regions, or the international community. The second part of the paper discusses the relational securities of women and men in the fisheries, including increasing vulnerabilities to economic and identity insecurities, and vulnerabilities to recent trends in fisheries development such as towards crustacean stocks, and the processing of crab.

Introduction

This chapter discusses the merits of approaching gender and fisheries communities from a security-based perspective, particularly using the concept of human security. The rationale behind 'securitizing' gender and the fisheries is to contribute to efforts to highlight

---

[1] In the context of this chapter, the use of the word 'northern' is used to connote the global North, that is the largely Western, post-industrialized, wealthy countries consisting of North America and Europe. This is, generally speaking, as opposed to the 'global South,' referring to less-wealthy (if not poverty-stricken) countries experiencing various stages of industrialization. Of course, there are always exceptions to such broad categories, not least Australia and New Zealand which are located in the 'south' but have economies and life-styles consistent with that of the 'north.' This chapter makes use of these broad categories, not with the intent of reifying them, but for building bridges across these constructed divides.

the dynamics, both challenges and successes, of fisheries from a gender perspective. It is also to contribute to the internationalization or 'globalization' of this dynamic—highlighting shared experiences in the fisheries for women around the world, recognizing the specificity of context, but noting shared perspectives and experiences across global spaces, from the global South or North. Not only could the subject of gender and fisheries benefit from a security analysis, looking at the situation in a different light as well as present the situation to a different policy community (that of human security), but the concept of security, particularly human security, can likewise benefit in its own development, as increased knowledge of gender and fisheries can illuminate new ways in which the concept of security can be informed as well as understood.

The purpose of using a security approach to examining and evaluating the dynamics of coastal community life (with a particular focus on fishery lives in these communities) is to connect these dynamics with a larger global picture of human well-being. My intention in this chapter is to suggest ways in which to make use of the tool of human security in the context of gender and the fisheries. I do not profess an expertise in the field of fisheries itself, but as human security can be informed and understood through gender, indigenous, and political economy approaches, I would like to explore its relevance with regard to a thus far under-considered issue. As well, human security is a tool that has policy impacts and implications; as such, there might be policy and political advantages toward exploring the dynamics of gender and the fisheries in the human security context.

I wish to therefore present two possible and complementary research agendas: 1) demonstrate that a broadened notion of security, such as human security, can make visible insecurities as well as capabilities in a gender analysis of Northern (global North) fisheries; and, 2) that this analysis can serve as a bridge to Southern (global South) fisheries communities, demonstrating shared interests, securities and goals instead of polarizing differences between the global South and North. I will first discuss notions of security and how a gender perspective opens up the concept of human security. This will largely be a theoretical presentation and as such does not directly address the case of gender and fisheries at first. However, the second half of the paper examines the theory in relation to the fisheries context, first presenting a discussion of the Canadian fisheries, and then exploring how this example can be made to highlight human security issues in the context of Northern fisheries in general.

What is Human Security?

Traditionally, security is the sole purview of the state, guided by an anarchic international system and balance of power, and largely military in character (Kahler 1997; Buzan *et al.* 1998). Mainstream or dominant security definitions are very exclusive, isolating the notion of security as a matter of 'high politics' (Robertson 1997), including as little as possible except obvious threats to territory or the integrity/survival of the state; in other words, national security. The parameters of security are constructed on the basis of élite interests, and it is in the name of national security that the military are employed. Preserving the state and maintaining a focus on the military caters to ensuring and maintaining security for those interests that are most secure in the first place, largely, the state apparatus and élites within. As much as this approach to security has validity for the state, it clearly does not articulate all the dynamics of security, nor expresses all the interests and values of diverse security actors outside of the state. Nevertheless, its preeminence has remained due to a certain logic connected to the definition—the state itself is expected to provide security to 'the people,' to individuals.[2]

'Human security' was popularized in the 1994 United Nations *Human Development Report*, expanding the notion of security to include dimensions of food, health, community, environment, economy and personal and political security with the intention to, in part, address some of the glaring weaknesses of security theory and practice, not least the problem that often the state was not always capable of providing security to individuals and communities. Human security focuses on the individual as its referent. In the UNDP Report, human security is defined as "freedom from fear, freedom from want," and consists of four essential characteristics: universal, interdependent, easier to ensure through early prevention, and people-centred (UNDP 1994). No agreement exists about the definition of human security, and positions range from the Canadian Department of Foreign Affairs view of 'freedom from fear' that is considered prior to development (McRae and Hubert 2001), to human security applied to migration using the definition parameters of the UNDP Report going beyond the Canadian application (Poku *et al.* 2000), or those like Caroline Thomas who argues for a convergence of the development and security agendas (Thomas 2001).

The notion of security has developed in terms of levels (the international system, the state, the society or non-state collectivity, and the individual) and sectors (economic, military, political, and societal, etc.). The widening of the security concept pertains to both. Security can therefore initially be thought of as a grid, with the levels revealing themselves vertically and sectors crossing the levels laterally. The permutations and combinations of securities are extensive even when we restrict ourselves to only a few levels or sectors, and the nature of security is clearly dynamic and in flux. Given the different levels and sectors and the fact that security can therefore become overwhelming, choices and priorities of securities need to be made. As such, the concept of security is not neutral but a political concept that is subject to struggles of power. The relations of power behind the dominant security discourses and actors are the focus of gender security analyses, and by recognizing and responding to the articulations of security by the unheard, marginalized or non-dominant voices coming from a variety of regions around the world, we can begin to establish human security connections between the global South and North.

Dominant Security Discourses Perpetuate the 'Us–Them' Polarization

Dominant human security definitions contribute to an 'us–them' polarization or the secure *versus* the victim. It not only categorizes people and regions but also does not recognize the enabling features of security—how do people cope and adjust to vulnerabilities and establish their own security? It is assumed that fear and want does not play a role in, for example, the average Canadian or North American's life and even if it did, it is assumed that the state is well situated to address these fears. A critical and gender-informed definition of human security allows for a greater recognition of security/insecurity origins and solutions, and does not exclude 'democratic, secure' states from the formula. Insecurity originates 'inside' North America as much as from the 'outside.' Human security issues

---

[2] The role of the state in becoming a, if not the, sole provider of security is reflected in the influential works of Thomas Hobbes and the mythical agreement between individual and sovereign, whereby the individual relinquishes his/her sovereignty in return for security from the ruler of the state, who holds the monopoly of force.

such as economic insecurity, political and social isolation, the sex trade, the proliferation of disease, the arms trade and poverty and famine occur within the global North as well as the global South (albeit to differing degrees and within differing contexts), as well as between the North and South. Economic insecurity, political and social isolation and the degradation if not the elimination of identities are relevant in the case of fisheries (see discussion further in chapter), but continue to be ignored from the Canadian human security agenda. For example, in Canada, insecurity is perceived as originating from outside Canada, not the other way around (Axworthy 2001; Werthes/Härig 2007). According to the Canadian government, human security is not an issue inside Canada because insecurity on the individual level is assumed to be addressed by and through the state apparatus; thus 'human security' is only relevant to foreign policy, identifying when 'the other' experiences in/security. State-centric security definitions are assumed to be appropriate and functional in democratic settings because it is assumed that 'the people' can dictate their needs/securities to government and that these will be/are executed at the people's behest. Of course, this only addresses the fear or negative side of security. What about the positive side including ways of creating security for ourselves? There is minimal knowledge-gathering about what people do for themselves and each other to establish their own security outside of state mechanisms, at least within the human security context. It must be acknowledged that people are not perpetually helpless without the state. The field of human security could learn from, as well as make visible, the capabilities and enabling strategies employed, thereby providing more about what it means to be secure, as well as finding complementary approaches to these strategies. I think gender analysis provides us with some of the tools for uncovering both vulnerabilities and adaptations or enabling strategies from those most affected, from those who are 'below.'

What Does Gender Mean to Security?

There is a difference between 'applying' the human security agenda so that it suits the needs of women, and understanding human security through a gender perspective. To accomplish the latter means that we must include the views, experiences and perspectives of women, but it does not exclude men either. Instead, it informs the theory about structural relations that go largely unrecognized and render invisible relations of dominance and non-dominance. Like gender, security is a relational concept (McSweeney 1999; Bigo 2001; Hoogensen and Rottem 2004). It is in relation to 'the other' that we can determine what and how securities and insecurities exist. Such an evaluation is also rooted in context—the context determines the discourse, content and practices of security (McDonald 2002). Gender research makes visible the discriminatory and unequal relationships between the sexes, and has focused primarily on women as the subordinate subject in gender relationships. As such, gender research is usually best known through the development of feminist theories. The recognition of power relationships between dominant and non-dominant groups plays a central role. Although these power relations are based in gender, they speak to relations of dominance and non-dominance with regard to all relationships of identity. Identity assumes that insecurities and securities from the position of the 'individual' are dependent upon and related to such identities constructed on the basis of sex, race, class, or ethnicity thereby affecting the individual's perception and reception of security.

Using a gender perspective means that articulations of security can be identified through relations of dominance and non-dominance in a variety of contexts, and these articulations will necessarily differ according to context. A gender perspective which recognizes the dominant/non-dominant relationships as well as context dependence inherent to the creation of securities exposes securities and insecurities from large-scale, 'traditional' conflicts to those 'behind the closed doors' of the private sphere including domestic and sexual violences. It allows for the articulations of security and insecurity from the relational and power positions of men and women, from East Timor to the Arctic. These securities and insecurities do not always necessitate a response from state-based security mechanisms or departments of foreign affairs, but that does not de-legitimate their claims to being issues of security.

Gender and Security

A gender approach has the ability to transcend and integrate many of the levels and sectors of security that scholars have otherwise chosen to analyze separately, as well as make visible enabling strategies and capabilities. Instead of playing into the dominant approaches to security studies that focus on a very small portion of the security grid and from the top down, gender analysis often takes its starting point from the bottom up, demonstrating the relevance of the individual and community to broader institutions such as the state, international organizations, and world orders. Gender analysis acknowledges that the personal is political, and therefore, even the individual's experience is relevant. At the same time, it is recognized that individuals are part of communities and states, and that gender is a significant feature of individual identity in relation to others (community) and the state itself, and is, therefore, a part of societal and state security (Hoogensen and Rottem 2004). The social constructions of gender come into play in the analysis, as well as the ways in which humans have constructed their societies on the basis of gender roles, i.e., who has the 'right' to play which roles in society, and how people are supposed to relate to one another. Gender analysis is social, as it relies on the construction of relationships in society as the point of analysis. Globalized gender analysis has demonstrated not only the dominance of male or patriarchally-based societies, but culturally dominant societies and states, where the gendered demands (for example, American militarism) of one society are imposed upon other, less dominant societies. As such, gender analysis integrates the individual, national and global levels through acknowledgement of the social and relational interactions between peoples and constructions of societies.

Gender analysis, particularly from the many variations of feminisms (Western, non-Western, indigenous, etc.) that have permeated gender discourses, have much to offer in this respect and contribute to the identification of security needs from the margins, be they states, communities or individuals. It is often women who find themselves at the forefront of security challenges, but also women who find themselves marginalized when attempting to arrive at solutions for these same challenges (Karamé 2001). Feminist approaches to security generally embrace a global and multidimensional concept that includes ecological, political, economic and social facets in addition to the usually assumed military component. International relations and security in particular, has sustained and maintained a militarized, élite actor focus. "An IR lens focused exclusively on élite interstate actors and narrow definitions of security keeps us from seeing many other important realities" (Peterson and Runyan 1999:51). Often the dichotomy between the narrow conception of security and a broader conception can be illustrated by the differences between militarization and structural violence; the former focused on military defence and the removal of a physical, often institutional threat such as war or large scale violent conflict, and the latter representing reduced life expectancy as a consequence of oppressive political and economic structures that especially affects the lives of women and other subordinated groups (Peterson and Runyan 1999).

Christine Sylvester suggests that security can only be understood as complex, emanating from multiple nuanced 'standpoints' that generally go unacknowledged (if not purposefully ignored) by mainstream security approaches (Sylvester 1994).

> To work with this elusiveness requires a recognition that [security] is intensified in the late or postmodern era by the number and types of insecuring actions that rivet our attention. This is a time of simultaneous struggles, of storms with many centers unfolding on many fronts at once. Some simultaneous struggles are relatively easy to see, as in South Africa, where efforts to homestead the acrid terrain of apartheid move in cross-cutting directions; there are similar struggles, it seems, in Peru, erstwhile Yugoslavia, Liberia, Canada, Angola. Other types of struggles are more difficult to follow, as in the see-sawing efforts to 'secure' the international environment or to secure reproductive rights or religious identities (Sylvester 1994:183).

Security is therefore context-dependent and logically changes according to context rather than being a one-dimensional concept applied to all contexts. The problem of isolating one dimension of security is further illustrated in the divide and discourses of non-Western feminisms, highlighting the 'us-them' polarization inherent within Western thinking overall, and Western feminisms themselves. Western feminisms may aptly identify the lack of acknowledgement of diverse insecurities by the traditional security framework, but non-Western feminisms wage similar arguments against Western feminisms themselves: "Western feminist scholarship cannot avoid the challenge of situating itself and examining its role in ... a global economic and political framework" (Witt 1999:17). Gayatri Spivak (1999) writes that indigenous and non-Western feminisms increasingly become marginalized within mainstream feminism. Thus, gender analysis, from both the global North and South, highlights features of the security dynamic which have been isolated, ignored, and made invisible because the realities of gender and 'other' have not been acknowledged. Some gendered indigenous approaches give voice to the personal, economic, environmental community, and physical security that are intertwined. Joyce Green and Cora Voyageur (1999) demonstrate the dynamics of the non-dominant and insecure position of indigenous women in the Canadian context, noting the multiple insecurities such as poverty, hunger, social and identity marginalization, political isolation and domestic and societal violence faced by many aboriginal Canadian women. This is not the picture of Canada that dominates the human security literature, but an invisible and unwanted view that, due to the marginalization of aboriginal women, is all too easy to ignore.

The recognition of human security in non-military contexts, such as in domestic violence, or in social, political and economic contexts such as fisheries, is a subject of interest for the Global Environmental Change and Human Security (GECHS) project, and is additionally relevant here (GECHS 2005). Power relations of dominance and non-dominance can be recognized through vulnerability assessments, which measure not only the impact of climate and environmental change upon communities, but assess a variety of stress factors upon different groups within given communities, noting that some groups are more vulnerable to climate and environmental change than are others (Vogel and O'Brien 2004). The relationship between social and ecosystems are complex and affect people and their contexts in different ways. As Vogel and O'Brien state: "within any society, wealthy or poor, some members are likely to be more vulnerable that others" (Vogel and O'Brien 2004:2). The ability to respond to change by both the ecosystem and the social group is assessed, which differs from an 'impact' assessment which addresses the impact, but not the ability to respond or cope with, the stress in question, hence the vulnerability (Vogel and O'Brien 2004). Unequal distribution of coping mechanisms and adaptability makes different groups more or less vulnerable to the effects of environmental or climate change. Both the recognition of the power dynamics between dominant and non-dominant groups, and the variability of vulnerability to (human-induced) environmental and climate change are relevant to a human security analysis of fisheries.

Human Security and the North-South Issue

The implementation of the human security agenda, adopted by many Northern countries as a 'new' way of providing assistance to Southern countries, was intended to eradicate both the traditional and obsolete notion of security, and the problems associated with it, such as continuing poverty and human strife.[3] However, a common critique of the human security agenda is its perpetuation of the superior-subordinate/dominance-non-dominance relationship such that the concept entrenches linear and élitist thinking with regard to security. Thus far, the general assumption has been that the North has eradicated its own human security issues, and is well placed to assist the South in eradicating the same. The result becomes an imbalance in perceptions and explanations of what occurs within and across regions. Assuming that human security has no application in the global North serves to disguise and prevent shared human security concerns and experiences. Instead of broadly brushing one area of the world as 'secure' and another as 'insecure,' we can look at securities of non-dominance. Relations of dominance/non-dominance exist in all parts of the world, albeit distributed in different ways. Here, our discussion turns to the fisheries, which provides such an example of different contexts and shared experiences. In the global North where it is broadly understood that the entire area is 'secure,' a securities approach recognizing a dominance/non-dominance relationship would recognize non-dominant insecurities otherwise invisible to the dominant discourse, such as those experienced by women and men in the fisheries. This approach allows for learning about security through the strategies of people in fisheries communities. What this means is that we need to learn about the fisheries context. Not only from 'above,' where governments decide and regulate fishing activity, or where and how firms distribute fish processing plants and the workers within them, but also from 'below,' from the people who are located and relocated, employed and unemployed, exposed to new vulnerabilities and adapting new strategies to meet and overcome these vulnerabilities.

---

[3] See, for example, the Human Security Program implemented by the Canadian government since the late 1990s (DFAIT 2005). This particular program focuses primarily on the protection of civilians from violent conflict (freedom from fear) and does not address issues of 'freedom from want.' The program is directed towards countries of the global South.

Common Local Fisheries Experiences Through Trade-From-Above (International Trade Agreements)

One example of the many concerns expressed around fisheries relates to the impact trade agreements have on the efficacy and benefit of these agreements to developing countries, as the potential impacts more greatly stress a developing economy than one considered 'developed.' According to a 2003 report published by the International Food Policy Research Institute and World Fish Center, there are already many studies that focus on the ways in which policies for poverty reduction, economic growth, and environmental sustainability and security are affected by the changes in the fisheries sector in developing countries (Delgado et al. 2003). However, impacts can also be noted in the economic, environmental and social development of developed nations, especially those that have underdeveloped or over-exploited natural resource industries. The impact of changes in the fisheries sector in developed countries additionally affects developing nations when, for example, fishers from the global North seek out employment off the shores of the global South. In a climate where increasing and justifiable fears surround the inefficient and unsustainable fishery management practices of virtually all states, the relationship between environmental security in the form of conservation and sustainability, and economic and community security, have profound relevance for current international fishery strategies. More significantly, at least from a human security perspective, these agreements impact greatly the experiences of those 'on the ground'—the people who are dependent upon fisheries as a source of income but also as an identity and way of life. I would argue that a number of their experiences and concerns are shared across borders and regions and that these experiences should not only receive attention from the human security debate, but should also teach us about the ways in which in/securities manifest themselves within different contexts.

Trade, the fisheries, and the people 'on the ground'

Continued demand, previously by developed countries but now outpaced by that of developing countries, spurs on the depletion of stocks already threatened, and prevents renewal for those stocks decimated by improper fishing practices, improper nets, excessive by-catch, IUU (illegal, unreported and unregulated) fishing, habitat loss, pollution and climate change. Fish is money, and thus, it will be harvested, even under the most stressful of circumstances (such as depleting stocks).

Many have already argued that there exists a link and/or an inverse relationship between trade (here seen as a function of economic security), and the environment and this is seen in the case of the fisheries (Munn 1999; Hackett 2001; Delgado et al. 2003; Leal 2004). According to the World Wildlife Fund (WWF), as a result of accelerated tariff reduction, in part instituted by APEC leaders in November 1997, fish and fish product exports have increased between 12 and 14 percent (WWF 2005). The establishment of the 200 mile exclusive economic zones has been in part blamed for the decline in food fish production in developed countries, along with overfishing and declining stocks (Delgado et al. 2003). Economic policy often takes the approach of short term solutions to save jobs, but these have been implemented at the expense of the long-term sustainability of ecosystems, and eventually at the expense of the long-term sustainability of jobs dependent upon those ecosystems (Lino Grima 1999). Delgado et al. (2003) list some of the impacts the fisheries have had on the environment. Overexploitation due to advanced technology, inflexibility in labour and assets, over-capacity, and ecological destruction are cited as key features of the fish trade that wreak havoc on the development of fish stocks (Delgado *et al.* 2003). Bycatch, or the non-target species that are caught amongst the targeted fish, are often discarded for 'economic or regulatory reasons' (Delgado *et al.* 2003). Regulating fisheries on purely economic grounds is found to be grossly insufficient and will result in a collapsed industry if there is no concurrent public or community responsibility and management (Lino Grima 1999). For those countries dependent on natural resources, and hence the environment, it is a monumental task to stimulate and drive their economies while at the same time responding positively to environmental and conservationist agreements that could curtail, if not eliminate, excessive, abusive, and destructive exploitation of the environment. Eliciting the support of economic stakeholders is very difficult: "This requires the backing businesses, consumers, and other constituencies, which may not be easily secured" (Mastny and French 2002:1). This applies to fishing as it does to other natural resource sectors, and to Canada and other fish-exporting nations around the world. Evidence of the problems surrounding the trade/environment debate can be found, among other things, in the increase in environmental crime that abounds within these industries, such as IUU fishing (*see* below).

Canada has often dealt domestically and internationally with this problem, with mixed results. Canada's 'turbot wars' experience with Spanish and Portuguese trawlers fishing near Canada's EEZ is illustrative of the controversies and debates surrounding the perceptions and realities of IUU fishing (Mastny and French 2002). Although international agreements such as the 1993 *Compliance Agreement*, the 1995 *UN Agreement Relating to the Conservation and Management of Straddling Fish Stocks and Highly Migratory Fish Stocks* and the 2001 agreement to combat "illicit activity including that by flag-of-convenience vessels" (Mastny and French 2002), as well as increased regional patrols by nations such as Canada, IUU fishing continues to be a significant source leading to overfished and depleted stocks around the world. It is apparent that numerous trade agreements (if not all) allowing foreign vessels to fish in designated national waters have produced immediate financial gain through royalties, but have also threatened the long-term livelihood of local fisherpeople, depleted fish stocks, and contributed to the non-sustainability of the fishing industry (Mutume 2002). Trade agreements are encouraging greater exploitation of fish resources (Neis and Williams 1997), without adequate concurrent public or community monitoring and management of the resource (Lino Grima 1999). In other words, state and internationally-based economic securities, expressed through trade agreements, lead to human insecurities or insecurities experienced 'below' the state or international levels, such as threatened community identities (as indigenous peoples, fisherpeople, etc.), threatened economic and food securities, and political insecurity demonstrated by the political isolation of these communities from the decision-making processes that determine their fate, or even threaten their survival (Duhaime *et al.* 2004).

Relations Between Economic and Societal Impacts and Human Insecurities

Varying 'interests' are inevitably pitted against one another as social, economic, and sometimes personal security are found to be at odds with environmental and health security. With the closure of, and moratorium on, the cod fishery in Atlantic Canada in 1993, fishing communities were devastated. What used to be the 'largest and most productive cod fishery in the world' became a wasteland, and "by 1995, Newfoundland fisheries employed about 10% of the labour they had during the late 1980s and generated only 20%

of previous export earnings. The Task Force on Incomes and Adjustment in the Atlantic Fishery observed that 'the groundfish resource failure means a total or at least major economic collapse for hundreds of communities in Atlantic Canada' (Hutchings and Myers 1995:20). The impacts were not just economic, but political, social, cultural, gendered, ethnic, ethical and environmental. Initial reports from the United Nations Environment Programme (UNEP) are showing that under certain conditions, trade agreements and trade liberalization do not mix well with fisheries sustainability and the economic and social well-being of fishers (Ahmad 2002).

Despite these reports, and from an economic point of view, the Canadian fisheries appear to be in good stead according to the Organization for Economic Cooperation and Development (OECD). In its *Review of the Fisheries in OECD Countries* reports, the OECD presents an economic and trade based analysis of the fisheries. I would argue that the strength behind initiating trade agreements, which expose local fisher communities to various social, economic and political vulnerabilities and insecurities, come from such narrow economic analyses as OECD reports. I wish to examine a number of the points the 2001 OECD Review raised with relation to Canada, and offer an example of the tensions between economic and environmental and other human insecurities.

The 2001 OECD Review illuminated a number of significant economic developments in the Canadian fishing industry, including the move of the 1998 Asia Pacific Economic Cooperation (APEC) 'fish and seafood trade liberalization initiative' to the WTO to allow for broader participation, the Canadian ratification of the *United Nations Fish Agreement* (UNFA) in 1999, and the 1999 Supreme Court of Canada ruling in favour of supporting treaty rights, including the right to earn a 'moderate livelihood' from fishing, hunting and gathering to the Mi'kmaq, Maliseet, and Passamaquody First Nations (OECD 2001).

The OECD Review noted the low volume of landings in the Canadian fishing industry in relation to historical levels, but increased crustacean landings and an improvement in aquaculture secured Canada a record overall volume in 1999 of 1.1 million tonnes, or CAD 1.9 billion in value (OECD 2001). Recognition was accorded to the efforts of the Canadian government to increase conservation through bilateral and multilateral fishing agreements such as the *Pacific Salmon Treaty* and the *International Plan of Action* (IPOA) adopted by the Food and Agriculture Organization of the United Nations (FAO), but Canada nonetheless appears to be encouraging these efforts from a deficit position. Historically low levels of Canada's significant fish stocks such as cod and salmon make the conservation efforts appear too little too late, as the industry shifts its reliance onto currently thriving stocks such as crustaceans.

Although Canada has managed to sustain, and even increase, its overall fish/seafood volumes (exports in 1998 were at their highest value thus far), this has been entirely dependent upon new sources of product, rather than from the traditional stocks that previously defined the Canadian fisheries market. Cod stocks continue to be dangerously low, with Pacific salmon stocks (coho in particular) following close on the heels of the cod if not surpassing it with regard to high risk of extinction (OECD 2001). Overcapacity in the processing sector has equally placed undue pressure on the fish stocks, with the result that the government of Canada is presently pursuing initiatives to re-orient the focus of the processing sector, including value-added secondary processing, aquaculture, and 'rationalization' of the industry (OECD 2001). However, pressure to export fish products continues to mount, as Canada is the number one supplier of seafood to the US market: "The United States is the second largest importer of fish products in the world, after Japan" (Anania 1994:518) and the European Union opened its tariff rate quotas on cooked and peeled shrimp from Canada, allowing greater export potential.

What cannot be overemphasized in such a report as the OECD's is the overall positive spin it still places on developments within the Canadian fishing industry. Financially, Canadian fisheries are doing rather nicely, albeit on the basis of new stocks, such as crustaceans. The acknowledgement of overfishing and depleted fish stocks is recognized but overshadowed by the numerical successes coming from other stocks (dollar amount). To the extent that the survival of fishing communities is recognized, it is only with regard to a very small number of indigenous groups, and only insofar as a 'moderate livelihood' is provided—these communities are not sharing in the multi-billion dollar industry praised in the report. This overall 'success story' (economically speaking, that is), does not take the local picture into account, and therefore ignores or undervalues the vulnerabilities to which people and communities are exposed. These would be the communities that were dependent upon those now depleted fish stocks, or those who are now working in the crustacean industry and who are exposed to yet new vulnerabilities (SafetyNet 2005). As well, the economic success story reduces the significance of overfishing as focus is turned towards new fish stocks that generate even more wealth than the 'older' stocks (OECD 2001). But the over-exploitation and depletion of a natural resource is the crux of the issue. As Bente Aasjord (2002:37) states: "Overfishing has to do with unsustainable development. Overfishing is to take risks. Overfishing has to do with stealing others' livelihoods. Overfishing is giving rights to some, and marginalizing others. Overfishing creates winners and losers."

Overfishing has the effect of driving fishing communities, already suffering significant economic insecurity, into great poverty. This is, however, only a part of the picture. Economic insecurity does not exist in isolation of other insecurities and vulnerabilities. Reduced economic power reduces access to food, education (particularly post-secondary education), and health care (Neis and Williams 1997). These insecurities need to be seen together and as relationships. An economic assessment such as that provided by the OECD does not make these relationships visible, demonstrating that economic assessments that in turn contribute to the development of trade agreements, are woefully inadequate, and actually a threat to communities that depend on natural resources for their livelihoods and identities. Local fishing communities are often family-based, labour intensive fisheries that have social and political as well as economic significance (Aasjord 2002; Duhaime et al. 2004). Economic slowdowns force many nations, developing ones in particular, to allow continued overfishing in their waters to generate some level of income for the nation. Solutions may lie in quotas and increased costs for foreign/international fishing fleets (Duhaime et al. 2004) but against the strength of subsidy reduction heralded at the Doha (2001) WTO talks, these recommendations could fall on deaf ears.

Politically, fishers and fisher communities are isolated, "subject to external influences that may even threaten its survival" (Duhaime et al. 2004:73). The depletion of fish stocks, the negotiation of trade agreements pertaining to these same stocks, and the regulations upon which fishers must rely are beyond the control of fishers and their families (Duhaime et al. 2004). As trade regimes further liberalize, the fishing communities are further marginalized (Aasjord 2002). Individual fishing quotas, instituted as a measure toward better fisheries management, have the additional effect of concentrating economic dependence upon fewer and fewer owners and processors, raising fears about the privatization of the fisheries and reduced access by traditional fisher communities (Caulfield 2004). Women face even further marginalization in these processes as their work in fisheries is largely unrecognized and invisible, and they are not represented on those few political bodies that fishers can participate in (as owners of production, as members of fishers associations and unions, or on international management bodies such as the Northwest Atlantic Fisheries Organization or NAFO) (Aasjord 2002).

An important social factor of security involves health—physical, mental and spiritual. Overall reduction in social services ensures that health security becomes an increasing concern. Access to adequate health care is a challenge for remote locations catering to the needs of Indigenous and non-Indigenous citizens based on predominantly natural resource industries (be they traditionally-oriented or otherwise). Advances in distance delivery systems such as telemedicine are extremely important to the health security of the Arctic region (Hild et al. 2004; Linstad forthcoming). Overfishing has the effect of reducing or eliminating fish from many diets, contributing to greater risks of cardiovascular diseases as people in particularly developing countries no longer have access to essential protective fats available in fish (Anania 1994). As such, environmental securities have significant impacts on women of the Arctic connected with economic securities (for those economically dependent upon the land) as well as with health and cultural securities, among others. Persistent organic pollutants have been linked to increases in breast cancer, impacts on reproductive health, and transfer to infants through breastmilk (Kafarowski 2002). The health of the ecosystem is also integrally linked to the culture of many indigenous peoples in the North, with much of the protection and promotion of this culture being conducted by women in the homeplace (Frink et al. 2002). Another prime example of a work-related health concern particularly relevant to the fisheries, and to women who dominate in the processing industry, is the increasing incidence of work-related asthma found in snow crab processing plants (WHSCC 2005). As noted previously, the crustacean fisheries are on the increase in some regions, providing a substantial amount of the fisheries revenue today. As such, increasing demand is and will continue to be placed on the harvesting and processing of this resource. Finally, one can look to identities and the ways in which traditional lifestyles are embedded into how people and communities define themselves and each other. One Northern health issue, not explicitly related to but neither completely divorced from lives in the fisheries, is the high rates of suicide particularly amongst young men who do not see a future for themselves as contributors to their communities (Hild et al. 2004). The destruction of traditional and/or long-standing community, family, and individual identities by liberalization of fisheries and concurrent marginalization of fishers, need to be examined in light of health implications. Environmental exploitation without any sense of sustainability threatens Northern lifestyles and well-being by wreaking havoc on cultural, ethnic and national identities and security.

Why human security and the fisheries?

As the notion of security expands, recognizing a variety of sectors of security such as health, cultural, identity, community, economic, political, personal, food, and environmental securities, it is becoming increasingly apparent that gender plays a significant role in understanding the meanings of, as well as the challenges within, the concept and its operationalization. Gender roles and perspectives, whether oppressive, neutral, or liberating in nature, infuse our understandings of what it means to be secure.

As such, meanings of security to women and men in Northern fisheries become a part of the security dynamic. Gender views of security are additionally important as gender perspectives and experiences can also transcend some of the divisions we otherwise place between each other through race, class, culture and ethnicity. As such, we can examine and respond to articulations of security from all women in the North, both indigenous and non-indigenous, as well as share and understand securities from other global contexts, such as the global South and the fisheries in these areas. Resources play a major role in the economy, settlement, history, and culture of fisheries communities, and therefore, constitute a focal point for diverse securities. Women are often relegated to processing industries, and more importantly are depended upon as a community and family support and resource (Neis and Williams 1997). With the fish stocks close to collapse or already collapsed, communities are in crisis and the burden often falls to women; families are moving out of the communities, birth rates are falling, communities are in poverty, and the lack of self-esteem and erosion of fisher identities amongst the men increase the tensions and stress borne by women (Neis and Williams 1997; Aasjord 2002). Economic security is relevant for women in general, but especially where, among the prominent resource extraction industries in the North such as oil, gas, and mining that offer some of the highest average earnings, the vast majority of employees are male. Higher rates of female employment occur within the care-giving, education and/or social services sectors. In many of the Northern regions, economic cutbacks by national governments have often had a negative impact on rural or remote locations, reducing the standard of living and quality of life in these areas through limited employment opportunities, low wage levels, and underdeveloped social services (Chamberlain 2002; Rumyantseva 2003; Winberg 2002). Access to services and often a source of employment is therefore limited for these women. Work opportunities for women are becoming more and more important, however, they are often the ones families fall back upon to ensure the survival of the family; women try to subsidize if not replace a major decrease in the family income (Sloan and Kafarowski 2004); they support others in the family (fathers, sons) who have lost their livelihoods and in many respects their identities; they work to keep together local social institutions, and they try to retain family and community pride and dignity (Aasjord 2002).

There is also a direct relationship of dominance and non-dominance between both fishers in general and women in particular, and the directions trade has been taking through trade agreements. Barbara Neis and Susan Williams (1997) note that women have in fact been blamed for the lack of industrialization in the inshore fisheries, as well as for the increasing dependence upon fisheries-related employment. Women are considered a significant impetus behind the increased use of employment insurance, which in turn supports and inflates the labour market (Neis and Williams 1997). What goes unrecognized by the OECD and other bodies which evaluate the efforts of individual countries to comply with acceptable trade practices is that: 1) The statistics on women fishers and processing workers are misleading as they have not adequately represented or understood women's formal and informal roles in the fishing industry over time; 2) Fewer employment opportunities and higher costs of living force many people to rely more heavily on the fisheries; and 3) Ever increasing costs include fishing costs raise greater incentives toward fisheries incomes within each household to offset these higher costs borne by these households (Neis and Williams 1997).

Security, particularly throughout the twentieth century and the Cold War, has been dominated by the interests of the state. We were told that the concept ought only to be employed in situations demanding extraordinary measures against existential threats against the state (Walt 1991; Buzan *et al.* 1998). Security, however, is no longer just a matter for 'high politics.' Human security identifies a myriad of sectors of security that demand attention in the interest of establishing human well-being. What can we learn from Northern and coastal communities, including the fisheries? By integrating a gender perspective into the concept of human security, thereby acknowledging the dynamics and relations between dominant and non-dominant, as well as interpreting human security from these relations of dominance and non-dominance in the Northern and fisheries contexts, securities can be understood as not mutually dependent categories especially within a gender context. Insecurities such as sexual violence, domestic violence, economic deprivation, environmental degradation and political isolation are relevant to the relationships between Northern women and men and their environments, including within the fisheries. Women and men in coastal communities, regardless of North/South and assumptions of their 'developedness,' can share security experiences, learn from each other, and teach academics and policymakers about their meaning of security.

References

Aasjord, B. (2003). 'Where have all the fishes gone?,' pp. 36-44 in *Taking Wing: Conference on Gender Equality and Women in the Arctic*. Helsinki: Ministry of Social Affairs and Health.

Ahmad, K. (2002). UN agency concerned by exploitation of poor nations' fish stocks, *The Lancet* 359(9301):144.

Anania, G., ed. (1994). *Agricultural Trade Conflicts and GATT: New Dimensions in U.S.-European Agricultural Trade Relations*. Boulder, CO; Westview Press.

Axworthy, L. (2001). 'Introduction,' in R. McRae and D. Hubert, eds. *Human Security and the New Diplomacy: Protecting People, Promoting Peace*. Montreal and Kingston: McGill-Queen's University Press.

Bigo, D. (2001). 'The Möbius Ribbon of Internal and External Security(ies),' in M. Albert, D. Jacobsen and Y. Lapid, eds. *Identities, Borders Orders: Rethinking International Relations Theory*. Minneapolis, London: University of Minnesota Press.

Buzan, B., O. Wæver, and J. de Wilde (1998). *Security: A New Framework For Analysis*. Boulder, CO and London: Lynne Rienner Publishers.

Caulfield, R. (2004). 'Resource Governance,' pp. 121-138 in N. Einarsson et al, eds. *Arctic Human Development Report*. Akureyri, Iceland: Stefansson Arctic Institute.

Chamberlain, L. (2002). 'Arctic Inspirations: Women Creating Small Businesses and Personal Success in Small Communities,' pp. 65-67 in *Taking Wing: Conference on Gender Equality and Women in the Arctic*. Helsinki: Ministry of Social Affairs and Health.

Delgado, C.L., N. Wada, M.W. Rosegrant, S. Meijer, and M. Ahmed (2003). *Fish to 2020: Supply and Demand in Changing Global Markets*. Washington, D.C. and Penang, Malaysia: International Food Policy Research Institute and WorldFish Center.

DFAIT (2005). *Government of Canada (DFAIT)*. Ottawa: Department of Foreign Affairs and International Trade. Electronic resource: http://www.humansecurity.gc.ca/. Last accessed March 3, 2009.

Duhaime, G. *et al.* (2004). 'Economic Systems,' pp. 69-84 in N. Einarsson et al, eds. *Arctic Human Development Report*. Akureyri, Iceland: Stefansson Arctic Institute.

Frink, L., R. Shepard and G. Reinhardt, eds. (2002). *Many Faces of Gender: Roles and Relationships Through Time in Indigenous Northern Communities*. Boulder: University Press of Colorado.

GECHS (2005). *Global Environmental Change and Human Security*. Electronic document: www.gechs.org. Last accessed: March 3. 2009.

Green, J and C. Voyageur. (1999). 'Globalization and Development at the Bottom,' pp. 142-157 in M. Porter and E. Judd, eds. *Feminists doing Development.* London and New York: Zed Books.

Hackett, S. (2001). *Environmental and Natural Resources Economics: Theory, Policy, and the Sustainable Society, 2nd edition.* New York: ME Sharpe Publishers.

Hild, C.M, *et al.* (2004). 'Human Health and Well-being,' pp. 155-168 in N. Einarsson et al., eds. *Arctic Human Development Report.* Akureyri: Stefansson Arctic Institute.

Hoogensen, G. and S.V. Rottem (2004). Gender Identity and the Subject of Security. *Security Dialogue* 35(2): 154-171.

Hunter, E. and D. Harvey (2002). Indigenous suicide in Australia, New Zealand, Canada and the United States. *Emergency Medicine* 14: 14-23.

Hutchings, J. and Myers, R. (1995). 'The biological collapse of Atlantic cod off Newfoundland and Labrador,' pp. 37-93 in R. Arnason & L. Felt , eds. *The North Atlantic fisheries: Successes, failures and challenges.* Charlettetown, PEI: Institute of Island Studies.

Kafarowski, J. (2002). 'Women and Natural Resources in the Circumpolar North: Striving for Sustainable Development Through Leadership,' pp. 73-78 in *Taking Wing: Conference on Gender Equality and Women in the Arctic.* Helsinki: Ministry of Social Affairs and Health.

Kahler, M. (1997). 'Inventing International Relations: International Relations: Theory After 1945,' pp. 20-53 in M.W. Doyle, and G.J. Ikenberry, eds. *New Thinking in International Relations Theory.* Boulder, CO: Westview Press.

Karamé, K. (2001). *Gendering Human Security: From Marginalization to the Integration of Women in Peace-Building.* Fafo Report 352/NUPI report no. 261, Oslo: Norwegian Insitute for International Affairs and Fafo Institute for Applied Social Science.

Leal, D.R., ed. (2004). *Evolving Property Rights in Marine Fisheries.* Lanham: Rowman and Little field Publishers, Inc.

Lino Grima, A.P. (1999). 'Economic Growth vs Environmental Protection: The Flight to the Economic Market,' pp. 30-32 in R.E. Munn, ed. *Emerging Environmental Issues in Ontario.* Environmental Monograph No.15. Toronto: Institute for Environmental Studies.

Linstad, L. (forthcoming). 'Telemedicine as a human security issue in the Arctic,' in G. Hoogensen and D. Bazely, eds. *Human Security in the Arctic.* London: Earthscan.

Mastny, L. and H. French (2002). 'Crimes of (a) global nature' *World Watch* 15(5): 12-22.

McDonald, M. (2002). 'Human Security and the Construction of Security' *Global Society* 16(3): 277-295.

McRae, R. and D. Hubert, eds. (2001). *Human Security and the New Diplomacy: Protecting People, Promoting Peace.* Montreal: McGill-Queen's University Press.

McSweeney, B. (1999). *Security, Identity and Interests: A Sociology of International Relations.* Cambridge: Cambridge University Press.

Munn, R.E., ed. (1999). *Emerging Environmental Issues in Ontario.* Environmental Monograph No.15. Toronto: Institute for Environmental Studies.

Mutume, G. (2002). Fish and Empire. *Multinational Monitor* 23(7/8): 7-8.

Neis, B. and S. Williams (1997). The New Right, Gender and the Fisheries Crisis: Local and Global Dimensions. *Atlantis* (21)2: 47-63.

Nussbaum, M. and J. Glover, eds. (1996). *Women, Culture, and Development: A Study of Human Capabilities.* Oxford: Oxford University Press.

Organization for Economic Cooperation and Development (2001). Canada. Pp. 17-28 in *Review of the Fisheries in OECD Countries 2001.* Paris: OECD.

Peterson, V. S. and A.S. Runyan (1999). *Global Gender Issues: Dilemmas in World Politics. 2nd edition.* Boulder, CO: Westview Press, Inc.

Poku, N.K., N. Renwick, and J. Glenn (2000). 'Human Security in a Globalising World,' pp. 9-22 in D.T. Graham and N.K. Poku, eds. *Migration, Globalization, and Human Security.* London: Routledge.

Robertson, C.L. (1997). *International Politics since World War II: A Short History.* Armonk, NY: M.E. Sharpe.

Rumyantseva, T. (2003). 'Women's Living Conditions and Employment Opportunities,' pp. 51-59 in *Taking Wing: Conference on Gender Equality and Women in the Arctic.* Helsinki: Ministry of Social Affairs and Health.

SafetyNet (2005). *Internet.* Electronic document: www.safeteynet.mun.ca. Last accessed: March 3, 2009.

Sloan, L. and J. Kafarowski (2004). 'Gender Issues: Focus on Fisheries,' pp. 200-201 in N. Einarsson et al., eds. *Arctic Human Development Report.* Akureyri: Stefansson Arctic Institute.

Solheim, B.O. (1994). *The Nordic Nexus: A Lesson in Peaceful Security.* Westport: Praeger Publishers.

Sørensen, B.W. (1999). *'Men in Transition': The Representation of Men's Violence against Women in the Arctic.* Strasbourg: European Council of Europe.

Spivak, G. (1999). *A critique of post-colonial reason: toward a history of the vanishing present.* Cambridge, MA: Harvard University Press.

Sylvester, C. (1994). *Feminist Theory and International Relations in a Postmodern Era.* Cambridge: Cambridge University Press.

Thomas, C. (2001). *Global Governance, Development & Human Security: The Challenge of Poverty & Inequality.* London, GBR: Pluto Press.

Thompson, A. (2000). Canadian foreign policy and straddling stocks: Sustainability in an interdependent world. *Policy Studies Journal* 28(1): 219-235.

UNDP (1994). *Human Development Report 1994.* New York and Oxford: Oxford University Press.

Gender, Globalization and Fisheries Network (2002). Statement from the gender, globalization and fisheries network. *Women and Environments* 54/55: 23.

Verhaag, M.A. (2003). It Is Not Too Late: The Need for a Comprehensive International Treaty to Protect the Arctic Environment. *Georgetown International Environmental Law Review* 15(3): 555-579.

Vogel, C. and K. O'Brien (2004). Vulnerability and Global Environmental Change: Rhetoric and Reality. *Aviso* 13. Electronic document: http://www.gechs.org/aviso/13/ Last accessed: March 3, 2009

Walt, S.M. (1991). The renaissance of Security Studies. *International Studies Quarterly* 35(2): 211-239.

Werthes, S. and S. Härtig (2007). Human Security: New Threats, New Responsibilities. *Military Technology* 31(6): 147-149.

Winberg, M. (2002). 'Prostitution and Trafficking are Male Violence Against Women,' pp. 191-194 in *Taking Wing: Conference on Gender Equality and Women in the Arctic.* Helsinki: Ministry of Social Affairs and Health.

Witt, D. (1999). *Black Hunger: Food and the Politics of U.S. Identity.* New York: Oxford University Press.

WHSCC (2005). *Crab Asthma Brochure.* Electronic document: http://www.safetynet.mun.ca/login1.htm. Last accessed: March 3, 2009.

APPENDICES

LIST OF ACRONYMS

ACIA	Arctic Climate Impacts Assessment
ADFG	Alaska Department of Fish and Game
AKTEA	p.v. European Network of Women in Fisheries and Aquaculture or Womens' FishNet
APEC	Asia-Pacific Economic Cooperation
CARC	Canadian Arctic Resources Committee
CDN$	Canadian dollar
DFAIT	Department of Foreign Affairs and International Trade (Canada)
DFO	Department of Fisheries and Oceans (Canada)
EEA	European Economic Area
EEZ	Exclusive Economic Zone
EFTA	European Free Trade Association
EU	European Union
FAO	Food and Agriculture Association (United Nations)
FIFS	Fiskeriverkets Författningsamling (Sweden)
FJMC	Fisheries Joint Management Committee (Inuvialuit, NWT, Canada)
GECHS	Global Environmental Change and Human Security
HACCP	Hazard Analysis and Critical Control Points
HTA	Hunters' and Trappers' Association
HTC	Hunters' and Trappers' Committee (NWT, Canada)
HTO	Hunters' and Trappers' Organization (Nunavut, Canada)
ICFWS	International Conference on Fisheries and their Supporters
ICSF	International Collective in Support of Fishworkers
ILO	International Labour Organization
IPOA	International Plan of Action
ITK	Inuit Tapiriit Kanatami (Canada)
IUU	Illegal, Unreported, Unregulated
NAFO	North Atlantic Fisheries Organization
NOK	Norwegian Kroner
NWMB	Nunavut Wildlife Management Board (Nunavut, Canada)
NWT	Northwest Territories (Canada)
OECD	Organization for Economic Cooperation and Development
SDF	Sami Development Fund (Finmark, Norway)
SEK	Swedish Kroner
TAC	Total Allowable Catch
TIAA	The Icelandic Aquaculture Association
UAF	University of Alaska, Fairbanks
UN	United Nations
UNDP	United Nations Development Programme
UNEP	United Nations Environmental Programme
UNFA	United Nations Fisheries Agreement
WHMIS	Workplace Hazardous Materials Information System
WTO	World Trade Organizations
WWF	World Wildlife Fund

CONTRIBUTORS

Sine Anahita is Associate Professor of Sociology at the University of Alaska Fairbanks (UAF). Her research and teaching interests center on organized inequalities—how organizations, states, and institutions organize markers of difference such as gender, sexuality, ethnicity, and dis/ability, into systems of inequality. Currently, she is Chair of the Department of Sociology at UAF and also Coordinator of the Women's Studies Program. Three other collaborative projects Anahita is involved with include a study of water inequalities in Alaska (with Nicole Grewe); an institutional ethnography of UAF women faculty in science, technology, engineering, and math disciplines (with Joy Morrison); and an analysis of how masculinities shape Alaska's wolf control policies (with Tamara Mix.)

Elisabeth Angell received a degree in Economics from the University of Oslo, Norway in 1993. She is now a researcher at Norut NIBR Finnmark, a regional research institute in northern Norway. Her main research interests address regional economic development and regional policy. Much of her work is based in regions in which the primary sector plays an important role (especially fisheries and reindeer hunting) and where the petroleum industry is increasing in significance. Furthermore, the area in which Angell works is the main geographical area of the Sámi people. She has been particularly interested in the effects of national policy (political measures) on the relationship between the primary sector and other industries. Consequences for women and for the Sámi people are an integrated part of most projects. E-mail: elisabeth.angell@finnmark.norut.no Website: www.finnmark.norut.no

Siri Gerrard trained in social anthropology and ethnography and is Senior Lecturer at the Department of Planning and Community Studies, University of Tromsø, Norway. Gerrard's major research interests are in the field of social and cultural processes in fishing villages and in the fishing industry with special emphasis on women's and men's roles, relations and positions. She has more than 30 years' experience of research especially in northern Norway, but also in northern Tanzania and northern Cameroon. She has written articles and co-edited books, mostly focusing on women and/or gender relations, but also on feminist theories.

Elina Helander-Renvall has studied languages, ethnography and political science and received her Ph.D from the University of Umeå in Sweden. She is a senior social scientist who lives in Utsjoki, Northern Finland where she has a reindeer farm. Currently, she works as director for the Arctic Indigenous Peoples and Sami Research Office at the Arctic Centre, University of Lapland in Rovaniemi, Finland. Over the last few years, Helander-Renvall has conducted research on sustainable development, customary law, indigenous knowledge, women´s knowledge and climate change. Her recent publications include: *Rätt i Deanodat. Norges Offentlige utredninger. Samiske sedvaner og rettsoppfatninger*. NOU 2001(34):459-496; 'A marginalised minority remains marginalized? On the management of fjord resources,' pp. 203-221 in K. Karppi and J. Eriksson, eds. *Conflict and Cooperation in the North* (2002). *Umeå: Kulturens Frontlinjer; Snowscapes, Dreamscapes. Snowchange Book on Community Voices of Change*. E. Helander and T.

Mustonen, eds. Tampere: Tampere Polytechnic; 'The Nature of Sami Customary law,' pp. 88-96 in Arctic Governance. Juridica Lapponica. Arctic Centre (2004); and *Biological Diversity in the Arctic* (2005) Draft Report to SCBD.

Gunhild Hoogensen is Associate Professor in the Department of Political Science at the University of Tromsø. She holds a Ph.D. in Political Science specializing in International Relations and Comparative Politics from the University of Alberta. Her main research interest is the application of the human security concept, informed by gender and indigenous perspectives, to the Arctic context. Dr. Hoogensen leads the "Human Security in the Arctic" project along with her Norwegian and Canadian colleagues Dr. Dawn Bazely (York University), Dr. David Malcolm (Arctic Energy Alliance), Dr. David DeWitt (York University), and Dr. Geir Wing Gabrielsen (Norwegian Polar Institute). Her book *International Relations, Security, and Jeremy Bentham* (Routledge) was released July 2005, with other articles most recently appearing in *Security Dialogue, Canadian Foreign Policy*, and *International Studies Review*. For more information, visit her project website at: http://uit.no/statsvitenskap/humansecurity/

Joanna Kafarowski holds a Bachelor's degree in English from Carleton University, Ottawa, Ontario, a Master's degree in Geography from the University of Victoria, British Columbia and a Ph.D in the Natural Resources and Environmental Studies program at the University of Northern British Columbia. Her doctoral thesis focused on the intersections between gender, decision-making and contaminants in Nunavik, Canada. Current research interests include human dimensions of environmental change; Indigenous peoples with a focus on gender; natural resource management; environmental and social justice; human security and protected areas. Much of her work is conducted in the circumpolar North. She is currently a Research Associate at the Canadian Circumpolar Institute. She can be reached at gypsy_four@hotmail.com

Anna Karlsdóttir graduated from the University of Roskilde in Denmark in 1996 with an M.Sc. in Human Geography and a minor in Public Administration. Anna moved back to Iceland in 2000, where she established a new educational program in environmental management at the Agricultural College of Iceland. In 2002, Anna became Assistant Professor at the University of Iceland in Human Geography and Tourism Studies. She has been involved in research on the transformation of the fishery and seafood sector and its impact in the North Atlantic region, as well as studying women in fisheries and aquaculture. Her research activities also focus on cruise tourism in the Arctic and North Atlantic. Anna has two children aged 15 and 8. She is also completing her PhD studies at the University of Roskilde.

Phyllis Morrow is the current Dean of the College of Liberal Arts and a professor of Anthropology and Women's Studies at the University of Alaska Fairbanks. She received her Ph.D. from Cornell University (1987) in Social and Cultural Anthropology. She has been an expert legal witness on Yup'ik cultural issues and a consultant on the development of cross cultural materials and approaches for educational and social service institutions. She is a member of the American Anthropological Association and the Alaska Anthropological Association, as well as being on the editorial board for *Arctic Anthropology*. Phyllis' research interests include linguistic anthropology, socio-legal studies, folklore, and Inuit/Yup'ik ethnography.

Virginia Mulle is Associate Professor of Sociology and Associate Dean of the College of Arts and Science at the University of Alaska Southeast. Originally from New York, she received her B.A. in Sociology from the State University of New York, an M.S. in Psychology and a Ph.D. in Sociology from the University of Florida. She joined the faculty at the University of Alaska Southeast in 1994. Teaching areas of specialization are research methods, gender, stratification, and race and ethnicity. Current research projects include a study of the relationship between child maltreatment and substance abuse in Alaska, the changing roles of Native women in contemporary society, and the social construction of identity of Native Americans in national parks. She resides in Juneau with Mike Turek, a research associate in the Division of Subsistence, Alaska Department of Fish and Game, and six cats. The cats are unemployed.

Darlene Northway was born and raised in the Upper Tanana drainage of Alaska, specifically in an area called Scottie Creek, near the Alaska-Canada border. While growing up, she helped take care of her family and learned to hunt, trap, fish, cook, and sew, which she continues to do. Darlene is a well respected Elder in Northway village and throughout Alaska. As a speaker of Upper Tanana Athabascan, Darlene is a frequent consultant on Native language issues. For the past 17 years she has worked as an interpretive ranger for the Tetlin National Wildlife Refuge (USFWS). Throughout the Northway Whitefish Project, Darlene served on the Project's Whitefish Advisory Board. Darlene is the mother of six girls and one boy, grandmother to nineteen, and great-grandmother to three.

Katherine Reedy-Maschner earned a Ph.D. in Social Anthropology from the University of Cambridge, U.K., in 2004. Dr. Reedy-Maschner is Assistant Professor in the Department of Anthropology at Idaho State University, Pocatello. She has worked in the Eastern Aleutians since 1995, completing a dissertation on "Aleut Identity and Indigenous Commercial Fisheries." Her research interests range from identity and representations of arctic peoples to fisheries policy, ocean management and ecology. She has received multiple grants from the National Science Foundation, and is currently working on the Sanak Islands Biocomplexity Project, documenting traditional knowledge of the Steller sea lion, and analyzing Alaska Native testimony in fisheries policymaking. She lives in Pocatello with her husband and two sons.

Melissa A. Robinson grew up in Utah and has moved steadily north since 1998. She holds a Bachelor's and a Master's of Science degree in Wildlife Biology from the University of Montana (2002) and the University of Alaska Fairbanks (2005), respectively. At the University of Montana, Melissa conducted research with the Salish and Kootenai tribes living on the Flathead Reservation. Since 2001, she has worked for the U.S. Fish and Wildlife Service (USFWS) in Alaska, conducting wildlife and fisheries studies and working with Native people across the state. While working for the USFWS, she completed her Master's research on linking local knowledge and fisheries science regarding humpback whitefish. At the University of Alaska Fairbanks, Melissa was part of the Regional Resilience and Adaptation Program, an interdisciplinary program examining issues related to social and ecological sustainability. Melissa also worked as a Refuge Operations Specialist for the Koyukuk and Nowitna National Wildlife Refuges in Galena, Alaska.

Kerrie-Ann Shannon is currently Assistant Professor in the Department of Anthropology at University of Alaska Fairbanks. She has conducted research in Nunavut, Canada since 1995. Her current research interests include procurement, human-environment relations, kinship, gender, exchange, games, and sled dogs. She holds a Ph.D in Anthropology from the University of Aberdeen, Scotland and an M.A. from the University of Alberta, Canada.

Martina Tyrrell is a British Academy Postdoctoral Fellow at the Department of Anthropology, University of Aberdeen, Scotland. She received her doctorate from the University of Aberdeen in 2005. She has previously worked on perceptions of the environment, both in Ireland and in the Canadian Arctic. In Ireland, she explored changing cultural perceptions of the boglands over the past two and a half centuries, and on the northwest coast of Hudson Bay she has explored Inuit sensory perception of the sea. This work on perception of the sea in Arviat, Nunavut, has led to her most recent research into human-animal relations. She is currently undertaking research in two Nunavut communities on the impact of hunting quotas and contaminants on social relationships and on the relationships between Inuit and beluga whales.

Maria Kristina Udén earned a M.Sc. degree in Mineral Processing and Metallurgy at Luleå University of Technology, Sweden in 1988. She returned to the university and the Department of Human Work Sciences in the early 1990s and defended her doctoral thesis entitled: Women technically speaking in 2000. Areas of interest include feminist approaches within the engineering sciences and the advancement of the Arctic economy particularly regarding the condition of women. Publications include 'The impact of women on engineering' published in *The International Journal of Engineering Education*, and 'Providing Internet Connectivity to the Saami Nomadic Community' in the *Proceedings of 2nd International Conference on Open Collaborative Design for Sustainable Innovation.* http://www.thinkcycle.org (with A. Doria and D.P. Pandey).